Advanced Physics:
Materials and Mechanics

Advanced Physics:
Materials and Mechanics

T. DUNCAN BSc M Inst P

Senior Lecturer in Education, University of Liverpool
Formerly Senior Physics Master, King George V School, Southport

JOHN MURRAY · LONDON

Printed and bound in Great Britain at
The Camelot Press Ltd, Southampton

0 7195 2844 5

Preface

This book provides a course which, it is hoped, will be suitable for students in sixth forms and colleges. In selecting and treating topics recent 'A' level syllabus revisions and current teaching trends have been kept in mind. Thus, an attempt has been made to regroup subjects so that greater emphasis is placed on explanations of the macroscopic behaviour of matter using atomic models. Also, detailed accounts of experimental methods of measuring quantities such as the gravitational constant, moments of inertia, expansivities, etc., have either been omitted or dealt with briefly. Many recent physics textbooks have been influenced by Nuffield and other courses and have drawn on their materials; this one is no exception.

The comparatively new subject of Materials Science—the modern version of Properties of Matter—is introduced and given, especially in Chapters 1 and 2, the rather fuller consideration required by those studying Nuffield 'A' level physics, engineering science, ONC physics and physical science.

To be in line with what is taught in more advanced courses a distinction is drawn between heat and internal energy. Heat, like work, is regarded as involving energy transfer. By internal energy is meant molecular kinetic *and* potential energy. The term thermal energy is avoided because in the past it has been used to mean heat or internal energy or both.

To cater for the individual preferences of teachers and to allow for the different 'O' level courses students may have taken, the treatment permits considerable flexibility in the order in which topics are followed. The chapters in Part 1 (Materials) may be taken in any sequence and need not all be covered before Part 2 (Mechanics) is started. If desired, some or all of Part 2 can precede Part 1.

An effort has been made to present numerical data consistently, a point which has been somewhat neglected in the past in examination and other questions. In general, answers to problems are given only to the number of significant figures justified by the data. (Occasionally the next figure is quoted in brackets, e.g. 2.4(3) kg.) When appropriate, powers of ten notation is used; thus, a length may be written as 1.53×10^4 m (rather than as 15 300 m), indicating three-figure accuracy. Similarly $1.80 \times 10^{-5} \, {}^{\circ}C^{-1}$ is used in preference to $0.000\,018\,0 \, {}^{\circ}C^{-1}$.

In examples worked out in the text, numbers *and* units are substituted for symbols in expressions. The *unit* of the required quantity as well as its value is then found (for example, see pp. 32 and 59). This procedure forms a useful check and is recommended.

I am much indebted to Dr J. W. Warren of Brunel University and to Mr J. Dawber of Wade Deacon Grammar School for Boys, Widnes, both of whom read the manuscript. They corrected numerous small errors and made many helpful suggestions. I must also thank Professor J. Stringer and Dr J. C. Gibbings of the University of Liverpool. The former very kindly commented on Chapters 1 and 2 and useful discussions were held with the latter on Chapter 10. Mr B. Baker of Chesterfield High School, Crosby, undertook the task of constructing and testing the objective-type questions at the end of the book. I am most grateful to him and also to my son-in-law and daughter, Mr and Mrs B. L. N. Kennett, who checked the numerical answers. Thanks are also due to my wife for preparing the typescript.

For permission to use questions from recent examinations grateful acknowledgement is made to the various examining boards, indicated by the following abbreviations: *A.E.B.* (Associated Examining Board); *C* (Cambridge Local Examination Syndicate); *J.M.B.* (Joint Matriculation Board); *L* (University of London); *O* (Oxford Local Examinations); *O and C* (Oxford and Cambridge Schools Examination Board); *S* (Southern Universities Joint Board); *W* (Welsh Joint Education Committee).

T.D.

Contents

Acknowledgements

Thanks are due to the following who have kindly permitted the reproduction of copyright photographs:

Figs. 1.1*a, b*, Cambridge Scientific Instruments Ltd; 1.1*c*, 1968 The Plessey Co Ltd; 1.1*d*, A. Dinsdale, The British Ceramic Research Association; 1.7, Dr R. T. Southin; 1.9, Cenco Instrumenten; 1.10, J. W. Martin (from *Elementary Science of Metals*, Wykeham Publications (London) Ltd); 1.16*b, c, d, e*, The Royal Society (*Proceedings*, 1947, A, 190); 1.19*a*, G. Bell & Sons Ltd (from *An Approach to Modern Physics* by E. Andrade); 1.19*b*, Dr C. Henderson, Aberdeen University; 2.1, 2.2, Avery-Denison Ltd; 2.13, V. A. Phillips and J. A. Hugo; 2.14, The Royal Society (*Proceedings*, 1947, A, 190); 2.15, Joseph V. Laukonis, Research Laboratories, General Motors Corporation; 2.17*b* and cover, Penguin Books Ltd (from *Revolution in Optics* by S. Tolansky); 2.22, British Engine Boiler & Electrical Insurance Company Ltd; 2.24, Vosper Thornycroft, Southampton; 2.25*a*, Central Office of Information, from *Project*; 2.25*c*, Fulmer Research Institute Ltd; 2.28*a*, Professor E. H. Andrews, Queen Mary College; 2.28*b*, Malayan Rubber Fund Board (London) Inc.; 2.20, Eidenbenz and Eglin, used in *Science*, ed. J. Bronowski (Aldus Books); 3.20, Avro Limited, Dover; 3.22, Central Electricity Generating Board; 3.24, Royal Aircraft Establishment, Farnborough; 3.25, Mullard Ltd; 3.30, Educational Measurements Ltd; 4.4*a, b, c*, U.S.I.S.; 5.37*a, b*, Penguin Books Ltd (from *Revolution in Optics* by S. Tolansky); 5.37*c*, Rank Precision Industries; 5.76*a*, Penguin Books Ltd (from *Revolution in Optics* by S. Tolansky); 5.87, Hale Observatories; 5.97, Times Newspapers Ltd; 6.1, Shell Chemicals U.K. Ltd; 6.17, reprinted by permission of the publishers, D. C. Heath & Co, Lexington (from *PSSC Physics*); 6.25*a, b*, 6.34*a, b*, J. T. Jardine, George Watson's College; 6.35*b*, Lord Blackett; 7.9, Popperfoto; 7.29, 7.30, 7.31, NASA; 8.1*a, b*, Mrs F. B. Farquharson; 8.25, Royal Aircraft Establishment, Farnborough; 9.1, Popperfoto; 9.3*b*, British Leyland (Austin-Morris) Ltd; 9.3*c*, Pressure Dynamics Ltd; 9.14*a*, L. H. Newman (photo by W. J. C. Murray); 9.14*b*, Focal Press (photo by Oskar Kreisal from *Focal Encyclopedia of Photography*); 10.9, Dr J. C. Gibbings, Department of Mechanical Engineering, University of Liverpool, and the Editor, *Physics Bulletin* – © J. C. Gibbings.

Part 1 | MATERIALS

1 Structure of materials

Materials science

Advances in technology depend increasingly on the development of better materials. This is especially true of those industries engaged in aircraft production, space projects, telecommunications, computer manufacture and nuclear power engineering. Structural materials are required to be stronger, stiffer and lighter than existing ones. In some cases they may have to withstand high temperatures or exposure to intense radioactivity. Materials with very precise electrical, magnetic, thermal, optical or chemical properties are also demanded.

A great deal has been known for many years about materials that are useful in everyday life and industry. For example, the metallurgist has long appreciated that alloys can be made by adding one metal to another or that heating, cooling or hammering metals changes their mechanical behaviour. Materials *technology* is a long-established subject. The comparatively new subject of materials *science* is concerned with the study of materials as a whole and not just with their physical, chemical or engineering properties. As well as asking *how* materials behave, the materials scientist also wants to know *why* they behave as they do. Why is steel strong, glass brittle and rubber extensible? To begin to find answers to such questions has required the drawing together of ideas from physics, chemistry, metallurgy and other disciplines.

The deeper understanding of materials which we now have has come from realizing that the properties of matter in bulk depend largely on the way the atoms are arranged when they are close together. Progress has been possible because of the invention of instruments for 'seeing' finer and finer details. The electron microscope, which uses beams of electrons instead of beams of light as in the optical microscope, reveals structure just above the atomic level. The field ion microscope and X-ray apparatus allow investigation at that level.

The scanning electron microscope, Fig. 1.1*a*, is a development from the electron microscope and 'scans' a surface with electrons in the way that a television screen is scanned. It gives higher magnifications and much greater depth of focus than optical microscopes using reflected light. It is useful for examining the surfaces of semiconductors, the hairlike fibres and 'whiskers' that are so important in the manufacture of the new generation of composite materials, man-made fibres, and corroded and fractured surfaces. A view of the end of a torn wire (× 75) is shown in Fig. 1.1*b* and of lead–tin telluride crystals (× 30) in Fig. 1.1*c*.

Materials science is a rapidly advancing subject with exciting prospects for the future. Its importance lies in the help it can give with the selection of materials for particular applications, with the design of new materials and with the improvement of existing ones. The strength of even a tea cup has been improved by research into ceramics, as Fig. 1.1*d* shows.

Atoms, molecules and Brownian motion

The modern atomic theory was proposed in 1803 by John Dalton, an English schoolmaster. He thought of atoms as tiny, indivisible particles, all the atoms of a given element being exactly alike and different from those of other elements in behaviour and mass. By making simple assumptions he explained the gravimetric (i.e. by weight) laws of chemical combination but failed to account satisfactorily for the volume relationships which exist between combining gases. This required the introduction in 1811 by the Italian scientist, Amedeo Avogadro, of the molecule as the smallest particle of an element or compound capable of existing independently and consisting of two or more atoms, not necessarily identical. Thus, whilst we could only have atoms of elements, molecules of both elements and compounds were possible.

At the end of the nineteenth century some scientists felt that evidence, more direct than that provided by the chemist, was needed to justify the basic assumption that atoms and molecules exist. In 1827 the Scottish botanist, Robert Brown, discovered that fine pollen grains suspended in water were in a state of constant movement, describing small, irregular paths but never stopping. The effect, which has been observed with many kinds of small particles suspended in both liquids and gases, is called *Brownian motion*. It is now considered to be due to the unequal bombardment of the suspended particles by the molecules of the surrounding medium.

Very small particles are essential. If the particle is fairly large, the impacts, occurring on every side and irregularly, will cancel out and there will be no average resultant force on the particle. Only if the particle is small will it suffer impacts with a few hundred molecules at any instant and the chances of these cancelling out are proportionately less. It is then likely that for a

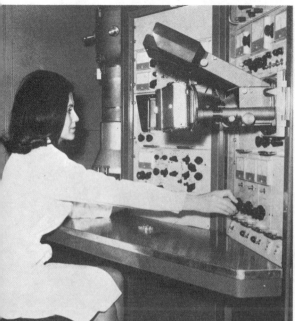

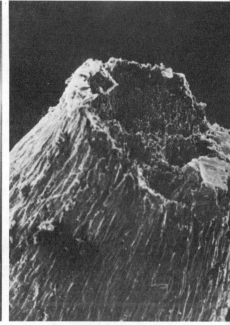

Fig. 1.1a

Fig. 1.1b

Fig. 1.1d

Fig. 1.1c

short time most of the impacts will be in one direction; shortly afterwards the direction will have changed. The phenomenon can be observed in smoke in a small glass cell which is illuminated strongly from one side and viewed from above with a low-power microscope, Fig. 1.2. How would the random motion be affected by (*i*) cooling the air to a low temperature, (*ii*) using smaller smoke particles?

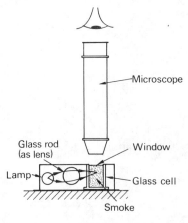

Fig. 1.2

The effect, on its own, does not offer conclusive proof for molecules but it clearly reveals that on the microscopic scale there is great activity in matter which macroscopically (on a large scale) appears to be at rest. The theory of the motion was worked out by Einstein and is found to correspond closely with observation. His basic assumption was that the suspended particles have the same mean kinetic energy as the molecules of the fluid and so behave just like very large molecules. Their motions should therefore be similar to those of the fluid molecules.

The Avogadro constant: mole

Atomic and *molecular masses* (previously called *atomic* and *molecular weights*) give the masses of atoms and molecules compared with the mass of another kind of atom. Originally the hydrogen atom was taken as the standard with atomic mass 1 since it has the smallest mass. In 1960 it was agreed internationally, for various reasons, to base atomic and molecular masses on the atom of carbon (more precisely, on the carbon 12 isotope $^{12}_{6}C$). On the carbon scale the atomic mass of carbon 12 is taken as exactly 12 making that of hydrogen 1.008 and of oxygen 16.00. Nowadays atomic masses are found very accurately using a *mass spectrometer*.

It follows from the definition of atomic mass that any number of atoms of carbon will have, near enough, 12 times the mass of the *same* number of atoms of hydrogen. Therefore any mass of hydrogen, say 1 g, will contain the same number of atoms as 12 g of carbon. In general, the atomic mass of any element expressed in grams, contains the same number of atoms as 12 g of carbon. This number is thus, by definition, a constant. It is called the *Avogadro constant* and is denoted by N_A. Its accepted experimental value is 6.02×10^{23}.

The number of molecules in the molecular mass in grams of a substance is also (because of the way molecular masses are defined) the same for all substances and equal to the Avogadro constant. There are, therefore, 6.02×10^{23} molecules in 2 g of hydrogen (molecular mass 2) and in 18 g of water (molecular mass 18). In fact, the Avogadro constant is useful when dealing with other particles besides atoms and molecules and a quantity which contains 6.02×10^{23} particles is called, especially by chemists, a *mole*. We can thus have a mole of atoms, a mole of molecules, a mole of ions, a mole of electrons, etc.—all contain 6.02×10^{23} particles. We must always have a mole of some kind of particle and so

$$N_A = 6.02 \times 10^{23} \text{ particles per mole}$$

It should be noted that the mole (abbreviation mol) is based on the gram and not the kilogram, which makes it an anomaly in the SI system of units. Sometimes, however, it is expressed in terms of the number of particles per kilogram-mole and its value then is 6.02×10^{26}.

The Avogadro constant has been measured in various ways. In an early method alpha particles emitted by a radioactive source were counted by allowing those within a small known angle to strike a fluorescent screen. Each particle produced one scintillation on the screen and if it is assumed that one particle is emitted by each radioactive atom an approximate value for N_A can be obtained (see question 6, p. 26). Other methods give more reliable results—one involves X-ray crystallography.

Size of a molecule

(a) *Monolayer experiments.* An experimental determination of the size of a molecule was made by Lord Rayleigh in 1899. He used the fact that certain organic substances, such as olive oil, spread out over a clean water surface to form very thin films.

A simple procedure for performing the experiment is to obtain a drop of olive oil by dipping the end of a loop of thin wire, mounted on a card, into olive oil, quickly withdrawing it and then estimating the diameter of the drop by holding it against a $\frac{1}{2}$ mm scale and viewing the drop and scale through a lens, Fig. 1.3a. If the drop is then transferred to the centre of a waxed tray overbrimming with water, the surface of which has been previously cleaned

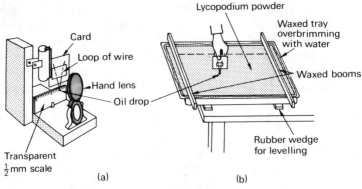

Fig. 1.3

by drawing two waxed booms across it and then lightly dusted with lyco-podium powder, Fig. 1.3*b*, it spreads out into a circular film pushing the powder before it. Assuming the drop is spherical, the thickness of the film can be calculated if its diameter is measured. It is found to be about 2×10^{-9} metre, i.e. 2 nanometres (2 nm).

Oil-film experiments do not necessarily prove that matter is particulate but from them we can infer that if molecules exist and if the film is one molecule thick, i.e. a monolayer, then in the case of olive oil one dimension of its molecule is 2 nm.

(*b*) *Predictions from kinetic theory of gases.* Information about the mole-cular world can sometimes be obtained from observations of the behaviour of matter in bulk, i.e. from macroscopic observations. Thus with the help of the kinetic theory of gases, expressions can be derived relating such pro-perties as rate of diffusion with the size of the gas molecules involved.

(*c*) *Using the Avogadro constant.* Consider copper which has atomic mass 64 and density 9.0 g cm^{-3}. One mole of copper atoms, therefore, has mass 64 g and volume $64/9$ cm^3; it contains 6.0×10^{23} atoms. The volume available to each atom is $64/(9 \times 6 \times 10^{23})$ cm^3 and the radius *r* of a sphere having this volume is given by

$$\frac{4}{3}\pi r^3 = \frac{64}{9 \times 6 \times 10^{23}}$$

$$\therefore \quad r = 0.14 \times 10^{-7} \text{ cm}$$

$$= 0.14 \times 10^{-9} \text{ m}$$

$$= 0.14 \text{ nm}$$

If copper atoms are spherical, would their radius be larger or smaller than this even if they were packed tightly? Why? A more accurate way of calcu-lating the size of a copper atom is indicated in questions 10 to 13 on pp. 27–8.

A word of caution is necessary regarding atomic dimensions. Nowadays atoms and molecules are no longer pictured as having hard, definite surfaces like a ball and there is, therefore, little point in trying to give their diameters too exact values; most are within the range 0.1 to 0.5 nm. Also, although we shall usually treat atoms and molecules as spheres, it is necessary on occasion to consider them as having other shapes.

Periodic Table

With the passage of time the early nineteenth-century picture of an indivisible atom came to be doubted in the light of fresh information. During the 1860s chemical knowledge increased sufficiently for it to be clear that there were elements with similar chemical properties. Moreover, atomic masses were being established with greater certainty and attempts were made to relate properties and atomic masses.

It was found that if the elements were arranged in order of increasing atomic masses then, at certain repeating intervals, elements occurred with similar chemical properties. Sometimes it was necessary to place an element of larger atomic mass before one of slightly smaller atomic mass to preserve the pattern. The first eighteen elements of this arrangement, called the Periodic Table, are shown in Table 1.1. The third and eleventh (3 + 8) elements are the alkali metals lithium and sodium; the ninth and seventeenth (9 + 8) are the halogens fluorine and chlorine—here the repeating interval is eight. The serial number of an element in the table is called its *atomic number*.

Table 1.1

Group 1	Group 2	Group 3	Group 4	Group 5	Group 6	Group 7	Group 0
1 Hydrogen							2 Helium
3 Lithium	4 Beryllium	5 Boron	6 Carbon	7 Nitrogen	8 Oxygen	9 Fluorine	10 Neon
11 Sodium	12 Magnesium	13 Aluminium	14 Silicon	15 Phosphorus	16 Sulphur	17 Chlorine	18 Argon

The Periodic Table suggests that the atoms of the elements may not be simple entities but are somehow related. There must be similarities between the atoms of similar elements and it would seem that the similarity might be due to the way they are built up.

We now believe that atoms are composed of three types of particles—protons, neutrons and electrons. (Many other subatomic particles, such as positrons, mesons and antiprotons, are known but most are short-lived and are not primary components.) Protons and neutrons are packed together into a very small nucleus which is surrounded by a cloud of electrons, the diameter of the atom as a whole being at least 10 000 times greater than that of the nucleus. The comparative masses and charges of the three basic particles are given in Table 1.2. The nucleus is positively charged and the electron cloud negatively charged but the number of protons equals the number of electrons so that the atom is electrically neutral.

Table 1.2

Particle	Mass	Charge
electron	1	$-e$
proton	1836	$+e$
neutron	1839	0

e = electronic charge

The number of protons in the nucleus of an atom has been found to be the same as its atomic number which therefore means that each element in the Periodic Table has one more proton and one more electron in its atom than the previous element. Hydrogen, the first element, has one proton and one electron. Helium, the second element, has two protons and two electrons. Lithium, with atomic number three, has three protons and three electrons. Neutrons are present in all nuclei except that of hydrogen.

Interatomic bonds

Materials consist of atoms held together by the attractive forces they exert on each other. These forces are electrical in nature and create interatomic bonds of various types. The type formed in any case depends on the outer electrons in the electron clouds of the atoms involved.

(a) *Ionic bond*. This is formed between the atoms of elements at opposite sides of the Periodic Table, for example between sodium (Group 1) and chlorine (Group 7) when they are brought together to form common salt. A sodium atom has a loosely held outer electron which is readily accepted by a chlorine atom. The sodium atom becomes a positive ion, i.e. an atom deficient of an electron, and the chlorine atom becomes a negative ion, i.e. an atom with a surplus electron. The two ions are then bonded by the electrostatic attraction between their unlike charges.

A sodium ion attracts *all* neighbouring chloride ions in other pairs of bonded ions and vice versa. Each ion becomes surrounded by ions of opposite sign and the resulting structure depends among other things on the relative sizes of the two kinds of ion.

The ionic bond is strong. Ionic compounds are usually solid at room temperature and have high melting points. They are good electrical insulators in the solid state since the electrons are nearly all firmly bound to particular ions and few are available for conduction.

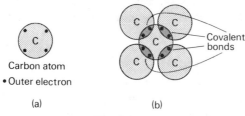

Carbon atom
• Outer electron

(a)

(b)

Fig. 1.4

(*b*) *Covalent bond.* In ionic bonding electron *transfer* occurs from one atom to another. In covalent bonding electron *sharing* occurs between two or more atoms. Thus the atoms of carbon can form covalent bonds with other carbon atoms. Each carbon atom has four outer electrons, Fig. 1.4*a*, and all can be shared with four other carbon atoms to make four bonds, Fig. 1.4*b*, each consisting of two interlocking electron clouds.

Covalent bonds are also strong and many covalent compounds have similar mechanical properties to ionic compounds. However, unlike the latter, they do not conduct electricity when molten.

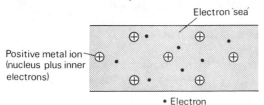

• Electron

Fig. 1.5

(*c*) *Metallic bond.* Metal atoms have one or two outer electrons that are in general loosely held and are readily lost. In a metal we picture many free electrons drifting around randomly, not attached to any particular atom as they are in covalent bonding. All atoms share *all* the free electrons. The atoms thus exist as positive ions in a 'sea' of free electrons, Fig. 1.5; the strong attraction between the ions and electrons constitutes the metallic bond.

The nature of the metallic bond has a profound influence on the various properties of metals, as we shall see later.

(*d*) *van der Waals bond*. van der Waals forces are very weak and are present in all atoms and molecules. They arise because, although the centres of negative and positive charges in an atom coincide over a period of time, they do not coincide at any instant—for reasons too advanced to be considered here. There is a little more of the electron cloud on one side of the nucleus than the other. A weak electric 'dipole' is produced giving rise to an attractive force between opposite ends of such dipoles in neighbouring atoms.

The condensation and solidification at low temperatures of oxygen, hydrogen and other gases is caused by van der Waals forces binding their molecules together. (In the molecules of such gases the atoms are held together covalently.) van der Waals forces are also important when considering polymers (p. 25).

Two further points: first, sometimes more than one of the previous four types of bonding is involved in a given case; second, information about the strength of interatomic bonds in solids is obtained from heat of sublimation measurements in which solid is converted directly to vapour and all atoms separated from their neighbours (see question 8, p. 27).

States of matter

The existence of three states or phases of matter is due to a struggle between interatomic (intermolecular) forces and the motion which atoms (molecules) have because of their internal energy (see p. 115).

(*a*) *Solids*. In the four types of interatomic force just discussed only attractions were considered but there must also be interatomic repulsions, otherwise matter would collapse. Evidence, both theoretical and experimental, suggests that at distances greater than one atomic diameter the

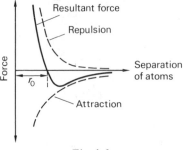

Fig. 1.6

attractive force exceeds the repulsive one, whilst for small distances, i.e. less than one atomic diameter, the reverse is true. In Fig. 1.6 the dotted graphs show how the *short-range attractive* force and the *very short-range repulsive* force between two atoms vary with the separation of the atoms; the total or resultant force is shown by the continuous graph.

It can be seen that for one value of the separation r_0, the resultant interatomic force is zero. This is the situation that normally exists in a solid, but if the atoms come closer together—for example, when the solid is compressed—they repel each other; they attract when they are pulled farther apart. We have only considered two atoms whereas in a solid each atom has interactions with many of its close neighbours. The conclusion about the existence of an equilibrium separation, however, will still hold good.

In an ionic bond the short-range attractive part of the interatomic force arises from the attraction between positive and negative ions which pulls them together until their electron clouds start to overlap, thus creating a very short-range repulsive force. The attractive and repulsive forces in the other types of bond also arise from the electric charges in atoms.

Now consider the motion of the atoms, the other contestant. In a solid the atoms vibrate about their equilibrium positions alternately attracting and repelling one another but the interatomic forces have the upper hand. The atoms are more or less locked in position and so solids have shape and appreciable stiffness.

(b) *Liquids.* As the temperature is increased the atoms have larger amplitudes of vibration and eventually they are able partly to overcome the interatomic forces of their immediate neighbours. For short spells they are within range of the forces exerted by other atoms not quite so near. There is less order and the solid melts. The atoms and molecules of a liquid are not much farther apart than in a solid but they have greater speeds, due to the increased temperature, and move randomly in the liquid while continuing to vibrate. The difference between solids and liquids involves a difference of structure rather than a difference of distance between atoms and molecules.

Although the forces between the molecules in a liquid do not enable it to have a definite shape, they must still exist otherwise the liquid would not hold together or exhibit surface tension (i.e. behave as if it had a skin on its surface) and viscosity nor would it have latent heat of vaporization.

(c) *Gases.* In a gas or vapour the atoms and molecules move randomly with high speeds through all the space available and are now comparatively far apart. On average their spacing at s.t.p. is about 10 molecular diameters and their mean free path (i.e. the distance travelled between collisions) is roughly 300 molecular diameters. Molecular interaction only occurs for those brief spells when molecules collide and large repulsive forces operate between them.

Conditions in gases and solids are, by comparison, simpler than those in liquids and in general their behaviour is better understood.

Types of solids

There are two main types of solids.

(*a*) *Crystalline*. Most solids, including all metals and many minerals, are crystalline. In substances such as sugar the crystal form is evident but less so in the case of metals, although large crystals of zinc are often visible on a freshly galvanized iron surface.

The crystalline structure of a metal can be revealed by polishing the surface, treating it with an etching chemical, sometimes a dilute acid, and then viewing it under an optical microscope. The metal is seen to consist of a mass of tiny crystals, called grains, at various angles to one another; it is said to be *polycrystalline*. Grain sizes are generally small, often about 0.25 mm across. Fig. 1.7 shows crystal grains in a cross-section of an aluminium–copper casting. The grains show up on the surface after etching

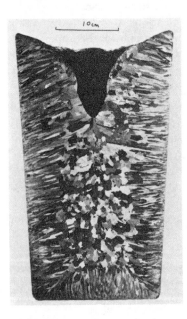

Fig. 1.7

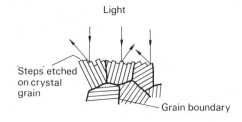

Fig. 1.8

because 'steps' are formed on each grain due to the rate of chemical action differing with different grain orientations. Light is then reflected in various directions by the different grains so that some appear light and others dark, Fig. 1.8.

Grain boundaries are revealed at the atomic level by the field ion microscope, Fig. 1.9, which uses beams of helium ions to 'illuminate' the object rather than electrons or light. The field ion micrograph in Fig. 1.10 shows the tip of an iridium needle, viewed from the point and magnified 2 500 000

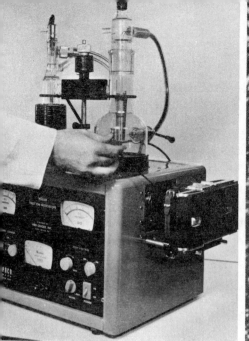

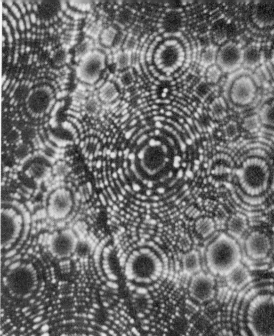

Fig. 1.9   Fig. 1.10

times. Each bright spot represents an iridium atom and the abrupt pattern change is clearly visible at the grain boundary.

The essence of the structure of a crystal, whether it be a large single crystal or a tiny grain in a polycrystalline specimen, is that the arrangement of atoms, ions or molecules repeats itself regularly many times, i.e. there is a long-range order.

(b) *Amorphous*. Here the particles are assembled in a more disordered way and only show order over short distances; there is no long-range order. The structure of an amorphous solid has been likened to that of an instantaneous photograph of a liquid. It is much more difficult to unravel but is the subject of considerable research. The many types of glass are the commonest of the amorphous solids; we can think of them as having a structure of groups of atoms (e.g. of silicon and oxygen) that would have been crystalline had it not been distorted.

Crystal structures

The structure adopted by a crystalline solid depends on various factors including the kind of bond(s) formed and the size and shape of the particles involved. For example, in metals where all the positive ions attract all electrons (p. 11), the bonding pulls equally in all directions, i.e. is non-directional, and every ion tends to surround itself by as many other ions as is geometrically possible. A close-packed structure results. On the other

15

hand, in covalent solids the bonding is directional, i.e. every shared electron, is localized between only two atoms. This does not encourage close-packing since the number of atoms immediately surrounding each atom is limited to the number of covalent bonds it forms. Would you describe the ionic bond as directional or non-directional?

Some common crystal structures will now be described.

(a) *Face-centred cubic* (FCC) packing is shown in Fig. 1.11a; there is a particle at the centre of each of the six faces of the cube in addition to the eight at the corners. Copper and aluminium have this structure. The sodium

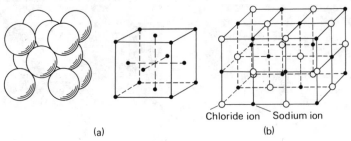

Chloride ion   Sodium ion

(a)   (b)

Fig. 1.11

crystal can be regarded as two interpenetrating FCC structures, one of sodium ions and the other of chloride ions, Fig. 1.11b; each sodium ion is surrounded by six chloride ions and vice versa.

(b) *Hexagonal close-packing* (HCP) is represented in Fig. 1.12; it is built up from layers of hexagons. Zinc and magnesium form HCP crystals.

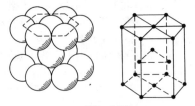

Fig. 1.12

These two structures give the closest possible packing and account for 60% of all metals. They are not very dissimilar if we consider how they can be assembled from successive layers. Fig. 1.13 shows a layer A of hexagonal close-packed spheres in which each sphere touches six others—this is the closest packing possible for spheres. A second hexagonal close-packed layer B can be placed on top and the packing of these two layers will be closest when the spheres of B sit in the hollows formed by three neighbouring spheres of A. A third hexagonal close-packed layer can be placed on top of B in two ways. If it rests in the hollows of B so that its spheres are

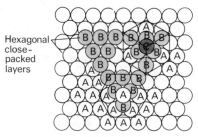

Hexagonal close-packed layers

Fig. 1.13

directly above the *spheres* in A, then an HCP crystal results, Fig. 1.13 (bottom), and the layer stacking is ABAB. However, if the third layer rests in other hollows in B, its spheres can be above *hollows* in A and the structure is FCC with layer stacking ABCABC.

(*c*) *Body-centred cubic* (BCC) packing has a particle at the centre of the cube and one at each corner, Fig. 1.14. Alkali metals have this less closely packed structure.

Fig. 1.14

(*d*) *Tetrahedral* structures have a particle at the centre of a regular tetrahedron and one at each of the four corners, Fig. 1.15*a*. This more open arrangement is found in carbon (as diamond), silicon and germanium—all substances which form covalent bonds. The hardness of diamond is partly due to the fact that its atoms are not in layers and so cannot slide over each other as they can in graphite, the other crystalline form of carbon. Graphite forms layers of six-membered rings of carbon atoms that are about two-and-a-half times farther apart than are the carbon atoms in the layers, Fig. 1.15*b*. The forces between the layers are weak, thus explaining why

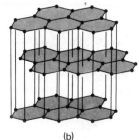

(a)   (b)

Fig. 1.15

graphite flakes easily and is soft and suitable for use in pencils and as a lubricant. Graphite and diamond provide a good example of the importance of structure in determining properties.

Two further points: first, there is in every crystal structure a typical cell, called the *unit cell*, which is repeated over and over again—Figs. 1.11a, 1.12 and 1.14 are examples of unit cells; second, the structures described are those of perfect crystals. In practice there are imperfections in crystals and these are important in determining the properties of a material, as we shall see later.

Bubble raft

Soap bubbles pack together in an orderly manner and provide a good representation, in two dimensions, of how atoms are packed in a crystal.

A bubble raft is made by attaching a glass jet (about 1 mm bore) or a 25 gauge hypodermic needle on a 1 cm^3 syringe barrel, to the gas tap, via a length of rubber tubing and a screw clip. The jet is held below

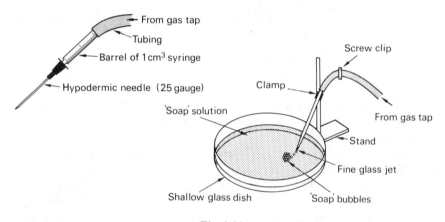

Fig. 1.16a

the surface of a 'soap' solution (1 Teepol, 8 glycerol and 32 water is satisfactory) in a shallow glass dish, at a constant depth which gives bubbles of about 2 mm diameter, Fig. 1.16a. If the dish is placed on an overhead projector a magnified image of the raft can be viewed. What pulls the bubbles together and what keeps them from getting too close?

A perfect, hexagonally close-packed array is shown in Fig. 1.16b. Grain boundaries are readily seen in Fig. 1.16c, 'vacancies' in Fig. 1.16d and a bubble of a different size in Fig. 1.16e. What might (d) and (e) represent in a real crystal and what effect do they have on the structure?

Fig. 1.16b

Fig. 1.16c

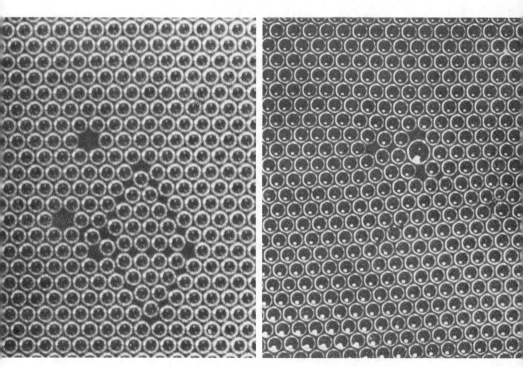

Fig. 1.16d

Fig. 1.16e

X-ray crystallography

Before the discovery which enabled the arrangement of atoms in a crystal to be determined experimentally, crystallographers had simply assumed that the regular external shapes of crystals were due to the atoms being arranged in regular, repeating patterns.

In 1912 three German physicists, Max von Laue, W. Friedrich and P. Knipping, found that a beam of X-rays, on passing through a crystal, formed a pattern of spots on a photographic plate (see Fig. 1.19b opposite). Shortly afterwards W. L. (later Sir Lawrence) Bragg and his father Sir William Bragg showed how the pattern could be used to reveal the positions of the atoms in a crystal. Together they proceeded to unravel the atomic structures of many substances and started the science of X-ray crystallography. In recent years the structures of many complex organic molecules, including some like DNA (deoxyribonucleic acid) that play a vital part in the life process, have been discovered by this technique.

The analysis of crystal structures by X-rays depends on the fact that X-rays, like light, have a wave-like nature and when they fall on a crystal they are scattered by the atoms so that in some directions the scattered beams reinforce each other while in others they cancel each other. X-rays are used because their wavelengths are of the same order as the atomic spacings in crystals—about 10 000 times less than those of light.

The simplest way of regarding what occurs when X-rays fall on a crystal is to consider the crystal as made up of regularly spaced layers of atoms each of which produces a weak 'reflected' beam such that the angle of incidence equals the angle of reflection, as for the reflection of light by a

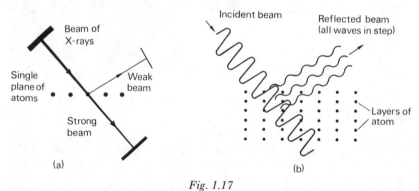

Fig. 1.17

mirror. In Fig. 1.17a reflection by a single layer of atoms is shown: most of the beam passes through. Fig. 1.17b shows a beam of X-rays falling on a set of parallel layers of atoms. If all the reflected waves are to combine to produce a strong reflected beam (and give an intense spot on a photograph) then they must all emerge in step. For this to happen the path difference between successive layers must be a whole number of wavelengths of the X-rays—as they are in this case. Otherwise crests of one wave may coincide with troughs from another and the two tend to cancel out.

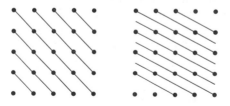

Fig. 1.18

The atoms in a crystal can be considered as arranged in several different sets of parallel planes, from all of which strong reflections may be obtained to give a pattern of spots characteristic of the particular structure. Two other possible sets of planes are shown for the array in Fig. 1.18. In a poly-crystalline sample many planes are involved at once and thousands of spots are produced resulting in circles or circular arcs on the photograph. Fig. 1.19a is due to a polycrystalline sample of gold and Fig. 1.19b to a single crystal of the same material.

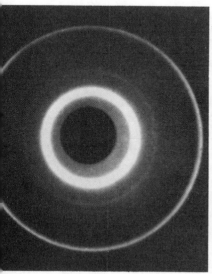

Fig. 1.19a   Fig. 1.19b

Microwave analogue

Microwaves are very short radio waves with wavelengths extending from 1 cm or so to about 1 m and are used for radar and satellite communication. A large-scale demonstration with microwaves shows that regularly spaced polystyrene tiles or layers of spheres give strong reflections at certain angles, depending on the tile or layer separation and that this can be explained as being due to interference between waves reflected from successive layers— as it can with X-rays and layers of atoms.

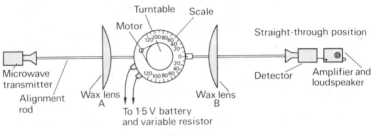

Fig. 1.20a

The apparatus is shown in Fig. 1.20a; wax lens A produces a parallel beam from the 3 cm microwave transmitter and B focuses it on the detector. First it should be shown, with the transmitter and detector in line, that microwaves can largely penetrate a polystyrene ceiling tile but not a metal sheet.

(a) *One tile.* A single tile held vertically on the turntable reflects (partially) the beam at *any* angle. Through what angle from the straight-through position must the detector be turned when the *glancing* angle (i.e. the angle between the tile and the incident beam) is θ (see pp. 149–50)?

(b) *Two tiles.* When a second tile is brought up behind and parallel to the first (already positioned for the reflected beam to be detected), the intensity of the reflection rises and falls as the extra path to the second tile varies between an even and an odd number of half-wavelengths. Interference is occurring between the waves reflected from each tile when they come together. Will this occur for all glancing angles?

(c) *Ten tiles at 3 cm centre-to-centre spacing*, Fig. 1.20b. If the detector is swung round as the array of tiles revolves on the turntable, a strong reflection is obtained only when the detector makes an angle of 60° with the straight-through position. The tiles then bisect the angle between the detector and the straight-through position, i.e. they make an angle of 30° with the straight-through position, and give a glancing angle of 30°, Fig. 1.20c.

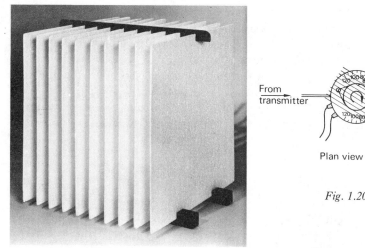

Fig. 1.20b

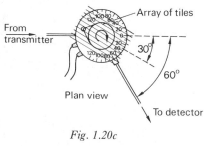

Fig. 1.20c

(d) *Polystyrene ball 'crystal'*. This is made from seven hexagonal close-packed layers of 5 cm diameter spheres glued together to give a face-centred cubic (FCC) structure (see Appendix 6, p. 402). As the crystal rotates on the turntable, Fig. 1.20d, *two pairs* of strong reflections are obtained per revolution when the detector is at 44° from the straight-through position ($\theta = 22°$), the time between each pair being greater than that between the two signals in each. At 50° ($\theta = 25°$) there are *two single* strong reflections per revolution and similarly at 74° ($\theta = 37°$).

Fig. 1.20d

The first pair of 44° reflections are produced when two easily identifiable sets of vertical 'hexagonal' layers (i.e. with the packing of spheres in each layer the closest possible) bisect *in turn* the angle between the straight-through position and the detector; the second pair arises half a revolution later when the 'backs' of the two sets of the same layers are in the bisecting position. At 50°, a set of 'square' layers (i.e. with less closely packed spheres —see question 9, p. 27) is responsible for the strong signals when, twice in each revolution, it bisects the angle between the beam and the detector.

Polymers

Polymers are materials with giant molecules each containing anything from 1000 to 100 000 atoms and are usually carbon (organic) compounds. An example of a natural polymer is cellulose whose long, tough fibres give strength and stiffness to the roots, stems and leaves of plants and trees. Rubber, wool, proteins, resins and silk are others. Man-made polymers include plastics such as polythene, Perspex and polystyrene, fibres like nylon and Terylene, synthetic rubbers and the epoxy resins which are well known for their strong bonding properties and toughness.

The unravelling of the intricacies of nature's polymers required X-ray apparatus, the electron microscope and other instruments. Their molecules were each found to consist of a large number of repeating units, called *monomers*, arranged in a long flexible chain. Thus every molecule of cellulose comprises a long chain of from a few hundred to several thousand glucose sugar ($C_6H_{12}O_6$) molecules.

Artificial polymers are made by a chemical reaction known as *polymerization*, in which large numbers of small molecules join together to form a large one. Polythene or polyethylene (to give it its full name) is made by polymerizing ethylene (C_2H_4), a gas obtained when petroleum is 'cracked'. In one process the ethylene molecules, heated to 100 °C–300 °C under a pressure several thousand times greater than atmospheric, link with one another to give the long chain molecules of polythene, Fig. 1.21.

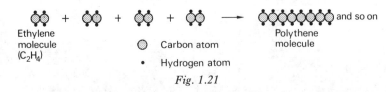

Ethylene molecule (C_2H_4)

◎ Carbon atom

• Hydrogen atom

Polythene molecule

Fig. 1.21

If the chains run parallel to each other, like wires in a cable, the structure shows a certain amount of order and is said to be 'crystalline', Fig. 1.22*a*. This contrasts with the disorder of tangled chains in an 'amorphous' structure, Fig. 1.22*b*. Many polymers have both crystalline and amorphous

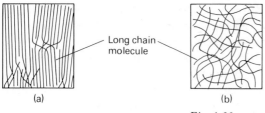

(a)   (b)   (c)

Fig. 1.22

regions, Fig. 1.22*c*. If crystallinity predominates an X-ray photograph shows sharp spots (but the pattern is never as sharp as for wholly crystalline materials) and the polymer is fairly strong and rigid. A polymer with a largely amorphous structure is soft and flexible and gives diffuse rings in an X-ray photograph. The proportion of crystalline to amorphous regions in a polymer depends on its chemical composition, molecular arrangement and how it has been processed. The intermolecular forces between chains are of the weak van der Waals type, but in crystalline structures the chains are close together over comparatively large distances and so the total effect of these forces is to produce a stiff material.

Crystallization is one of two principles that have been used to produce strong, stiff polymers (e.g. polythene, nylon); the other is the formation of strong covalent bonds between chains—a process known as cross-linking. In vulcanizing raw rubber, i.e. heating it with a controlled amount of sulphur, a certain number of sulphur atoms form cross-links between adjacent rubber molecules to give a more solid material than raw rubber which is too soft for use, Fig. 1.23. As more cross-links are added to rubber it stiffens and ultimately becomes the hard material called ebonite.

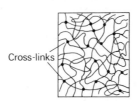

Cross-links

Fig. 1.23

Polymers such as ebonite and bakelite (the first plastic to be made) with many strong cross-links do not soften with increased temperature but set once and for all after their initial moulding. They are called *thermosetting plastics* and remain comparatively strong until excessive heating leads to breakdown of the cross-links and chemical decomposition. By contrast in *thermoplastic* polymers only the weak van der Waals forces hold the chains together and these materials can be softened by heating and if necessary remoulded. On cooling they recover their original properties and retain any new shape. This treatment can be repeated almost indefinitely so long as temperatures are below those causing decomposition, i.e. breakdown of the covalent bonds that hold together the atoms in the long chain.

The possibility of using man-made polymers in the future as load-bearing structural materials for houses, buildings, cars, boats and aircraft depends largely on how far their strength and stiffness can be increased. As we have seen, two methods are used to do this at present. One is by having 'crystallized' long chains—a physical feature—and the other requires cross-links to be formed between chains—a chemical feature. Current research is directed towards producing molecular chains which are themselves stiff (most existing man-made polymer chains are inherently flexible) by polymerizing monomers which have a ring-shaped structure. It is then hoped to achieve the desired strength and stiffness by crystallizing and cross-linking those chains.

QUESTIONS

1. Experiment shows that 3 g of carbon combine with 8 g of oxygen to form 11 g of carbon dioxide. If 1 atom of carbon reacts with 2 atoms of oxygen to give 1 molecule of carbon dioxide (i.e. $C + O_2 = CO_2$)

(a) compare the mass of an oxygen atom with that of a carbon atom. (*Hint :* start by supposing that 1 g of C contains x atoms.)

(b) what is the atomic mass of oxygen on the carbon 12 scale?

(c) what mass of oxygen contains the same number of atoms as 12 g of C?

2. (a) If the atomic mass of nitrogen is 14, what mass of nitrogen contains the same number of atoms as 12 g of carbon?

(b) What mass of chlorine contains the same number of atoms as 32 g of oxygen? (Atomic masses of chlorine and oxygen are 35.5 and 16 respectively.)

3. Taking the value of the Avogadro constant as 6.0×10^{23} how many atoms are there in (a) 14 g of iron (at. mass 56), (b) 81 g of aluminium (at. mass 27), (c) 6.0 g of carbon (at. mass 12)?

4. What is the mass of (a) one atom of magnesium (at. mass 24), (b) three atoms of uranium (at. mass 238), (c) one molecule of water (mol. mass 18)? Take the value of the Avogadro constant as 6.0×10^{23}.

5. (a) A *mole* is the name given to the *quantity of substance* which contains a certain number of particles. What is this number?

(b) What is the mass of 1 mole of hydrogen molecules?

(c) If the density of hydrogen at s.t.p. is 9.0×10^{-5} g cm^{-3} what volume does 1 mole of hydrogen molecules occupy at s.t.p.?

(d) How many molecules are there in 1 cm^3 of hydrogen at s.t.p.?

6. By counting scintillations it is found that 1.00 mg of polonium in decaying completely emits approximately 2.90×10^{18} alpha particles. If one particle is emitted by each atom and the atomic mass of polonium is 210, what is the Avogadro constant?

7. Estimate (a) the mass and (b) the diameter of a water molecule (assumed spherical) if water has molecular mass 18 and the Avogadro constant is 6.0×10^{23} per mole.

8. (a) Suggest an *approximate* but reasonable value for the heat of sublimation of copper (in $J g^{-1}$) from the following data.

$$\text{Specific latent heat of fusion} = 2.0 \times 10^2 \text{ J g}^{-1}$$
$$\text{Specific latent heat of vaporization} = 4.8 \times 10^3 \text{ J g}^{-1}$$

What additional information would enable a better estimate to be made?

(b) If the Avogadro constant is 6.0×10^{23} per mole and the atomic mass of copper is 64, what is the heat of sublimation of copper in J/atom? Why is a knowledge of this quantity useful?

9. (a) 'Square' and 'hexagonal' methods of packing spheres are shown in Figs. 1.24a and b respectively. How many other spheres are touched by (i) A, (ii) B? In which arrangement is the packing closest?

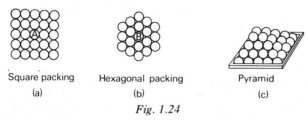

Square packing   Hexagonal packing   Pyramid

(a)   (b)   (c)

Fig. 1.24

(b) Fig. 1.24c is a pyramid of spheres in which the second and successive layers are formed by placing balls in the hollows of the layer below it. How are the balls packed in (i) the sloping sides of the pyramid, (ii) the horizontal layers?

Fig. 1.25

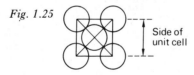

Side of unit cell

10. One face of the unit cell of an FCC crystal is shown in Fig. 1.25, atoms being represented by spheres. If r is the atomic radius in cm, calculate (a) the length of a side of the unit cell, (b) the volume of a unit cell, (c) the number of unit cells in 1 cm³.

11. In a crystal built up from a large number of similar unit cells the atoms at the corners and on the faces of individual cells are shared with neighbouring cells. In an FCC unit cell

(a) how many corner atoms are there?

(b) how many neighbouring cells share each corner atom?

(c) what is the effective number of corner atoms per cell?

(d) how many *face* atoms are there?

(e) how many neighbouring cells share each face atom?

(f) what is the effective number of face atoms per cell?

(g) what is the total effective number of atoms per cell?

(h) what is the total effective number of atoms in 1 cm³ of unit cells (use your answer from 10(c)).

12. X-ray diffraction shows that copper, atomic mass 64 and density 9.0 g cm⁻³, has an FCC structure. If the Avogadro constant is 6.0×10^{23} per mole, how many atoms are there in 1.0 cm³ of copper?

13. Using your answers to 11(*h*) and 12, calculate the atomic radius of copper.

14. For crystalline sodium chloride, draw the unit cell which is repeated throughout the lattice. Label precisely the two kinds of particle at the lattice sites. What are the forces maintaining them in their relative positions?

Calculate the distance between adjacent particles in crystalline sodium chloride, given that its formula weight is 58.5 and its density is 2.16 g cm⁻³ (2.16×10^3 kg m⁻³). (The Avogadro constant $= 6.03 \times 10^{23}$ mole⁻¹.)

Discuss the effect of a small stress on a crystalline lattice. (*C. Phys. Sc.*)

2 Mechanical properties

Stress and strain

The mechanical properties of a material are concerned with its behaviour under the action of external forces—a matter of importance to engineers when selecting a material for a particular job. Four important mechanical properties are strength, stiffness, ductility and toughness.

Strength deals with how great an applied force a material can withstand before breaking. *Stiffness* tells us about the opposition a material sets up to being distorted by having its shape or size, or both, changed. A stiff material is not very flexible. There is no such thing as a perfectly stiff or rigid (unyielding) material; all 'give' in some degree although the deformation may often be very small. *Ductility* or workability relates to the ability of the material to be hammered, pressed, bent, rolled, cut or stretched into useful shapes. A *tough* material is one which is not brittle, i.e. it does not crack readily. Steel has all four properties, putty has none of them. Glass is strong and stiff but not tough or ductile. Which properties would you ascribe to rubber, nylon and diamond?

Fig. 2.1   *Fig. 2.2*

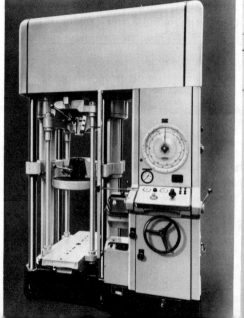
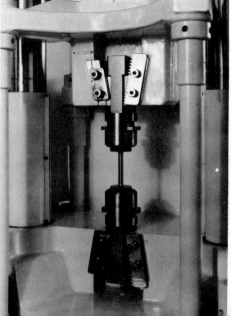

MECHANICAL PROPERTIES

Information about mechanical properties may be obtained by observing the behaviour of a wire or strip of material when it is stretched. The stretching of short rods or 'test-pieces' is done using a machine like that in Figs. 2.1 and 2.2.

The extension produced in a sample of material depends on (*i*) the nature of the material, (*ii*) the stretching force, (*iii*) the cross-section area of the sample and (*iv*) its original length. What effect would you expect (*iii*) and (*iv*) to have? To enable fair comparisons to be made between samples having different sizes the terms stress and strain are used when referring to the deforming force and the deformation it produces. *Stress is the force acting on unit cross-section area* and for a force F and area A it equals F/A, Fig. 2.3. The unit of stress is the *pascal* (Pa) which equals one newton per square metre (N m^{-2}). *Strain is the extension of unit length.* If e = extension and l = original length, the strain is e/l. Strain is a ratio and has no units. A stress which causes an increase of length puts the sample in tension, and so we talk about a tensile stress and a tensile strain.

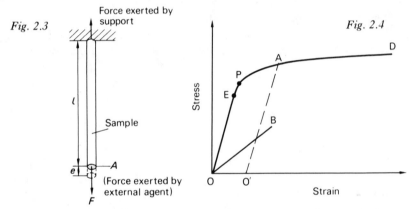

Fig. 2.3　　　　　　　　　　　　　　　　　　　　　　　　*Fig. 2.4*

The shape of the stress–strain graph for the stretching of a sample (e.g. a wire) depends not only on the material but also on its previous treatment and method of manufacture. For a ductile material, i.e. a metal, it has the *general* form shown by OEPAD in Fig. 2.4. There are two main parts.

(*a*) *Elastic deformation.* The first part of the graph from O to E is a straight line through the origin showing that strain is directly proportional to stress, i.e. doubling the stress doubles the strain. Over this range the material suffers elastic deformation, i.e. it returns to its original length when the stress is removed and none of the extension remains.

(*b*) *Plastic deformation.* As the stress is increased the graph becomes non-linear but the deformation remains elastic until at a certain stress corresponding to point P, and called the *yield point*, permanent or plastic deformation starts. Henceforth the material behaves rather like Plasticine and retains some of its extension if the stress is removed. Recovery is no longer complete and on reducing the stress at A, for example, the specimen recovers along AO′ where AO′ is almost parallel to OE. OO′ is the permanent plastic extension produced. If the stress is reapplied, the curve O′AD is followed. At D the specimen develops one or more 'waists' and *ductile fracture* occurs at one of them. The stress at D is the greatest the material can bear and is called the *breaking stress* or *ultimate tensile strength*; it is a useful measure of the strength of a material.

The specimen appears to 'give' at P, and over the plastic region a given stress increase produces a greater increase of strain than previously. None the less it still opposes deformation and any increase of strain requires increased stress. Beyond P the material is said to *work-harden* or *strain-harden*.

Few ductile materials behave elastically for strains as great as $\frac{1}{2}\%$ (i.e. for an extension of $\frac{1}{2}$ cm in a 100-cm-long wire) but they may bear large plastic strains, up to 50%, before fracture. The breaking stress for steels may occur at stresses half as great again as that at the yield point, while for very ductile metals with low yield points it may be several times greater than the stress at the yield point. It is desirable that metals used in engineering structures should carry loads which only deform them elastically. On the other hand the fabrication of metals into objects of various shapes requires them to withstand considerable plastic deformation before fracture, i.e. be very ductile. The dominant position of metals in modern technology arises from their strength and ductility.

A brittle material such as a glass may give a curve like OB in Fig. 2.4 and fractures almost immediately after the elastic stage; little or no plastic deformation occurs and the glass is non-ductile.

Young's modulus

The stress–strain curve for the stretching of metals and some other materials (e.g. glasses), over almost all the elastic region, is a straight line through the origin. That is, *tensile strain is directly proportional to tensile stress during elastic deformation.* This statement is known as Hooke's law and in more elementary work it is often stated in the form: extension varies as the load. In mathematical terms it can be written

$$tensile\ strain \propto tensile\ stress$$

or $$\frac{tensile\ stress}{tensile\ strain} = a\ constant$$

This constant is called *Young's modulus* and is denoted by E. Its value depends on the nature of the material and *not* on the dimensions of the sample. If a material has large E, it resists elastic deformation strongly and a large stress is required to produce a small strain. E is thus a measure of the opposition of a material to change of length strains such as occur when a wire or rod is stretched elastically, i.e. it measures *elastic stiffness*.

Young's modulus is of great importance in engineering. In the early days of iron railway bridge construction, engineers relied heavily on 'rule of thumb' methods. It required a series of disasters like that of the Tay Bridge in Scotland in 1879 and a reputed collapse rate of twenty-five bridges per year at about the same time in the U.S.A. before it was accepted that reliable strength calculations were necessary for safety and the economical use of materials. The value of E is one of the pieces of information which must be known to calculate accurately the deformation (deflections) that will occur in a loaded structure and its parts. When a beam bends, one surface is compressed and the other stretched, Fig. 2.5, so that E is involved; 'beam theory' is one of the foundation stones of engineering.

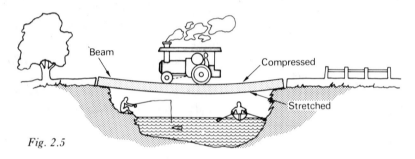

Fig. 2.5

If a stretching force F acting on a wire of cross-section area A and original length l causes an extension e we can write

$$E = \frac{tensile\ stress}{tensile\ strain} = \frac{F/A}{e/l} = \frac{Fl}{Ae}$$

Like stress, E is expressed in pascals (Pa) since strain is a ratio. Suppose a load of 1.5 kg attached to the end of a wire 3.0 m long of diameter 0.46 mm stretches it by 2.0 mm then $F = ma = mg = 1.5 \times 9.8$ N ($g = 9.8$ m s^{-2})

$$E = \frac{Fl}{Ae}$$

$$= \frac{(1.5 \times 9.8\ \text{N}) \times (3.0\ \text{m})}{(\pi \times 0.23^2 \times 10^{-6}\ \text{m}^2) \times (2.0 \times 10^{-3}\ \text{m})}$$

$$= \frac{1.5 \times 9.8 \times 3.0}{\pi \times 0.23^2 \times 10^{-6} \times 2.0 \times 10^{-3}} \frac{\text{N} \times \text{m}}{\text{m}^2 \times \text{m}}$$

$$= 1.3 \times 10^{11}\ \text{Pa}$$

Approximate values of E for some common materials are given in Table 2.1.

Table 2.1

Material	Young's modulus ($\times 10^{10}$ Pa)
steel	21
copper	13
glasses	7
polythene	about 0.5
rubber	about 0.005

Glasses are surprisingly stiff (and strong). The high elasticity of rubber (i.e. its ability to regain its original shape after a very large deformation) is not to be confused with its low elastic modulus. For steel a large stress gives a small strain while the same stress applied to rubber will give a very much larger strain. This means that by our scientific definition, steel has a greater modulus of elasticity than rubber.

Stretching experiments

1. Copper. Using an arrangement like that in Fig. 2.6 the extensions produced in a 2 metre length of copper wire (SWG 32) are found as it is loaded to breaking with 100 gram slotted weights.

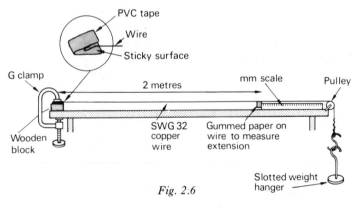

Fig. 2:6

A load–extension graph is plotted to see if Hooke's law is obeyed and to find the percentage strain copper can withstand before its elastic limit is exceeded.

The breaking stress of copper can be determined and, if wire of another gauge is used (say SWG 26), whether the breaking stress depends on wire thickness. The percentage plastic strain borne by copper before it breaks may also be found.

2. *Steel.* If *1* is repeated for 2 metres of steel wire (SWG 44) the same kind of information can be obtained.

3. *Rubber.* A load–extension graph can be plotted for a strip of rubber (5 cm long and 2 mm wide cut from a rubber band), suspended vertically and loaded with 100 g slotted weights and then unloaded. Information may be obtained about Hooke's law, and also the number of times its original length that rubber can be extended. The breaking stress of rubber should be found.

4. *Polythene.* A strip 15 cm long and 1 cm wide, cut *cleanly* from a piece of polythene, can be investigated as in *3*.

5. *Glass.* The breaking stress of glass may be found by hanging weights from a glass thread which has been freshly drawn from a length of 3 mm diameter soda glass rod. If parts of the rod are left at the top and bottom, the thread can be supported by clamping one end and a hook made at the other end for the weights.

6. *Young's modulus for a wire.* Using the apparatus of Fig. 2.7a or b the extensions of a wire can be measured with greater accuracy.

In Fig. 2.7a the right-hand wire is under test and carries a vernier scale (see Appendix 3, p. 396) which, when the right-hand wire is loaded, moves

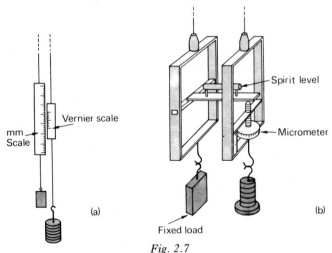

Fig. 2.7

over a millimetre scale attached to the left-hand wire and enables the extension to be measured. The alternative and more accurate arrangement in Fig. 2.7b is known as Searle's apparatus. In this case the micrometer screw (see Appendix 3, p. 397) is adjusted, after the addition of a load to the right-hand wire, so that the bubble of the spirit level is centralized. The extension is then found from the scale readings. By having two wires of the same material suspended from the same support, errors are eliminated if there is a change of temperature or if the support yields, since both wires will be affected equally.

Initially both wires should have loads that keep them taut and free from kinks. Readings are then taken as the load on the right-hand wire is increased by equal steps, without exceeding the elastic limit. The strain should therefore not be more than 0.1%, i.e. the wire should not be stretched much beyond 1/1000th of its original length. The length l of the wire to the top of the vernier or micrometer is measured with a metre rule and the diameter ($2r$) found at various points along its length with a micrometer screw gauge.

From a graph of 'load' against 'extension', an average value of 'load'/'extension' in kg mm^{-1} is given by PQ/OQ, Fig. 2.8. Young's modulus can then be calculated from

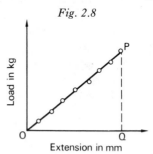

Fig. 2.8

$$E = \frac{Fl}{Ae}$$

where F/e is expressed in N m^{-1}, i.e. $F/e = \text{PQ} \times g/(\text{OQ} \times 10^{-3})$, $A(=\pi r^2)$ in m^2 and l in m.

Deformation and dislocations

The deformation behaviour of materials can be explained at the atomic level.

(a) *Elastic strain.* This is due to the stretching of the interatomic bonds that hold atoms together. The atoms are pulled apart very slightly, each is displaced a tiny distance from its equilibrium position and the material lengthens. Hooke's law is a result of the fact that the 'interatomic force–separation' graph, Fig. 2.9a, is a straight line for atomic separations close to the equilibrium separation r_0, Fig. 2.9b.

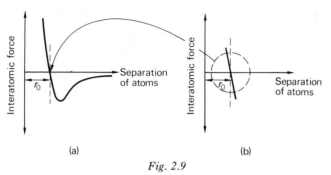

Fig. 2.9

35

Young's modulus E, the measure of elastic stiffness (i.e. resistance to elastic deformation), is high for materials with strong interatomic bonds. Covalent and ionic solids, and to a lesser extent metals, are in this category. Diamond (pure carbon), the hardest known natural substance, has a large number of very strong covalent bonds per unit volume and a very high value of E.

As well as determining the stiffness of a material, Young's modulus also governs ultimately, in theory, its strength since this too depends on the forces between atoms. However, we shall see later (pp. 40–2) that there are other factors which prevent solids displaying their theoretical strengths.

(b) *Plastic strain.* The ability to undergo plastic strain (and be ductile) is a property of crystalline materials. The yielding which occurs could therefore be attributed to the slipping of layers of atoms (or ions) over one another. With close-packed layers like those in Fig. 2.10a, the atoms would have to be moved farther apart, Fig. 2.10b. This would be resisted by the interatomic bonds, many of which will have to be broken simultaneously.

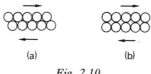

(a)   (b)

Fig. 2.10

Calculations based on the known strength of bonds show that the stresses needed to produce slip in this way are many times greater than those which actually cause plastic strain. The problem is, therefore, not so much to explain the strength of metals as their weakness. This led to a search for defects in crystal structures and in 1934 G. I. Taylor of Cambridge University proposed the *dislocation* as one such defect.

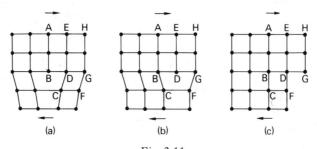

(a)   (b)   (c)

Fig. 2.11

The idea is that occasionally, due perhaps to growth faults during crystallization, there is an incomplete plane of atoms (or ions) in the crystal lattice, for example AB in Fig. 2.11a. We shall now see how the movement of the dislocation produces the same effect as a plane of atoms slipping over other planes, but much more easily.

If a stress is applied as shown by the arrows, atom B, whose bonds have already been weakened by the distortion of the structure, moves a small distance to the right and forms a bond with atom C, Fig. 2.11b. Plane of atoms DE is now incomplete. Then D flicks over and joins with F leaving GH as the incomplete plane, Fig. 2.11c. The result is just the same as if half-plane AB had slipped over the planes below BDG to the surface of the crystal. This process would have involved breaking a great many bonds at the same time. Instead, the dislocation, by moving a single line at a time has broken many fewer bonds and required a much smaller stress to do it. No atom has in fact moved more than a small fraction of the atomic spacing. Plastic deformation by this mechanism is clearly only possible in the well-ordered structure of a crystalline material.

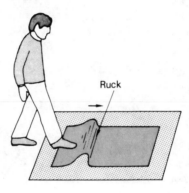

Ruck

Fig. 2.12

The passage of a dislocation in a crystal is like the movement of a ruck in a carpet. A greater force is needed to drag one carpet over another by pulling one end of it, than to make a ruck in the carpet and kick it along, Fig. 2.12. Calculations confirm that the stresses to make dislocations move in metals are in good agreement with their measured plastic flow stresses.

The first and most direct evidence for the existence of dislocations was obtained in 1956 by J. W. Menter, also of Cambridge University, using an electron microscope. An electron micrograph (an electron microscope photograph) is shown in Fig. 2.13 of an aluminium–copper alloy in which the planes of atoms are spaced about 0.20 nm apart. A dislocation can be seen; the extra plane of atoms ends in the white circle and distorts the arrangement of the surrounding planes.

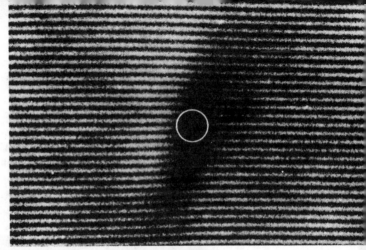

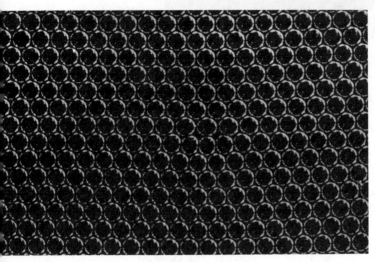

Fig. 2.13

Fig. 2.14

Dislocations can be obtained in a bubble raft (p. 18) and made to move if the raft is squeezed between two glass slides dipping into the 'soap' solution. There is one to be seen in Fig. 2.14.

Strengthening metals

Pure metals produced commercially are generally too weak or soft to be of much mechanical use—a rod of pure copper the thickness of a pencil is easily bent by hand. Their weakness can be attributed to the fact that they contain a moderate number of dislocations which can move about easily in the orderly crystal structure, thus allowing deformation under relatively small stresses. The traditional methods of making metals stronger and stiffer all involve obstructing dislocation movement by 'barriers', i.e. pockets of disorder in the lattice. Three barriers will be considered.

(a) *'Foreign' atoms.* In an alloy such as steel 'foreign' atoms (e.g. carbon) are introduced into the lattice of iron, disturbing its perfection and opposing dislocation motion. This makes for greater strength and stiffness.

(b) *Other dislocations.* One problem with the dislocation model is that when the dislocations have slipped out of the crystal, as in Fig. 2.11c, the crystal is then perfect and should have its theoretical strength. In general this is not observed and it would seem that further dislocations are generated whilst slip is occurring. This view can be justified in a more advanced treatment but here it is sufficient to say that as the metal is submitted to further stress, dislocations are created, move, meet and thereby obstruct each other's progress. A 'traffic jam' of dislocations builds up.

(c) *Grain boundaries.* In practice most metal samples are polycrystalline, i.e. consist of many small crystals or grains at different angles to each other (see Fig. 1.7, p. 14). The boundary between two grains is imperfect and can act as an obstacle to dislocation movement. In general, the smaller the grains the more difficult is it to deform the metal. Why?

An obvious way of strengthening metals would be to eliminate dislocations altogether and produce in effect perfect crystals. So far this has only been possible for tiny, hairlike single-crystal specimens called 'whiskers' that are only a few micrometres thick and are seldom more than a few millimetres long. Their strength, however, approaches the theoretical value and they can withstand elastic strains of 4 or 5% (compared with $\frac{1}{2}$% or less for most common engineering materials). Unfortunately, perhaps due to surface oxidation, dislocations soon develop and the 'whisker' weakens. At present they are the subject of much research. Fig. 2.15 shows 'whiskers' of pure iron magnified four times, each one virtually a perfect crystal, and having about fifty times the tensile strength of the same thickness of ordinary soft iron.

Fig. 2.15

Cracks and fracture

Cracks, both external and internal and however small, play an important part in the fracture of a material and prevent it displaying its theoretical strength. Different types of fracture usually occur in brittle and ductile materials.

(*a*) *Brittle fracture*. This happens after little or no plastic deformation (i.e. during elastic deformation) by the very rapid propagation of a crack. It takes place, for example, when a glass rod is cut by making a small but sharp notch on it with a glass knife or file and then 'bending' it, as in Fig. 2.16—with the notch on the far side of the rod. Why?

Fig. 2.16

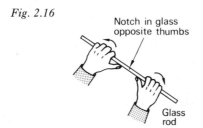

Notch in glass
opposite thumbs

Glass
rod

Round a scratch, notch or crack there is a concentration of stress which in general is greater the smaller the radius of curvature of the tip of the crack (i.e. the sharper the crack). Such stress concentrations may be seen by viewing a lamp through two 'crossed' Polaroid squares (i.e. one square is rotated to cut off most of the light coming from the other to give a dark field of view) having a strip of polythene between them, Fig. 2.17*a*. Pulling the strip causes colours to appear and is an indication that the polythene is under stress.

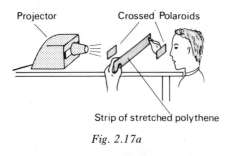

Projector   Crossed Polaroids

Strip of stretched polythene

Fig. 2.17a

Fig. 2.17b

(The phenomenon is called *photoelasticity* and is used to study stresses in plastic models of engineering structures. Fig. 2.17*b* shows these round a triangular hole in a plastic block under pressure.) When the strip is cut half-way across and again pulled, colours are seen at the tip of the cut showing the stress is high there. If it is sufficiently high, interatomic bonds are broken, the cut spreads and breaks a few more bonds at the new tip. Eventually complete fracture occurs.

Even tiny surface scratches, and they seem to arise inevitably on all materials, can lead to fracture. For example, a freshly-drawn $\frac{1}{2}$-metre-long glass fibre can be bent into an arc but it fractures if 'scratched' at A, Fig. 2.18, by gently stroking a few times with another glass fibre. It is less likely to break if scratched at B. Why?

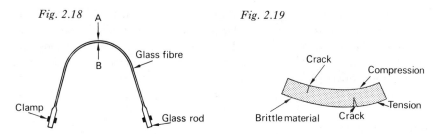

Fig. 2.18   A   B   Glass fibre   Clamp   Glass rod

Fig. 2.19   Crack   Compression   Tension   Brittle material   Crack

In brittle materials like concrete and glass, cracks spread more readily when the specimen is stretched or bent, i.e. is in tension. Crack propagation is much more difficult if such materials are used in compression (i.e. squeezed) so that any cracks close up, Fig. 2.19. Thus prestressed concrete contains steel rods that are in tension because they were stretched whilst the concrete was poured on them and set. As well as providing extra tensile strength these keep the concrete in compression even if the whole prestressed structure is in tension.

Glass can also be prestressed and its surface compressed, making it more resistant to crack propagation. In thermal toughening, jets of air are used to cool the hot glass and cause the outside to harden and contract whilst the inside is still soft. Later the inside contracts, pulls on the now reluctant-to-yield surface which is thus compressed. As we saw previously stresses in transparent materials are revealed by polarized light. The pattern of the air jets used for cooling the prestressed glass of a car windscreen can be seen through polarizing spectacles or sometimes by sunlight that has been partially polarized by reflection from the car bonnet. Prestressing glass in this way can increase its toughness 3000 times.

(*b*) *Ductile fracture.* In this case fracture follows appreciable plastic deformation, by *slow* crack propagation. After thinning uniformly along its length during the plastic stage, the specimen develops a 'waist' or 'neck' in

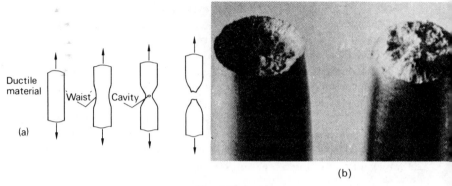

Ductile material

(a)

Waist

Cavity

(b)

Fig. 2.20

which cavities form. These join up into a crack and this travels out to the surface of the specimen, Fig. 2.20*a*, to give 'cup and cone' shaped, dull fracture surfaces, like those in Fig. 2.20*b* for an aluminium rod. It is possible that the internal cavities are formed during the later stages of plastic strain when stress concentrations arise in regions having a large number of inter-locking dislocations.

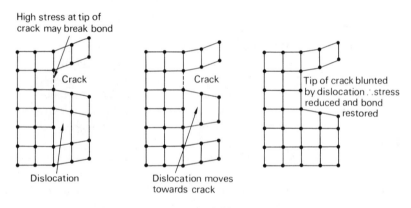

High stress at tip of crack may break bond

Crack

Dislocation

Crack

Dislocation moves towards crack

Tip of crack blunted by dislocation ∴ stress reduced and bond restored

Fig. 2.21

Cracks propagate with much greater difficulty in metals due to the action of dislocations. These can move to places of high stress such as the tip of a crack, thus reducing the effective stress by causing it to be shared among a greater number of interatomic bonds, Fig. 2.21. The crack tip is thereby deformed plastically, blunted by the dislocation and further cracking poss-ibly stopped. For the same reason surface scratches on metals have practic-ally no effect. In brittle materials dislocation movement is impossible (why?) and high local stresses can build up at cracks under an applied force.

Fatigue and creep

These are two other important aspects of the mechanical behaviour of metals that are conveniently considered here.

(*a*) *Fatigue*. This may cause fracture, often with little or no warning and happens when a metal is subjected to a large number of cycles of *varying* stress even if the maximum value of the stress could be applied *steadily* with complete safety. It is estimated that about 90% of all metal failures are due to fatigue; it occurs in aircraft parts, in engine connecting rods, axles, etc.

A typical fatigue fracture in a steel shaft is shown in Fig. 2.22; starting as a fine crack, probably at a point of high stress, it has spread slowly, producing a smooth surface (as on right of photograph), until it was half-way across the shaft which then broke suddenly. Stress concentrations may be due to bad design (e.g. rapid changes of diameter), bad workmanship (e.g. a tool mark) and are common at holes—the Comet aircraft disasters of 1954 were caused by fatigue failure started round small rivet holes.

For many ferrous metals there is a safe stress variation below which failure will not occur even for an infinite number of cycles. With other materials 'limited-life' design only is possible. As yet no fully comprehensive theory of fatigue exists.

Fig. 2.22

(*b*) *Creep*. In general this occurs at high temperature and results in the metal continuing to deform as time passes, even under constant stress. The effect is thought to be associated with dislocation motion due to the vibratory motion of atoms and in the creep resistant alloys that have been developed this motion is restricted. Such alloys are used, for example, to make the turbine blades of jet engines where high stress at high temperature has to be withstood without change of dimensions.

Some low melting-point metals can creep at room temperature, thus unsupported lead pipes gradually sag and the lead sheeting on church roofs has to be replaced periodically.

Composite materials

Composites are produced by combining materials so that the combination has the most desirable features of the components. The idea is not new. Wattle and daub (interlaced twigs and mud) have been used to build homes for a long time; straw and clay are the ingredients of bricks; Eskimos freeze moss into ice to give a less brittle material for igloo construction; reinforced concrete contains steel rods or steel mesh. In all cases the composite has better mechanical properties than any of its components.

The production of composites is man's attempt to copy nature. Wood is a composite of cellulose fibres cemented together with lignin. Bone is another composite material. Many modern technological applications require materials that are strong and stiff but light and heat resistant. The development of composites to meet these requirements is at present a major concern of materials scientists throughout the world and offers exciting possibilities to engineers in the future.

The highest *strength-to-weight* and *stiffness-to-weight* ratios are possessed not by metals but by materials such as glass, carbon and boron whose atoms are linked by many strong covalent bonds. (The strength of covalent bonds and their number per unit volume in these covalent solids accounts for the high strength and stiffness; the directional nature of the bond explains the non-close-packed structure and consequently small density, see p. 16.) Unfortunately these materials are brittle, partly because their structures make dislocation motion difficult under an applied stress and partly because they usually have small surface scratches that develop into cracks.

In modern composite materials the desirable properties of covalent solids are exploited by incorporating them as *fibres* in a weaker, yielding material called the *matrix*. Freshly drawn fibres are fairly scratch-free and are therefore strong. The matrix has three functions: first, it has to bond with and hold the fibres together so that the applied load is transmitted to them; second, it must protect the surface of the fibres from scratches; third, if cracks do appear it should prevent them from spreading from one fibre to another

—it can do this by acting as a barrier to the crack and deflecting it harmlessly along the interface it forms with the fibre, Fig. 2.23. A plastic resin or a ductile metal makes a suitable matrix. Fibre-reinforced composites are strong to stresses applied along the fibres.

Fig. 2.23

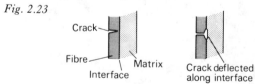

Crack — Fibre — Matrix — Interface — Crack deflected along interface

Fibre-glass was the first of the successful modern composites. It consists of high-strength glass fibres in a plastic resin (called glass-reinforced plastic —GRP) which is widely used where the stiffness and heat resistance required are not too great, e.g. for making boat hulls, storage tanks, pipes, car components. Fig. 2.24 shows the launching at Southampton in 1972 of the minesweeper H.M.S. *Wilton*, believed to be the world's first true ship (as distinct from boat) to have a hull made of GRP. Lightweight lift jet engines for vertical take-off also use GRP extensively for low-temperature parts and produce a thrust sixteen times their own weight. The best engines in commercial service today produce a thrust less than five times their weight.

Fig. 2.24

Fig. 2.25a

Carbon-fibre-reinforced plastics (CFRP) are similar but carbon fibres (about 6×10^5 per cm^2 of cross-section) of greater strength and stiffness replace glass fibres. These are stronger and stiffer than steel and much lighter (see question 7, p. 54). At present they are costly to produce; nevertheless CFRP are particularly attractive to the aircraft industry. They are not subject to fatigue failure or to high-stress concentrations round holes and cracks as metals are. They do not, however, flow plastically like metals but are elastic until failure, when extensive damage may occur. Good design is therefore necessary to avoid overloading. Their resistance to corrosion is also poor and prohibits certain applications. Fig. 2.25a shows a bundle of carbon fibres being assembled for tensile testing; Fig. 2.25b is a section of a piece of CFRP.

Fig. 2.25b

Fibre   Resin

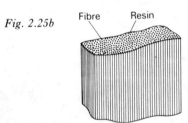

Much research is taking place on composites. The interface between a fibre and its matrix holds the key to the formation of a successful composite; at present a great deal is not understood about the properties of the interface.

Fig. 2.25c

If the fibres are not covered uniformly with the matrix, small holes form at the interface. In service these cause high-stress concentrations and premature failure. They may also allow liquids and gases to penetrate the composite and attack the fibres. The effect of coating the fibre with some other material to protect its surface from scratches, corrosion, etc., before combining it with the matrix is being studied. Fig. 2.25c is a scanning electron microscope photograph of a tungsten-coated carbon fibre (× 3000).

The fibre-reinforcement of metals and concrete is also being investigated.

Strain energy

Energy has to be supplied to stretch a wire. If the stretching force is provided by hanging weights, there is a loss of potential energy which is stored in the stretched wire as *strain energy*. Provided the elastic limit is not exceeded this energy can usually be recovered completely (rubber is a notable exception as we shall see in the next section). If it is exceeded, the part of the energy used to cause crystal slip (i.e. plastic strain) is retained by the wire.

We require to determine the strain energy stored in a wire stretched by a known amount. Consider a material with a force–extension graph like that in Fig. 2.26. (This is of the same form as its stress–strain graph.)

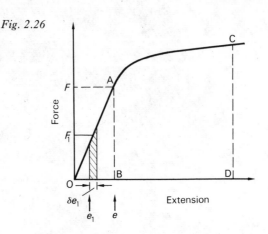

Fig. 2.26

Suppose the wire is already extended by e_1 and then suffers a further extension δe_1 which is so infinitesimally small that the shaded area is near enough a rectangle. If F_1 is the average but *nearly* constant value of the stretching force during the extension δe_1, then the energy transferred to the wire, i.e. the work done δW_1, is given by

$$\delta W_1 = \text{force} \times \text{distance}$$

$$= F_1 \times \delta e_1$$

$$= \text{area of shaded strip}$$

∴   Total work done during whole extension e = area OAB

i.e.   Strain energy stored in wire = area OAB

If Hooke's law is obeyed, OA is a straight line (as shown) and OAB is a triangle.

Therefore strain energy in wire for whole extension e and final stretching force F, is given by

$$\text{strain energy} = \text{area} \triangle \text{OAB}$$

$$= \tfrac{1}{2}\text{AB} \times \text{OB}$$

$$= \tfrac{1}{2}Fe$$

The expression $\frac{1}{2}Fe$ gives the strain energy in joules if F is in newtons and e in metres.

If l is the original length of the wire and A its cross-section area, then volume of wire $= Al$.

$$\therefore \quad \text{Strain energy per unit volume} = \frac{1}{2}Fe/(Al)$$

$$= \frac{1}{2}\left(\frac{F}{A} \times \frac{e}{l}\right)$$

But $F/A = \text{stress}$ and $e/l = \text{strain}$.

$$\therefore \quad \text{Strain energy per unit volume} = \frac{1}{2}(\text{stress} \times \text{strain})$$

If Hooke's law is not obeyed so that OA is not a straight line, what is the value of the strain energy in the wire for extension e? If the wire suffers plastic deformation, e.g. an extension OD, what represents the total strain energy stored?

Rubber

The two most striking mechanical properties of rubber are (*a*) its range of elasticity is great—some rubbers can be stretched to more than ten times their original length (i.e. 1000% strain) before the elastic limit is reached and (*b*) its value of Young's modulus is about 10^4 times smaller than most solids and *increases* as the temperature rises, an effect not shown by any other material.

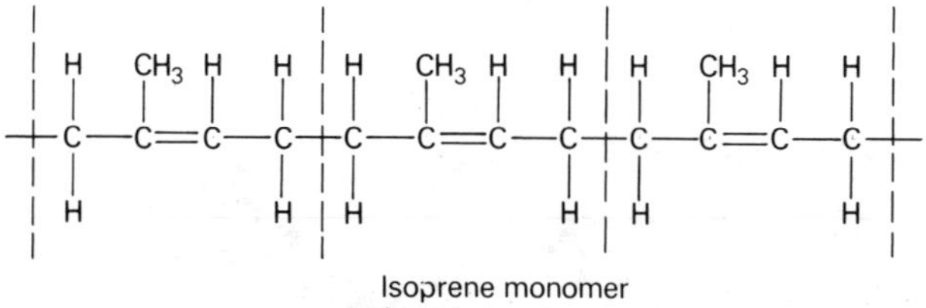

Isoprene monomer
(C_5H_8)

Fig. 2.27

Rubber is a polymer consisting of up to 10^4 isoprene molecules (C_5H_8) joined end-to-end into a long chain of carbon atoms, Fig. 2.27. The enormous extensibility and low value of E cannot be due to the stretching of the strong covalent bonds between atoms in the carbon chain.

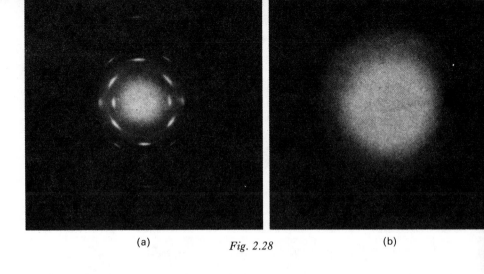

(a) Fig. 2.28 (b)

If a sample of stretched rubber is 'photographed' by a beam of high-energy electrons, sharp spots are obtained, Fig. 2.28a, like those produced by X-rays and a crystal. This suggests there is some order among the molecules in such a sample. Fig. 2.28b is a similar photograph of unstretched rubber. A plausible explanation of the behaviour of rubber might be that its long-chain molecules are intertwined and jumbled up like cooked spaghetti. Fig. 2.29 is a model of one rubber molecule. A stretching force would tend to make the chains uncoil and straighten out into more or less orderly lines alongside each other. When the force is removed they coil up again. There is also some cross-bonding between chains, achieved during manufacture by vulcanizing raw rubber (see p. 25); this cross-linking as well as causing stiffening also greatly increases the reversible strain possible by anchoring together the long molecules. Fully extended rubber is strong because the bonds are then stretched directly.

Fig. 2.29

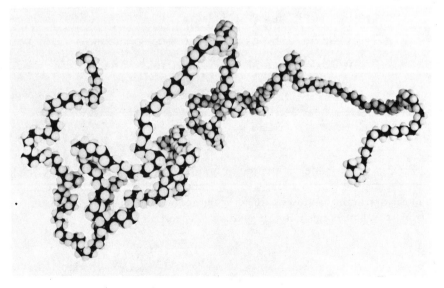

The rise in value of E with temperature can be attributed to the greater disorder among the chains when the material is heated; their resistance to alignment by a stretching force therefore increases.

If a stress–strain curve is plotted for the loading and unloading of a piece of rubber the two parts do not coincide, Fig. 2.30. OABC is for stretching

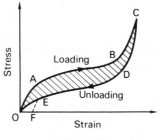

Fig. 2.30

and CDEO for contracting. The strain for a given stress is greater when unloading than when loading. The unloading strain can be considered to 'lag behind' the loading strain; the effect is called *elastic hysteresis*. It occurs with other substances, noticeably with polythene and glasses and to a small extent with metals.

We note that Hooke's law is not obeyed. This is typical of non-crystalline polymers and contrasts with the usually linear behaviour of crystalline materials, e.g. metals which obey the law during elastic strain. It is also evident that rubber stretches easily at first but is stiffer at large extensions. It has been shown (p. 48) that the area enclosed by OABC and the strain axis, represents the energy supplied to cause stretching; similarly the area under CDEO represents the energy given up by the rubber during contraction. The shaded area is called a *hysteresis loop*; it is a measure of the energy 'lost' as heat during one expansion–contraction cycle. The changes in temperature can be felt by placing a wide rubber band on the lips; if it is stretched quickly the temperature rises due to the transfer of mechanical energy to the rubber. When released under control the temperature falls. (The temperature change on free contraction is small.)

Rubber with a hysteresis loop of small area is said to have *resilience*. This is an important property where the rubber undergoes continual compression and relaxation as does each part of a car tyre when it touches the road and rotates on. If the rubber used in tyres does not have high resilience there is appreciable loss of energy resulting in increased petrol consumption or lower maximum speed. Should the heat build-up be large the tyre may disintegrate.

When rubber is stretched and released there may be a small permanent set as shown by the dotted line EF in Fig. 2.30.

Elastic moduli

All deformations of a body whether stretches, compressions, bends or twists can be regarded as consisting of one or more of three basic types of strain. For many materials experiment shows that *provided the elastic limit is not exceeded*

$$\frac{stress}{strain} = \text{a constant}$$

This is a more general statement of Hooke's law. The constant is called an *elastic modulus* of the material for the type of strain under consideration. There are three moduli, one for each kind of strain.

(*a*) *Young's modulus* (*E*). This has already been considered (p. 31) and is concerned with change of length strains. It is defined by

$$E = \frac{tensile\ stress}{tensile\ strain}$$

where stress is force per unit area (F/A) and strain is change of length per unit length (e/l).

(*b*) *Rigidity modulus* (*G*). In this case the strain involves a change of shape without change of volume. Thus if a tangential force F is applied along the top surface of area A of a rectangular block of material fixed to the bench, the block suffers a change of shape and is deformed so that the front and rear faces become parallelograms, Fig. 2.31. The shear stress is F/A and angle α

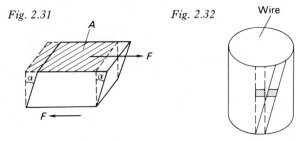

Fig. 2.31   *Fig. 2.32*   Wire

is taken as a measure of the strain produced. (The force F on the bottom surface of the block is exerted by the bench.) The rigidity modulus is defined as

$$G = \frac{shear\ stress}{shear\ strain} = \frac{F}{A\alpha}$$

When a wire is twisted, a small square on the surface becomes a rhombus, Fig. 2.32, and is an example of a shear strain. G can be found from experiments on the twisting of wires. If a spiral spring is stretched, the wire itself is not extended but is twisted, i.e. sheared. The extension thus depends on the rigidity modulus of the material as well as on the dimensions of the spring.

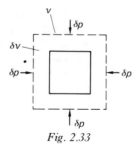

Fig. 2.33

(c) *Bulk modulus* (*K*). If a body of volume *V* is subjected to an *increase* of external pressure δp which changes its volume by δV, Fig. 2.33, the deformation is a change of volume without a change of shape. The bulk stress is δp, i.e. increase in force per unit area, and the bulk strain $\delta V/V$, i.e. change of volume/original volume; the bulk modulus *K* is defined by

$$K = \frac{bulk\ stress}{bulk\ strain} = \frac{-\delta p}{\delta V/V}$$

$$= -V\frac{\delta p}{\delta V}$$

The negative sign is introduced to make *K* positive since δV, being a decrease, is negative.

Note. δ (pronounced 'delta') is the Greek letter small *d* and when used as a prefix to the symbol for a quantity it indicates a change in that quantity is being considered.

Solids have all three moduli, liquids and gases only *K*. All moduli have the same units—pascals.

QUESTIONS

1. (*a*) Why are stresses and strains rather than forces and extensions generally considered when describing the deformation behaviour of solids?

(*b*) A length of copper of square cross-section measuring 1.0 mm by 1.0 mm is stretched by a tension of 40 N. What is the tensile stress in Pa?

(*c*) If the breaking stress of steel is 1.0×10^9 Pa will a wire of this material of cross-section area 4.0×10^{-4} cm² break when a 10 kg mass is hung from it? (*g* = 10 m s⁻²)

(*d*) A strip of rubber 6 cm long is stretched until it is 9 cm long. What is the tensile strain in the rubber as (*i*) a ratio, (*ii*) a percentage?

(*e*) A wire originally 2 m long suffers a 0.1% strain. What is its stretched length?

2. (a) Young's modulus for steel is greater than that for brass. Which would stretch more easily? Which is stiffer?

(b) How does a deformed body behave when the deforming force is removed if the strain is (i) elastic, (ii) plastic?

(c) A brass wire 2.5 m long of cross-section area 1.0×10^{-3} cm^2 is stretched 1.0 mm by a load of 0.40 kg. Calculate Young's modulus for brass. (Take $g = 10$ m s^{-2}.)

What percentage strain does the wire suffer? Use the value of E to calculate the force required to produce a 4.0% strain in the same wire. Is your answer for the force reliable? If it isn't, would it be greater or less than your answer? Explain.

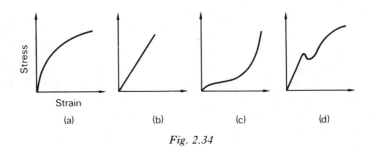

Fig. 2.34

3. Stress–strain curves for four different materials are shown in Fig. 2.34. Describe what you would feel if a specimen of each was pulled.

4. (a) A 0.50 kg mass is hung from the end of a wire 1.5 m long of diameter 0.30 mm. If Young's modulus for the material of the wire is 1.0×10^{11} Pa, calculate the extension produced. (Take $g = 10$ m s^{-2}.)

(b) Two wires, one of steel and one of phosphor bronze, each 1.5 metres long and of diameter 0.20 cm, are joined end to end to form a composite wire of length 3.0 metres. What tension in this wire will produce a total extension of 0.064 cm? (Young's modulus for steel $= 2.0 \times 10^{11}$ Pa and for phosphor bronze $= 1.2 \times 10^{11}$ Pa.)

(L. part qn.)

5. Write a short essay on 'strong' materials. Your essay might include discussion of such topics as: the need for strong materials, why some materials are weak, how the weaknesses can be avoided, composite materials such as plywood, reinforced concrete and fibre-glass, and the new technology of composite materials. These are only suggestions: you might write about a few of them, or, if you prefer, about all of them, and you may of course discuss other related issues if you wish.   (O. and C. Nuffield)

6. Define stress, strain, Young's modulus.

Describe in detail how Young's modulus for a steel wire may be determined by experiment.

A vertical steel wire 350 cm long, diameter 0.100 cm, has a load of 8.50 kg applied at its lower end. Find (a) the extension, (b) the energy stored in the wire. (Take Young's modulus for steel as 2.00×10^{11} Pa and $g = 9.81$ m s^{-2}.)   (L.)

7. Approximate values for Young's modulus, ultimate tensile strength and relative density of some materials are given in the table below.

MECHANICAL PROPERTIES

Material	Young's modulus ($\times 10^{10}$ Pa)	Ultimate tensile strength ($\times 10^8$ Pa)	Relative density
aluminium	7.0	1.0	2.7
carbon fibres	40	17	2.3
glasses	7.0	1.0	2.5
steel	21	10	7.8

(a) Explain the meaning and significance of each of these three terms.

(b) The tensile strength of glass equals that of aluminium. Why isn't glass used for the same kind of constructional applications as aluminium?

(c) In structures like aircraft the designer is concerned with obtaining as much strength and stiffness as possible for a given weight. Using the values in the table derive measures of 'stiffness-to-weight' and 'strength-to-weight' for the four materials listed.

3 Electrical properties

Conduction in solids

Materials exhibit a very wide range of electrical conductivities. The best conductors (silver and copper) are over 10^{23} times better than the worst conductors, i.e. the best insulators (e.g. polythene). Between these extreme cases is the now important group of semiconductors (e.g. germanium and silicon).

The first requirement for conduction is a supply of charge carriers that can wander freely through the material. In most solid conductors, notably metals, we believe that the carriers are loosely held outer electrons. With copper, for example, every atom has one 'free' electron (the one that helps to form the metallic bond, p. 11) which is not attached to any particular atom and so can participate in conduction. On the other hand if all electrons are required to form the bonds (covalent or ionic) that bind the atoms of the material together then that material will be an insulator. In semiconductors only a small proportion of the outer electrons are 'free' to move.

The 'free' electrons in a solid conductor are in a state of rapid motion, moving to and fro within the crystal lattice at speeds calculated to be about 1/1000 of the speed of light. This motion is normally completely haphazard (like that of gas molecules) and as many electrons with a given speed move in one direction as in the opposite direction with the same speed. There is, therefore, no net flow of charge and so no current.

If a battery is connected across the ends of the conductor, an electric field is created in the conductor which causes the electrons to accelerate and gain kinetic energy. Collisions occur between the accelerating electrons and atoms (really positive ions) of the conductor that are vibrating about their mean position in the crystal lattice but are not free to undergo translational motion. As a result the electrons lose kinetic energy and slow down whilst the ions gain vibrational energy. The net effect is to transfer chemical energy from the battery, via the electrons, to internal energy (see p. 115) of the ions. This shows itself on the macroscopic scale as a temperature rise in the conductor and subsequently energy may pass from the conductor to the surroundings as heat. The electrons are again accelerated and the process is repeated.

The *overall* acceleration of the electrons is zero on account of their collisions. They acquire a *constant average drift velocity* in the direction from negative to positive of the battery and it is this resultant drift of charge that is believed to constitute an electric current. An analogous situation may arise when a ball rolls down a long flight of steps. The acceleration caused by the earth's gravitational field when the ball drops can be cancelled by the force it experiences on 'colliding' with the steps. The ball may roll down the stairs with zero average acceleration, i.e. at constant average speed.

The 'free' electron theory is able to account in a general way for many of the facts of conduction and although in more advanced work it has been extended by the 'band' theory it will be adequate for our present purposes.

Current and charge

In metals, current is the movement of negative charge, i.e. electrons; in gases and electrolytes both positive and negative charges may be involved. Under the action of a battery, charges of opposite sign move in opposite directions and so a convention for current direction has to be chosen. As far as most external effects are concerned, positive charge moving in one direction is the same as negative charge moving in the opposite direction. By agreement all current is assumed to be due to the motion of positive charges and when current arrows are marked on circuits they are directed from the positive to the negative of the supply. If the charge carriers are negative they move in the opposite direction to that of the arrow.

The basic electrical unit is the unit of current—the *ampere* (abbreviated to A); it is defined in terms of the magnetic effect of a current. The unit of electric charge, the *coulomb* (C), is defined in terms of the ampere.

One coulomb is the quantity of electric charge carried past a given point in a circuit when a steady current of 1 ampere flows for one second.

If 2 amperes flow for 1 second, 2 × 1 coulombs (ampere-seconds) pass; if 2 amperes flow for 3 seconds then 2 × 3 coulombs pass. In general if a steady current I (in amperes) flows for time t (in seconds) then the quantity Q (in coulombs) of charge that passes is given by

$$Q = It$$

The flow of charge in a conductor is often compared with the flow of water in a pipe. The flow of water in litres per second say, corresponds to the flow of charges in coulombs per second, i.e. amperes.

The charge on an electron, i.e. the electronic charge, is 1.60×10^{-19} C and is much too small as a practical unit. In 1 C there are therefore $1/(1.60 \times$

10^{-19}), i.e. 6.24×10^{18} electronic charges. A current of 1 A is thus equivalent to a drift of 6.24×10^{18} electrons past each point in a conductor every second.

Smaller units of current are the milliampere (10^{-3} A), abbreviated to mA and the microampere (10^{-6} A), abbreviated to μA.

Drift velocity of electrons

On the basis of the 'free' electron theory an expression can be derived for the drift velocity of electrons in a current and an estimate made of its value. The results are surprising.

Consider a conductor of length l and cross-section area A having n 'free' electrons per unit volume each carrying a charge e, Fig. 3.1.

Fig. 3.1 n free electrons per unit volume

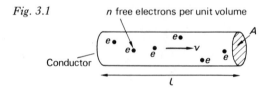

$$
\begin{aligned}
\text{Volume of conductor} &= Al \\
\text{Number of 'free' electrons} &= nAl \\
\text{Total charge } Q \text{ of 'free' electrons} &= nAle
\end{aligned}
$$

Suppose that a battery across the ends of the conductor causes the charge Q to pass through length l in time t with average drift velocity v. The resulting steady current I is given by

$$I = \frac{Q}{t}$$

$$= \frac{nAle}{t}$$

But $v = l/t$, therefore $t = l/v$.

$$\therefore \quad I = \frac{nAle}{l/v} = nAev$$

$$\therefore \quad v = \frac{I}{nAe}$$

To obtain a value for v consider a current of 1.0 ampere in SWG 28 copper wire of cross-section area 1.1×10^{-7} square metre. If we assume that each copper atom contributes one 'free' electron, it can be shown (see question 2a, p. 104) that $n \simeq 10^{29}$ electrons per cubic metre. Then, since $e = 1.6 \times 10^{-19}$ coulomb (the charge on an electron),

$$v = \frac{I}{nAe}$$

$$= \frac{(1.0\ \text{A})}{(10^{29}\ \text{m}^{-3}) \times (1.1 \times 10^{-7}\ \text{m}^2) \times (1.6 \times 10^{-19}\ \text{C})}$$

$$= \frac{1.0}{10^{29} \times 1.1 \times 10^{-7} \times 1.6 \times 10^{-19}} \cdot \frac{A}{\text{m}^{-3} \times \text{m}^2 \times \text{C}}$$

$$\simeq 6 \times 10^{-4} \frac{\text{C s}^{-1}}{\text{m}^{-1}\ \text{C}} \qquad\qquad (1\ \text{A} = 1\ \text{C s}^{-1})$$

$$\simeq 6 \times 10^{-4}\ \text{m s}^{-1}$$

$$\simeq 0.6\ \text{mm s}^{-1}$$

This is a remarkably small velocity and means that it takes electrons about half an hour to drift 1 m when a current of 1 A flows in this wire. The tiny drift velocity of electrons contrasts with their random speeds due to their vibrational motion (about 1/1000 of the speed of light) and is not to be confused with the speed at which the electric field causing their drift motion travels along a conductor. This is very great and is nearly equal to the speed of light, i.e. 3×10^8 m s^{-1} (See Appendix 7, p.404). Current therefore starts to flow almost simultaneously at all points in a circuit.

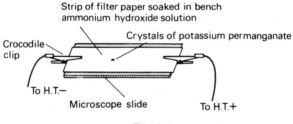

Strip of filter paper soaked in bench ammonium hydroxide solution

Crystals of potassium permanganate

Crocodile clip

To H.T.–

Microscope slide

To H.T.+

Fig. 3.2

The same expression for drift velocity holds for charge carriers other than electrons (in fact, it holds for the transport of other things as well as electric charge). In an electrolyte, conduction is due to ions and using the arrangement of Fig. 3.2 information can be obtained about their motion. On applying an electric field (from a 250 volt d.c. supply), the purple stains from the permanganate crystals travel very slowly towards the positive of the supply, and if we make the not unreasonable assumption that the stain travels with the charge carriers then ions too would appear to have tiny drift velocities of a similar value to those calculated for electrons.

Despite the slow movement of the carriers in conductors and their very small charge, large currents are possible. Why?

Potential difference

In an electric circuit electrical energy is converted into other forms of energy. A lamp converts electrical energy into heat and light and an electric motor converts electrical energy into mechanical energy. Such energy conversions, produced by suitable devices, are a useful feature of electrical circuits and form the basis of the definition of the term *potential difference* (p.d.)—an idea that helps us to make sense of circuits.

The potential difference between two points in a circuit is the amount of electrical energy changed to other forms of energy when unit charge passes from one point to the other.

The unit of potential difference is the *volt* and equals the p.d. between two points in a circuit in which 1 joule of electrical energy is converted when 1 coulomb passes from one point to the other. If 2 joules are converted per coulomb then the p.d. is 2 volts. If the passage of 3 coulombs is accompanied by the conversion of 9 joules of energy, the p.d. is 9/3 joules per coulomb, i.e. 3 volts.

It therefore follows that if the p.d. between A and B in Fig. 3.3 (we more commonly talk about the p.d. *across* AB) is 5 volts then when 4 coulombs pass

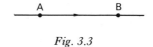

Fig. 3.3

from A to B, the electrical energy changed will be 5 joules per coulomb, i.e. 5×4 joules. In general if a charge of Q (in coulombs) flows in a part of a circuit across which there is a p.d. of V (in volts) then the energy change W (in joules) is given by

$$W = QV$$

If Q is in the form of a steady current I (in amperes) flowing for time t (in seconds) then $Q = It$ and

$$W = ItV$$

Although it is always the p.d. between two points which is important in electric circuits there are some occasions when it is helpful to consider what is called the *potential at a point*. This involves selecting a convenient point in the circuit and saying it has zero potential. The potentials of all other points are then stated with reference to it, i.e. the potential at any point is then the p.d. between the point and the point of zero potential. In practice one part of a piece of electrical equipment (e.g. a power supply) is often connected to earth; the earth and all points in the circuit joined to it are then taken as having zero potential.

If in Fig. 3.3 positive charge moves (i.e. conventional current flows) from A to B then A is regarded as being at a higher potential than B. Negative charge flow is therefore from a lower to a higher potential, i.e. from B to A. We can look upon p.d. as a kind of electrical 'pressure' that drives conventional current from a point at a higher potential to one at a lower potential.

Resistance

When the same p.d. is applied across different conductors different currents flow. Some conductors offer more opposition or *resistance* to the passage of current than others.

The resistance R of a conductor is defined as the ratio of the potential difference V across it to the current I flowing through it. That is,

$$R = \frac{V}{I}$$

The unit of resistance is the *ohm* (symbol Ω, the Greek letter omega) and is the resistance of a conductor in which the current is 1 ampere when a p.d. of 1 volt is applied across it. Larger units are the *kilohm* (10^3 ohm), symbol $k\Omega$, and the *megohm* (10^6 ohm), symbol $M\Omega$. The ratio V/I is a sensible measure of the resistance of a conductor since the smaller I is for a given V, the greater must be the opposition of the conductor, that is, the greater is R.

If the p.d. V across a *metallic* conductor is varied and the corresponding currents I measured, the ratio V/I is found to be constant so long as physical conditions, such as temperature, do not alter. Thus the resistance of such conductors is the same whatever the p.d. applied and we can write

$$\frac{V}{I} = \text{a constant}$$

This means $I \propto V$, i.e. *the current through a metallic conductor is directly proportional to the p.d. between its ends if the temperature and other physical conditions are constant.* This important result is known as Ohm's law and conductors, such as metals and alloys, which obey it are called *ohmic* conductors; a graph of V against I for them is a straight line through the origin. Many conductors do not obey Ohm's law, i.e. they are non-ohmic, and their resistance depends on the p.d. even though their temperature does not change during the measurements. The revolutionary advances in modern electronics are due largely to non-ohmic devices such as transistors.

The resistance of a metal can be regarded as arising from the interaction which occurs between the crystal lattice of the metal and the 'free' electrons as they drift through it under an applied p.d. This interaction is due mainly to collisions between electrons and the vibrating ions of the metal but collisions between defects in the crystal lattice (e.g. impurity atoms and dislocations) also play a part, especially at very low temperatures.

The *conductance* of a specimen is the reciprocal of its resistance and is measured in *siemens* (S).

Types of resistor

Conductors especially constructed to have resistance are called *resistors*, denoted by -\/\/\/- or -⊏⊐- ; they are required for many purposes in electrical circuits. Several types exist.

(*a*) *Carbon composition resistors*. These are made from mixtures of carbon black (a conductor), clay and resin binder (non-conductors) which are pressed and moulded into rods by heating. The resistivity of the mixture depends on the proportion of carbon. The stability of such resistors is poor and their values are usually only accurate to within $\pm 10\%$ but they are cheap, small and good enough for many jobs. Three sizes are available with power ratings of $\frac{1}{2}$, 1 and 2 watts.

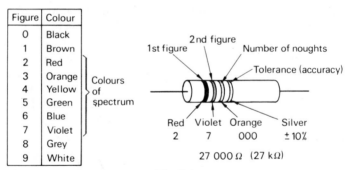

Figure	Colour
0	Black
1	Brown
2	Red
3	Orange
4	Yellow
5	Green
6	Blue
7	Violet
8	Grey
9	White

Fig. 3.4

Values are shown by colour markings as in Fig. 3.4. The tolerance colours are gold $\pm 5\%$, silver $\pm 10\%$, no colour $\pm 20\%$.

This colour code is now being replaced by a code with simpler markings.

Value	0.27 Ω	1 Ω	3.3 Ω	10 Ω	220 Ω	1 kΩ	68 kΩ	100 kΩ	1 MΩ	6.8 MΩ
Mark	R27	1R0	3R3	10R	K22	1K0	68K	M10	1M0	6M8

Tolerances are indicated by adding a letter; F = ±1%, G = ±2%, J = ±5%, K = ±10%, M = ±20%. Thus 5K6K = 5.6 kΩ ±10%.

(*b*) *Carbon film resistors.* Ceramic rods are heated to about 1000 °C in methane vapour which decomposes and deposits a uniform film of carbon on the rod. The resistance of the film depends on its thickness and can be increased by cutting a spiral groove in it, Fig. 3.5. The film is protected by an epoxy resin coating. The stability and accuracy of this type of resistor is commonly ±2% and the power rating $\frac{1}{8}$ to $\frac{1}{2}$ watt.

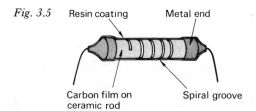

Fig. 3.5   Resin coating   Metal end

Carbon film on   Spiral groove
ceramic rod

(*c*) *Wire-wound resistors.* High-accuracy, high-stability resistors are always wire-wound, as are those required to have a large power rating (i.e. over 2 watts). They use the fact that the resistance of a wire increases with its length. Manganin (copper, manganese, nickel) wire is used for high-precision standard resistors because of its low temperature coefficient of resistance (p. 77); constantan or eureka (copper, nickel) wire is used for several purposes and nichrome (nickel, chromium) wire for commercial resistors.

Adjustable known resistances, called resistance boxes, are used for electrical measurements in the laboratory. They consist of a number of constantan coils which can be connected in series by switches or plugs to give the required value, Figs. 3.6 and 3.7. It is especially important with resistance

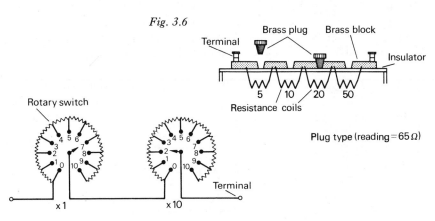

Fig. 3.6

Brass plug   Brass block

Terminal

Insulator

5   10   20   50
Resistance coils

Rotary switch

Plug type (reading = 65 Ω)

Terminal

×1   ×10

Switch type (reading = 27 Ω)   *Fig. 3.7*

boxes not to exceed the maximum safe current since overheating may change the resistance value or even burn out the coils. (The power limit is about 1 watt per coil and so a 1 ohm coil should not carry more than 1 ampere and a 100 ohm coil more than 0.1 ampere—from power = I^2R, see p. 83.)

(*d*) *Variable resistors.* Those used in electronic circuits, often as volume or other controls and sometimes called potentiometers, consist of an incomplete circular track of carbon composition or wire-wound card, with fixed connections to each end and a rotating arm contact which can slide over

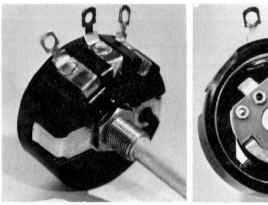

Fig. 3.8a   *Fig. 3.8b*

the track. Figs. 3.8a and b show the outside and inside respectively of a wire-wound potentiometer. If the track is 'linear', the resistance tapped off is proportional to the distance moved by the sliding contact; if it is 'logarithmic' it is proportional to the log of the distance and at the end of the track a small movement of the sliding contact causes a larger increase of resistance than at the start.

Larger current versions as used in many electrical experiments consist of constantan wire wound on a straight ceramic tube with the sliding contact carried on a metal bar above the tube.

There are two ways of using a variable resistor. It may be used as a *rheostat* for controlling the current in a circuit when only one end connection and the sliding contact are required, Fig. 3.9. It can also act as a *potential divider* for

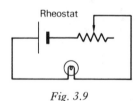

Fig. 3.9

ELECTRICAL PROPERTIES

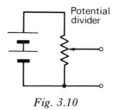

Fig. 3.10

controlling the p.d. applied to a device, all three connections then being used. In Fig. 3.10 any fraction of the total p.d. from the battery can be tapped off by varying the sliding contact.

Resistor networks

A network of resistors like that in Fig. 3.11 has a combined or equivalent resistance which can be found experimentally from the ratio of the voltmeter reading to the ammeter reading. Its value may also be calculated.

Fig. 3.11

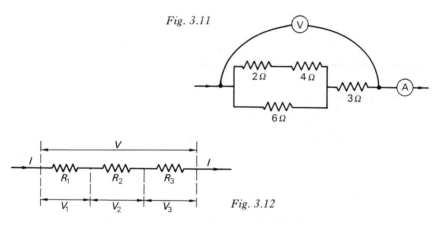

Fig. 3.12

(a) *Resistors in series.* Resistors are in series if the same current passes through each in turn. In Fig. 3.12 if the total p.d. across all three resistors is V and the current is I, the combined resistance R is given by

$$R = \frac{V}{I}$$

The electrical energy changed per coulomb in passing through all the resistors equals the sum of that changed in each resistor. Therefore if V_1, V_2 and V_3 are the p.d.s across R_1, R_2 and R_3 respectively then

$$V = V_1 + V_2 + V_3$$

65

By the definition of resistance, $R_1 = V_1/I$, $R_2 = V_2/I$ and $R_3 = V_3/I$, therefore $V_1 = IR_1$, $V_2 = IR_2$, $V_3 = IR_3$ and since $V = IR$ we have

$$IR = IR_1 + IR_2 + IR_3$$

$$\therefore \quad R = R_1 + R_2 + R_3$$

(b) *Resistors in parallel.* Here alternative routes are provided to the current which splits and we would expect the combined resistance to be less than the smallest individual resistance. In Fig. 3.13 if I is the total current through the network and I_1, I_2 and I_3 are the currents in the separate branches then since current is not used up in a circuit

$$I = I_1 + I_2 + I_3$$

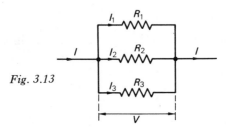

Fig. 3.13

In a parallel circuit the p.d. across each parallel branch is the same. If this is V then by the definition of resistance $R_1 = V/I_1$, $R_2 = V/I_2$, $R_3 = V/I_3$ and if R is the combined resistance then $R = V/I$. Therefore, $I = V/R$, $I_1 = V/R_1$, $I_2 = V/R_2$ and $I_3 = V/R_3$. Hence

$$\frac{V}{R} = \frac{V}{R_1} + \frac{V}{R_2} + \frac{V}{R_3}$$

$$\therefore \quad \frac{1}{R} = \frac{1}{R_1} + \frac{1}{R_2} + \frac{1}{R_3}$$

The single resistance R which would have the same resistance as the whole network can be calculated.

For the special case of two equal resistors in parallel we have $R_1 = R_2$ and

$$\frac{1}{R} = \frac{1}{R_1} + \frac{1}{R_1} = \frac{2}{R_1}$$

$$\therefore \quad R = \frac{R_1}{2}$$

In general for n equal resistances R_1, in parallel, the combined resistance is R_1/n.

The combined resistance of the network in Fig. 3.11 is 6 Ω; do you agree?

Using ammeters and voltmeters

Most ammeters and voltmeters are basically galvanometers (i.e. current detectors capable of measuring currents of the order of milliamperes or microamperes) of the moving-coil type which have been modified by connecting suitable resistors in parallel or in series with them as described in the next sections. Moving-coil instruments are accurate, sensitive and reasonably cheap and robust.

Connecting an ammeter or voltmeter should cause the minimum disturbance to the current or p.d. it has to measure. The current in a circuit is measured by breaking the circuit and inserting an ammeter in *series* so that the current passes through the meter. The resistance of an ammeter must therefore be *small* compared with the resistance of the rest of the circuit. Otherwise, inserting the ammeter changes the current to be measured. The perfect ammeter would have zero resistance, the p.d. across it would be zero and no energy would be absorbed by it.

The p.d. between two points A and B in a circuit is most readily found by connecting a voltmeter across the points, i.e. in *parallel* with AB. The resistance of the voltmeter must be *large* compared to the resistance of AB, otherwise the current drawn from the main circuit by the voltmeter (which is required to make it operate) becomes an appreciable fraction of the main current and the p.d. across AB changes. A voltmeter can be treated as a resistor which automatically records the p.d. between its terminals. The perfect voltmeter would have infinite resistance, take no current and absorb no energy.

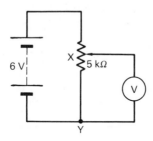

Fig. 3.14

It is instructive to set up the circuit in Fig. 3.14 for measuring the p.d. tapped off by the potential divider between X and Y, using first a high-resistance voltmeter (e.g. one with a 100 μA movement) for (V) and then a low-resistance voltmeter (e.g. one with a 10 mA movement). The reading in the second case is much lower. What will it be if both voltmeters are connected across XY?

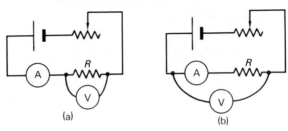

Fig. 3.15

The most straightforward method of measuring resistance uses an ammeter and a voltmeter as in Fig. 3.15a. The voltmeter records the p.d. across R but the ammeter gives the sum of the currents in R and in the voltmeter. If the voltmeter has a much higher resistance than R, the current through it will be small by comparison and the error in calculating R can be neglected. However, if the resistance of the voltmeter is not sufficiently high, perhaps because R is very high, the voltmeter should be connected across both R and the ammeter as in Fig. 3.15b. The ammeter now gives the true current in R. The voltmeter indicates the p.d. across R and the ammeter together, but the resistance of the latter is usually negligible compared with that of R and so the p.d. across it will be so small as to make the error in calculating R negligible.

Shunts, multipliers and multimeters

(*a*) *Conversion of a microammeter into an ammeter.* Consider a moving-coil meter which has a resistance (due largely to the coil) of 1000 Ω and which gives a full-scale deflection (f.s.d.) when 100 μA (0.0001 A) passes through it. If we wish to convert it to an ammeter reading 0–1 A this can be done by connecting a resistor (perhaps a misnomer) of very low value in parallel with it. Such a resistor is called a *shunt* and it must be chosen so that only 0.0001 A passes through the meter and the rest of the 1 A, namely 0.9999 A, passes through the shunt, Fig. 3.16. A full-scale deflection of the meter will then indicate a current of 1 A.

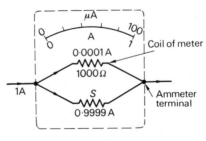

Fig. 3.16

To obtain the value S of the shunt, we use the fact that the meter and the shunt are in parallel. Therefore,

$$\text{p.d. across meter} = \text{p.d. across shunt}$$

Applying Ohm's law to both meter and shunt

$$0.0001 \times 1000 = 0.9999 \times S \qquad \text{(from } V = IR)$$

$$\therefore \quad S = \frac{0.0001 \times 1000}{0.9999}$$

$$= 0.1\Omega$$

The combined resistance of the meter and the shunt in parallel will now be very small (less than 0.1Ω) and the current in a circuit will be virtually undisturbed when the ammeter is inserted.

(*b*) *Conversion of a microammeter into a voltmeter.* To convert the same moving-coil meter of resistance 1000 Ω and f.s.d. 100 μA to a voltmeter reading 0–1 V, a resistor of high value must be connected in series with the meter. The resistor is called a *multiplier* and it must be chosen so that when a p.d. of 1 V is applied across the meter and resistor in series, only 0.0001 A flows through the meter and a full-scale deflection results, Fig. 3.17.

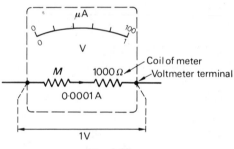

Fig. 3.17

To obtain the value M of the multiplier, we apply Ohm's law when there is an f.s.d. of 0.0001 A. Hence

$$\text{p.d. across multiplier and meter in series} = 0.0001 \, (M + 1000)$$

But the meter is to give an f.s.d. when the p.d. across it and the multiplier in series is 1 V. Therefore

$$0.0001(M + 1000) = 1$$

$$\therefore \quad M + 1000 = \frac{1}{0.0001} = 10\,000$$

$$\therefore \quad M = 9000 \, \Omega$$

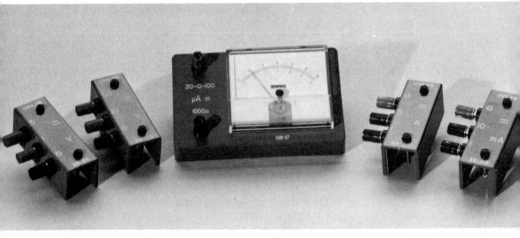

Fig. 3.18

In Fig. 3.18 a microammeter (20–0–100 μA) with its matching shunts and multipliers is shown.

Voltmeters are often graded according to their 'resistance per volt' at f.s.d. For the above voltmeter, 1 V applied across its terminals produces a full-scale deflection, i.e. a current of 100 μA, and so the resistance of the meter (coil + multiplier) must be 10 000 Ω (since $R = V/I = 1/0.0001 = 10\,000\ \Omega$). The 'resistance per volt' of the meter is thus 10 000 Ω/V. To be used as a voltmeter with an f.s.d. of 10 V it would need to have a total resistance of 100 000 Ω, i.e. a multiplier of 99 000 Ω to limit the full-scale current to 100 μA—but its resistance for every volt of deflection is still 10 000 Ω. A 100 Ω/V voltmeter has a resistance of 100 Ω for an f.s.d. of 1 V and draws a full-scale current of 10 mA ($I = V/R = \frac{1}{100} = 0.01\ \text{A} = 10\ \text{mA}$). Hence the higher the 'resistance per volt' of a voltmeter the smaller is the current it draws and the less will it disturb the circuit to which it is connected. A good voltmeter should have a resistance of at least 1000 Ω/V.

In electronic circuits resistances of 1 MΩ or higher are encountered and transistor or valve voltmeters which have very high resistances have to be used.

(*c*) *Multimeters.* A multi-range instrument or multimeter is a moving-coil galvanometer adapted to measure current, p.d. and resistance. There is a tapped shunt S across the meter and a tapped multiplier M in series with it, Fig. 3.19. A rotary switch allows the various ranges to be chosen.

One other position of the switch is marked 'ohms' and puts a dry cell B (usually 1.5 volts) and a rheostat R in series with the meter. To measure

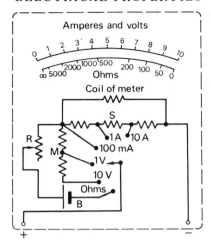

Fig. 3.19

resistance the terminals are short-circuited and R adjusted until the pointer gives a full-scale deflection, i.e. is on the zero of the ohms scale. The unknown resistance then replaces the short circuit across the terminals. The current falls and the pointer indicates the value in ohms. Fig. 3.20 shows a multimeter.

Fig. 3.20

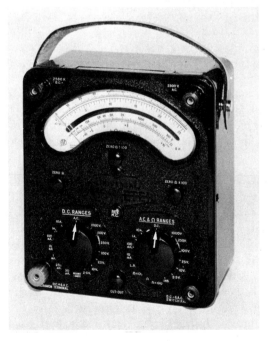

Rough calibration of a voltmeter

The calibration, which is done in two parts, is only approximate but uses the definition of the volt as a joule per coulomb. First, the temperature rise produced in a solid copper drum by a known force acting through a known distance is found and from it the amount of mechanical energy required to

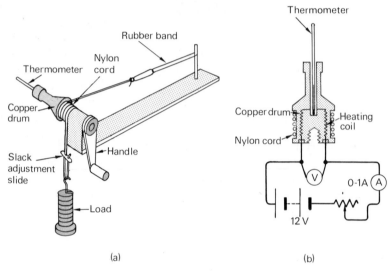

Rubber band

Nylon
Thermometer cord

Copper
drum

Slack
adjustment
slide

Handle

Load

(a)

Thermometer

Copper drum — Heating
coil
Nylon cord

0-1A (A)

12 V

(b)

Fig. 3.21

cause a 1 °C rise is calculated. Second, the drum is heated electrically for a certain time by a known current and the p.d. to be measured. Assuming that the amounts of electrical energy and mechanical energy needed to cause a 1 °C rise in the drum are the same, the p.d. can be calculated.

(a) *Mechanical experiment.* The apparatus is shown in Fig. 3.21a. The drum can be rotated about a horizontal axis by a handle and has a nylon cord wound round it five or six times. A heavy mass m (e.g. 8 kg) is hung from one end of the cord and a fixed rubber band attached to the other end. The initial temperature of the drum is noted. When the handle is turned the mass should lift slightly from the floor, the rubber band be *just* slack and the drum rotate steadily. The number of turns n to give a temperature rise of θ (about 10 °C) in time t is noted as is the diameter d (in metres) of the drum.

The mass m is supported by a frictional force F between the cord and the drum. If g is the acceleration due to gravity (9.8 m s^{-2}) then $F = mg$. During n revolutions, F acts over a total distance of πdn where πd is the circumference of the drum. The mechanical energy supplied by the person turning the handle is measured by the work and *work = force × distance in direction of force*. Here the work $= F \times \pi dn = mg \times \pi dn$. Therefore the mechanical energy W to produce a temperature rise of 1 °C in the drum is

$$W = \frac{mg\pi dn}{\theta}$$

(b) *Electrical experiment*. The copper drum is allowed to cool to room temperature and is connected into the electrical circuit of Fig. 3.21b with the nylon cord wound round it as before. The voltmeter to be calibrated is connected across the built-in heating coil in the drum and a steady current (0.7 A is suitable for some makes of apparatus) passed so that the temperature rises by about the same as in (a) (i.e. about 10 °C) in roughly the same time t. This ensures the heat losses in each experiment are similar and can be neglected. The electrical energy supplied to produce a temperature rise of 1 °C is W, the same as the mechanical energy required and is given by

$$W = \frac{ItV}{\theta_1}$$

where I is the ammeter reading in amperes, t the time in seconds for a temperature rise of θ_1 and V the unknown p.d. across the coil.
Hence

$$\frac{ItV}{\theta_1} = \frac{mg\pi dn}{\theta}$$

$$\therefore \quad V = \frac{mg\pi dn}{It} \cdot \frac{\theta_1}{\theta}$$

V can thus be calculated in joules per coulomb (volts) and compared with the reading on the voltmeter. In this experiment the same amount of internal energy (not heat as is often stated, see p. 115) is produced in the drum, first from mechanical energy and then from electrical energy.

Resistivity

The resistance of a conductor depends on its size as well as on the material of which it is made. To make fair comparisons of the abilities of different materials to conduct, the resistance of specimens of the same size must be considered.

Experiment shows that the resistance R of a uniform conductor of a given material is directly proportional to its length l and inversely proportional to its cross-section area A. Hence

$$R \propto \frac{l}{A}$$

or

$$R = \frac{\rho l}{A}$$

where ρ is a constant (for fixed temperature and other physical conditions), called the *resistivity* of the material of the conductor.

Hence since $\rho = AR/l$ we can say that the *resistivity of a material is numerically the resistance of a sample of unit length and unit cross-section area, at a certain temperature*. The unit of ρ is ohm metre (Ω m) since those of AR/ρ are metre^2 × ohm/metre, i.e. ohm metre.

Knowing the resistivity of a material the resistance of *any* specimen of that material may be calculated. For example, if the cross-section area of the live rail of an electric railway is 50 cm^2 and the resistivity of steel is 1.0×10^{-7} Ω m then neglecting the effect of joints, the resistance per kilometre of rail R follows—

$$R = \frac{\rho l}{A}$$

$$= \frac{(1.0 \times 10^{-7} \, \Omega \, \text{m}) \times (10^3 \, \text{m})}{(50 \times 10^{-4} \, \text{m}^2)}$$

$$= \frac{1.0 \times 10^{-7} \times 10^3}{50 \times 10^{-4}} \frac{\Omega \times \text{m} \times \text{m}}{\text{m}^2}$$

$$= 2.0 \times 10^{-2} \, \Omega$$

The resistivities at 20 °C of various materials are given in Table 3.1; their experimental determination is briefly described on p. 87.

The *conductivity* (σ) of a material is the reciprocal of its resistivity (ρ) i.e. $\sigma = 1/\rho$, and has unit ohm^{-1} metre^{-1} (Ω^{-1} m^{-1}).

Silver is the best conductor, i.e. has the lowest resistivity, and is followed closely by copper which, being much less expensive, is used for electrical connecting wire. Although the resistivity of aluminium is nearly twice that of copper, its density is only about one-third of copper's. The current-carrying-capacity-to-weight ratio of aluminium is therefore greater than that of copper. This accounts for its use in the overhead power cables of the Grid System where aluminium strands are wrapped round a core of steel

Table 3.1

Material		Resistivity (Ω m)	Use
CONDUCTORS			
Metals	Silver	1.6×10^{-8}	Contacts on small switches
	Copper	1.7×10^{-8}	Connecting wires
	Aluminium	2.7×10^{-8}	Power cables
	Tungsten	5.5×10^{-8}	Lamp filaments
Alloys	Manganin	44×10^{-8}	High-precision standard resistors
	Constantan or eureka	49×10^{-8}	Resistance boxes, variable resistors
	Nichrome (Ni, Cr)	110×10^{-8}	Heating elements
	Carbon	3000×10^{-8}	Radio resistors
SEMICONDUCTORS			
	Germanium	0.6	Transistors
	Silicon	2300	Transistors
INSULATORS			
	Glass	$10^{10}-10^{14}$	
	Polystyrene	10^{15}	

wires (54 aluminium strands to 7 steel wires for example, Fig. 3.22). The cable then has the strength it requires for suspension in long spans between pylons.

The resistivity of a pure metal is increased by small amounts of 'impurity' and alloys have resistivities appreciably greater than those of any of their

Fig. 3.22

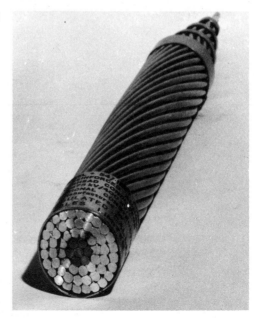

constituents. On the other hand, the addition of tiny traces of 'impurities' to pure semiconductors (a process known as 'doping' the semiconductor) reduces their resistivity. 'Impurity' atoms in a crystal lattice act as 'defects' and restrict the movement of charge carriers. When a semiconductor is 'doped' this is more than offset by the production of extra 'free' charges.

Electrical strain gauge

One device which engineers employ to obtain information about the size and distribution of strains in structures such as buildings, bridges and air-craft is the electrical strain gauge. It converts mechanical strain into a resistance change in itself by using the fact that the resistance of a wire depends on its length and cross-section area.

One type of gauge consists of a very fine wire (of an alloy containing mostly nickel, iron and chromium) cemented to a piece of thin paper as in Fig. 3.23. In use it is securely attached with a very strong adhesive to the component under test so that it experiences the same strain as the component. If, for example, an increase of length strain occurs, the gauge wire gets longer and

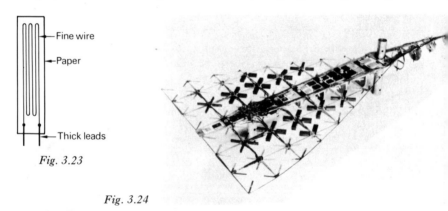

Fine wire

Paper

Thick leads

Fig. 3.23

Fig. 3.24

thinner and on both counts its resistance increases. Thick leads connect the gauge to a resistance measuring circuit (e.g. a Wheatstone bridge, p. 84) and previous calibration of the gauge enables the strain to be measured directly. What is the advantage of using a parallel-wire arrangement for the strain gauge?

Strain gauges are used to check theoretical calculations on new aircraft designs. Fig. 3.24 shows a model of a slender wing aircraft with weights and strain gauges attached at various places. The weights simulate the aero-dynamic loading when the aircraft is in flight.

Effects of temperature on resistance

(a) *Temperature coefficient.* The resistance of a material varies with temperature and the variation can be expressed by its *temperature coefficient of resistance* α. If a material has resistance R_0 at $0\,°C$ and its resistance increases by δR due to a temperature rise $\delta\theta$ then α for the material is defined by the equation

$$\alpha = \frac{\delta R}{R_0} \cdot \frac{1}{\delta\theta}$$

In words, α is *the fractional increase in the resistance at $0\,°C$* (i.e. $\delta R/R_0$) *per unit rise of temperature.* The unit of α is $°C^{-1}$ since δR and R_0 have the same units (ohms) and $\delta R/R_0$ is thus a ratio. For copper $\alpha \simeq 4 \times 10^{-3}\,°C^{-1}$, which means that a copper wire having a resistance of 1 ohm at $0\,°C$ increases in resistance by 4×10^{-3} ohm for every $1\,°C$ temperature rise.

Experiment shows that the value of α varies with the temperature at which $\delta\theta$ occurs but, to a good approximation, for metals and alloys we can generally assume it is constant in the range 0 to $100\,°C$. Thus if a specimen has resistances R_θ and R_0 at temperatures θ and $0\,°C$ respectively then replacing δR by $R_\theta - R_0$ and $\delta\theta$ by θ in the expression for α, we obtain

$$\alpha = \frac{R_\theta - R_0}{R_0\theta}$$

Rearranging gives $\qquad R_\theta - R_0 = R_0\alpha\theta$

and $\qquad\qquad\qquad R_\theta = R_0(1 + \alpha\theta)$

When using this equation where accuracy is important R_0 *should be the resistance at $0\,°C$.* A calculation shows the procedure when R_0 is not known.

Suppose a copper coil has a resistance of $30\,\Omega$ at $20\,°C$ and its resistance at $60\,°C$ is required. Taking α for copper as $4.0 \times 10^{-3}\,°C^{-1}$ we have

$$R_{20} = R_0(1 + 20\alpha)$$

and $\qquad\qquad\qquad R_{60} = R_0(1 + 60\alpha)$

Dividing $\qquad\qquad\qquad \dfrac{R_{60}}{R_{20}} = \dfrac{1 + 60\alpha}{1 + 20\alpha}$

$$\therefore \quad R_{60} = \frac{30(1 + 60 \times 4.0 \times 10^{-3})}{(1 + 20 \times 4.0 \times 10^{-3})}$$

$$= 34.5\,\Omega$$

If the calculation had *not* been based on the resistance at 0 °C and we had taken the original resistance (i.e. R_{20}) as R_0 then using $R_{60} = R_0(1 + \alpha\theta)$ where $\theta = (60 - 20)$ °C $= 40$ °C and $R_0 = 30$ Ω, we get

$$R_{60} = 30(1 + 4.0 \times 10^{-3} \times 40)$$

$$= 34.8 \ \Omega$$

This approximate method is quicker but in this example introduces an error of 0.3 in 34.5, i.e. about 1%, which is acceptable for many purposes.

The experimental determination of α is outlined on p. 87. Metals and alloys have *positive* *temperature* *coefficients* (they are p.t.c. materials), i.e. their resistance *increases* with temperature rise. The values for pure metals are of the order of 4×10^{-3} per °C or roughly $1/273$ per °C, the same as the cubic expansivity of a gas. In a tungsten-filament electric lamp the current raises the temperature of the filament to over 2730 °C when lit. The 'hot' resistance of the filament is, therefore, more than ten times the 'cold' resistance. Why doesn't a fuse blow every time lights are switched on? Alloys have much lower temperature coefficients of resistance than pure metals, that for manganin is about 2×10^{-5} per °C and a small temperature change has a small effect on its resistance.

Graphite, semiconductors and most non-metals have *negative* *temperature* *coefficients*, i.e. their resistance *decreases* with temperature rise (they are n.t.c. materials).

(b) *Superconductors.* When certain metals (e.g. tin, lead) and alloys are cooled to near -273 °C an *abrupt* decrease of resistance occurs. Below a definite temperature, different for each material, the resistance vanishes and a current once started seems to flow for ever. Such materials are called *superconductors* and their use in electrical power engineering and electronics is being explored.

(c) *Thermistors* (derived from *therm*al re*sistors*). These are devices whose resistance varies quite markedly with temperature. Depending on their composition they can have either n.t.c. or p.t.c. characteristics. The n.t.c. type consists of a mixture of oxides of iron, nickel and cobalt with small amounts of other substances and is used in electronic circuits to compensate for resistance increase in other components when the temperature rises and also as a thermometer for temperature measurement. The p.t.c. type, which is based on barium titanate, can show a resistance increase of 50 to 200 times for a temperature rise of a few degrees. It is useful as a temperature-controlled switch. Why? Fig. 3.25 shows a selection of thermistors.

(d) *Electrons, resistance and temperature.* The 'free' electron theory can account qualitatively for the variation of resistance with temperature of different materials. The increased average separation of the ions in a metal

Fig. 3.25
(a) disc n.t.c. type;
(b) plate n.t.c. type;
(c) p.t.c. type; (d) rod
n.t.c. type; (e) rod
voltage dependent re-
sistor; (f) bead-in-glass
type.

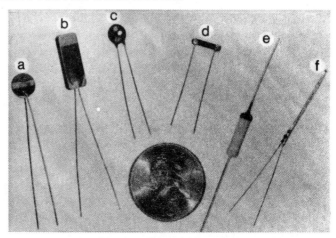

which accompanies a temperature rise (see p. 132) causes local distortion of the crystal lattice. As a result there is increased interaction between the lattice and the 'free' electrons when they drift under an applied p.d. The average drift speed is reduced and the resistance thus increases. In semiconductors this is more than compensated when greater vibration of the atoms 'frees' more electrons (an insignificant effect in metals) and thereby produces a marked decrease of resistance with temperature rise.

Electromotive force

Batteries and generators are able to maintain one terminal positive (i.e. deficient of electrons) and the other negative (i.e. with an excess of electrons). If we consider the motion of positive charges, then a battery, for example, moves positive charges from a place of low potential (the negative terminal) through the battery to a place of high potential (the positive terminal). The action may be compared with that of a pump causing water to move from a point of low gravitational potential to one of high potential.

A battery or generator therefore *does work on charges* and so energy must be changed within it. (Work is a measure of energy transfer, see p. 260.) In a battery chemical energy is transferred into electrical energy which we consider to be stored in the electric and magnetic fields produced. When current flows in an external circuit this stored electrical energy is changed, for example, to heat, but it is replenished at the same rate at which it is transferred. The electric and magnetic fields thus act as a temporary storage reservoir of electrical energy in the transfer of chemical energy to heat.

A battery or dynamo is said to produce an *electromotive force* (e.m.f.), defined in terms of energy transfer.

The electromotive force of a source (a battery, generator, etc.) *is the energy* (chemical, mechanical, etc.) *converted into electrical energy when unit charge passes through it.*

The unit of e.m.f., like the unit of p.d., is the *volt* and equals the e.m.f. of a source which changes 1 joule of chemical, mechanical or other form of energy into electrical energy when 1 coulomb passes through it. A car battery with an e.m.f. of 12 volts supplies 12 joules per coulomb passing through it; a power station generator with an e.m.f. of 25 000 volts is a much greater source of energy and supplies 25 000 joules per coulomb—2 coulombs would receive 50 000 joules and so on. In general, if a charge Q (in coulombs) passes through a source of e.m.f. E (in volts), the electrical energy supplied by the source W (in joules) is

$$W = QE$$

It should be noted that although e.m.f. and p.d. have the same unit, they deal with different aspects of an electric circuit. Whilst e.m.f. applies to a source supplying electrical energy, p.d. refers to the conversion of electrical energy in a circuit. The term e.m.f. is misleading to some extent, since it measures energy per unit charge and not force. It is true, however, that the source of e.m.f. is responsible for moving charges round the circuit.

A voltmeter measures p.d. and one connected across the terminals of an electrical supply such as a battery records what is called the *terminal p.d.* of the battery. If the battery is not connected to an external circuit and the voltmeter has a very high resistance then the current through the battery will be negligible. We can regard the voltmeter as measuring the number of joules of electrical energy the battery supplies per coulomb, i.e. its e.m.f. A working but less basic definition of e.m.f. is to say that it equals the terminal p.d. of a battery or generator *on open circuit*, i.e. when not maintaining current.

Internal resistance

A high-resistance voltmeter connected across a cell on open circuit records its e.m.f. (very nearly), Fig. 3.26a. Let this be E. If the cell is now connected to an external circuit in the form of a resistor R and maintains a steady current I in the circuit, the voltmeter reading falls; let it be V, Fig. 3.26b. V is the

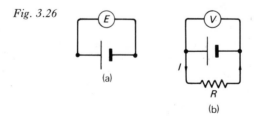

Fig. 3.26

(a)

(b)

terminal p.d. of the cell (but not on open circuit) and it is also the p.d. across *R* (assuming the connecting leads have zero resistance). Since *V* is less than *E*, then not all the energy supplied per coulomb by the cell (i.e. *E*) is changed in the external circuit to other forms of energy (often heat). What has happened to the 'lost' energy per coulomb?

The deficiency is due to the cell itself having some resistance. A certain amount of electrical energy per coulomb is wasted in getting through the cell and so less is available for the external circuit. The resistance of a cell is called its *internal resistance* (*r*) and taking stock of the energy changes in the complete circuit including the cell, we can say, assuming conservation of energy:

energy *supplied*   energy *changed*   energy *wasted* per
per coulomb by = per coulomb by + coulomb on internal
cell   external circuit   resistance of battery

Or, from the definitions of e.m.f. and p.d.,

$$\text{e.m.f.} = \text{p.d. across } R + \text{p.d. across } r$$

In symbols

$$E \quad = \quad V \quad + \quad v$$
e.m.f.   useful   'lost'
            volts   volts

where *v* is the p.d. across the internal resistance of the cell, a quantity which cannot be measured directly but only by subtracting *V* from *E*. From the equation $E = V + v$ we see that the *sum of the p.d.s across all the resistance in a circuit* (external and internal) *equals the e.m.f.*

Since $V = IR$ and $v = Ir$ we can rewrite the previous equation

$$E = IR + Ir$$

$$\therefore \quad E = I(R + r)$$

Suppose a high-resistance voltmeter reads 1.5 V when connected across a dry battery on open circuit, Fig. 3.27a, and 1.2 V when the same battery is supplying a current of 0.30 A through a lamp of resistance *R*, Fig. 3.27b.

Fig. 3.27

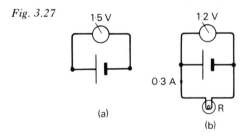

(a)

(b)

What is (a) the e.m.f. of the battery, (b) the internal resistance of the battery and (c) the value of R?

Using symbols with their previous meanings:

(a) Since the terminal p.d. on open circuit equals the e.m.f., we have $E = 1.5$ V.

(b) $E = V + v = V + Ir$ where $V = 1.2$ V and $I = 0.30$ A. Therefore

$$Ir = E - V \quad \text{and} \quad r = \frac{E - V}{I} = \frac{1.5 - 1.2}{0.30} = 1.0 \, \Omega$$

(c) From $V = IR$,

$$R = \frac{V}{I} = \frac{1.2}{0.30} = 4.0 \, \Omega$$

The internal resistance of an electrical supply depends on several factors and is seldom constant as is often assumed in calculations. However, it is sometimes useful to know its rough value and estimates can be made by taking p.d. and current measurements and proceeding as in the above example. Sources such as low-voltage supply units and car batteries from which large currents are required must have very low internal resistances. On the other hand if a 5000 V E.H.T. power supply does not have an internal resistance of the order of megohms to limit the current it supplies it will be dangerous.

The effect of internal resistance can be seen when a bus or car starts with the lights on. Suppose the starter motor requires a current of 100 A from a battery of e.m.f. 12 V and internal resistance 0.04 Ω to start the engine. How many volts are 'lost'? What is the terminal p.d. of the battery with the starter motor working? Why do the lights dim if they are designed to operate on a 12 V supply?

The terminal p.d. of a battery on open circuit as measured by even a very-high-resistance voltmeter is not quite equal to the e.m.f. because the voltmeter must take some current, however small, to give a reading. A small part of the e.m.f. is, therefore, 'lost' in driving current through the internal resistance of the battery. A potentiometer is used to measure e.m.f. to a very high accuracy (p. 90).

Power and heating effect

Current flow is accompanied by the conversion of electrical energy into other forms of energy and it is often necessary to know the *rate* at which a device brings about this conversion.

The power of a device is the rate at which it converts energy from one form into another.

If the p.d. across a device is V and the current through it is I, the electrical energy W converted by it in time t is (from the definition of p.d., p. 60).

$$W = ItV$$

The power P of the device will be

$$P = \frac{W}{t} = \frac{ItV}{t} = IV$$

The unit of power is the *watt* (W) and equals an energy conversion rate of 1 joule per second, i.e. $1 \text{ W} = 1 \text{ J s}^{-1}$. In the expression $P = IV$, P will be in watts if I is in amperes and V in volts. A larger unit is the *kilowatt* (kW) which equals 1000 watts.

Fig. 3.28

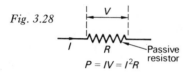

$$P = IV = I^2R$$

Passive resistor

If *all* the electrical energy is converted into heat by the device it is called a 'passive' resistor and the rate of production of heat will also be IV. If its resistance is R, Fig. 3.28, then since $R = V/I$ we have

$$P = IV$$

$$= \frac{V}{R} \cdot V = \frac{V^2}{R}$$

$$= I \cdot IR = I^2R$$

There are thus three alternative expressions for power but the last two are only true when all the electrical energy is changed to heat. The first, $P = IV$, gives the rate of production of all forms of energy. For example if the current in an electric motor is 5 A when the applied p.d. is 10 V then 50 W of electric power is supplied to it. However, it may only produce 40 W of mechanical power, the other 10 W being the rate of production of heat by the motor windings due to their resistance.

(*a*) *Heating elements and lamp filaments.* The expression $P = V^2/R$ shows that for a *fixed supply p.d.* of V, the rate of heat production by a resistor increases as R decreases. Now $R = \rho l/A$, therefore $P = V^2A/\rho l$ and so where a high rate of heat production at constant p.d. is required, as in an electric fire on the mains, the heating element should have a large cross-section area A, a small resistivity ρ and a short length l. It must also be able to withstand high temperatures without oxidizing in air (and becoming brittle). Nichrome is the material which best satisfies all these requirements.

Electric-lamp filaments have to operate at even higher temperatures if they are to emit light. In this case, tungsten, which has a very high melting point (3400 °C), is used either in a vacuum or more often in an inert gas (nitrogen or argon). The gas reduces evaporation of the tungsten (why?) and prevents the vapour condensing on the inside of the bulb and blackening it. In modern projector lamps there is a little iodine which forms tungsten iodide with the tungsten vapour and remains as vapour when the lamp is working, thereby preventing blackening.

(b) *Fuses.* When current flows in a wire its temperature rises until the rate of loss of heat to the surroundings equals the rate at which heat is produced. If this temperature exceeds the melting point of the material of the wire, the wire melts. A fuse is a short length of wire, often tinned copper, selected to melt when the current through it exceeds a certain value. It thereby protects a circuit from excessive currents.

It can be shown (see question 21, p. 108) that:

(*i*) the temperature reached by a given wire depends only on the current through it and is independent of its length (provided it is not so short for heat loss from the ends where it is supported, to matter); and

(*ii*) the current required to reach the melting point of the wire increases as the radius of the wire increases.

Fuses which melt at progressively higher temperatures can thus be made from the same material by using wires of increasing radius.

(c) *The kilowatt-hour* (kWh). For commercial purposes the kilowatt-hour is a more convenient unit of electrical energy than the joule.

The kilowatt-hour is the quantity of energy converted to other forms of energy by a device of power 1 kilowatt in 1 hour.

The energy converted by a device in kilowatt-hours is thus calculated by multiplying the power of the device in kilowatts by the time in hours for which it is used. Thus a 3 kW electric radiator working for 4 hours uses 12 kWh of electrical energy—often called 12 'units'. How many joules are there in 1 kWh?

Wheatstone bridge

(a) *Theory.* The Wheatstone bridge circuit enables resistance to be measured more accurately than by the ammeter–voltmeter method (p. 68). It involves making adjustments until a galvanometer is undeflected and so, being a 'null' method, it does not depend on the accuracy of an instrument. Other known resistors are, however, required.

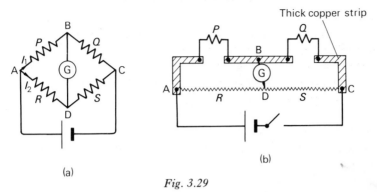

Fig. 3.29

Four resistors P, Q, R, S are joined as in Fig. 3.29a. If P is the unknown resistor, Q must be known as must the values of R and S or their ratio. A sensitive galvanometer G and a cell (dry or Leclanché) are connected as shown. One or more of Q, R and S are adjusted until there is no deflection on G. The bridge is then said to be balanced and it can be shown that

$$\frac{P}{Q} = \frac{R}{S}$$

whence P can be found. The proof of this expression follows.

At balance, no current flows through G, therefore the p.d. across BD is zero and so

$$\text{p.d. across AB} = \text{p.d. across AD}$$

Also, current through P = current through $Q = I_1$ and, current through R = current through $S = I_2$. Therefore

$$I_1 \times P = I_2 \times R$$

Similarly, p.d. across BC = p.d. across DC. Therefore

$$I_1 \times Q = I_2 \times S$$

Dividing,

$$\frac{P}{Q} = \frac{R}{S}$$

It can be shown that the same condition holds if the cell and G are inter-changed.

(b) *Metre bridge.* This is the simplest practical form of the Wheatstone bridge, Fig. 3.29b. The resistors R and S consist of a wire AC of uniform cross-section and 1 m long, made of an alloy such as constantan, with a resistance of several ohms. The ratio of R to S is altered by changing the position on the wire of the movable contact or 'jockey' D. The other arm of the bridge contains the unknown resistor P and a known resistor Q. Thick copper strips of low resistance connect the various parts. Figs. 3.29a and b have identical lettering to show their similarity.

The position of D is adjusted until there is no deflection on G, then

$$\frac{P}{Q} = \frac{R}{S} = \frac{\text{resistance of AD}}{\text{resistance of DC}}$$

Since the wire is uniform, resistance will be proportional to length and therefore

$$\frac{P}{Q} = \frac{AD}{DC}$$

Four practical points should be noted. Resistor Q should be chosen to give a balance point near to the centre of the wire, say between 30 and 70 cm, for three reasons. First, any errors in reading the balance lengths AD and DC are then small in comparison with their values. Second, where the resistors and metre wire are screwed or soldered to the copper strips there are 'connection' resistances which may, if desired, be determined experimentally and expressed in millimetres of bridge wire. However, when this is not done the 'end corrections', as they are called, will have least effect if neither AD nor DC is small. Third, the bridge is more sensitive near the middle since the unbalance current is larger per mm change of position.

In finding the balance point the cell key should be closed before the jockey makes contact with the wire. This is necessary because due to an effect known as 'self-induction', the currents in the circuit take a short time to grow to their steady values. During this time a momentary deflection of the galvanometer might be obtained even when the bridge is balanced for steady currents.

A high resistor should be joined in series with the galvanometer to protect it from damage whilst the balance point is being found. In the final adjustment it is shorted out and maximum sensitivity of the galvanometer obtained, Fig. 3.29c.

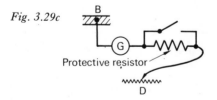

Fig. 3.29c

Protective resistor

Having obtained a balance point, P 'and Q should be interchanged and a second pair of values for AD and DC obtained and the means taken. This

Fig. 3.30

helps to compensate for errors arising from non-uniformity of the wire, from the wrong positioning of the millimetre scale in relation to the wire and from 'end corrections' provided the balance point is near the centre of the wire.

Wheatstone bridge methods are unreliable for finding resistances of less than 1 ohm, due to 'connection' resistance errors becoming appreciable, and are insensitive for resistances greater than 1 megohm unless a highly sensitive galvanometer is used. A modern dial-operated form of Wheatstone bridge for the use of maintenance engineers is shown in Fig. 3.30.

(c) *Measurement of resistivity using a metre bridge.* The resistivity ρ of a material can be determined by measuring the resistance R of a known length l of wire and also its average diameter d using a micrometer screw gauge. Then $\rho = AR/l$ where $A = \pi(d/2)^2$.

(d) *Measurement of temperature coefficient of resistance.* This may be found for, say, copper by measuring its resistance at different temperatures with the apparatus shown in Fig. 3.31a. A graph of resistance against temperature is plotted. Over small temperature ranges it is a straight line and from it the temperature coefficient α is calculated using $\alpha = (R_\theta - R_0)/R_0\theta$, Fig. 3.31b.

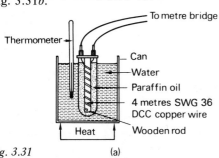

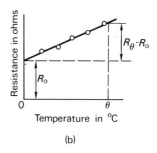

Fig. 3.31   (a)   (b)

Potentiometer

(a) *Theory.* A potentiometer is an arrangement which measures p.d. accurately. It can be adapted to measure current and resistance.

In its simplest form it consists of a length of resistance wire AB of *uniform* cross-section area, lying alongside a millimetre scale, and through which a *steady* current is maintained by a cell, called the driver cell, Fig. 3.32. (This is usually an accumulator because it gives a steady current for a long time.)

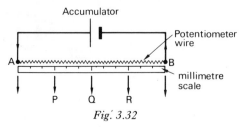

Fig. 3.32

As a result there is a p.d. between any two points on the wire which is proportional to their distance apart. Part of the p.d. across AB is tapped off and used to counterbalance the p.d. to be measured. If the p.d. across AB is 2 volts and if the wire is uniform and the current steady, what p.d.s can be tapped off between (*i*) A and P, (*ii*) A and Q, (*iii*) A and R, (*iv*) Q and B?

In practice the unknown p.d. V_1 is connected with its positive side to X in Fig. 3.33*a* if the positive terminal of the driver cell is joined to A, as shown.

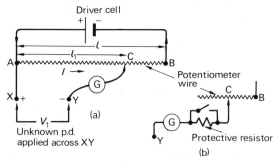

Fig. 3.33

The negative side of the unknown p.d. goes to a galvanometer G and a jockey C. In AX and YGC there are thus two p.d.s trying to cause current flow. The one tapped off between A and C from the potentiometer wire acts in an anticlockwise direction whilst the unknown p.d. V_1 tends to drive current in a clockwise direction. The direction of current flow through G therefore depends on whether V_1 is greater or less than the p.d. across AC. When the position of the jockey on the wire is such that there is *no current* through G, these two p.d.s are equal and the potentiometer is said to be balanced. The balance length l_1 is then measured.

If the resistance of AB per cm is r then the resistance of the balance length l_1 (in cm) is $l_1 r$ and if the steady current through AB from the driver cell is I, we have

$$\text{unknown p.d. } V_1 = \text{p.d. tapped off at balance}$$

$$\therefore \quad V_1 = I \times l_1 r$$

Since I and r are constants

$$V_1 \propto l_1$$

Let the p.d. across the whole potentiometer wire AB be V, then if $l = \text{AB}$

$$V = I \times lr$$

Therefore,

$$\frac{V_1}{V} = \frac{I l_1 r}{I l r} = \frac{l_1}{l}$$

$$\therefore \quad V_1 = \frac{l_1}{l} \cdot V$$

Knowing V, l and l_1, we can find V_1. When the driver cell is an accumulator of low internal resistance, V may be taken as its e.m.f. If this is 2.0 volts and if $l = 100$ cm and $l_1 = 80$ cm then $V_1 = (80/100) \times 2 = 1.6$ volts. Where higher accuracy is required a slightly different procedure is adopted as we shall see presently.

A potentiometer is a kind of voltmeter but is much more accurate than the best dial instrument since its 'scale' (i.e. the wire) may be made as long as we wish and its adjustment being a 'null' method does not depend on the calibration of the galvanometer. It has the further advantage of not altering the p.d. to be measured since at balance no current is drawn by it from the unknown p.d.; it behaves like a voltmeter of infinite resistance. On the other hand the wire form considered here is bulky and slow to use compared with an ordinary voltmeter. Modern potentiometers are dial-operated.

(b) *Practical points.* The following procedure should be noted.

(i) With a large protective resistor in series with the galvanometer, Fig. 3.33b, the circuit is tested by first placing the jockey on one end of the wire and then on the other; the deflections should be in opposite directions. If they are not then either the unknown p.d. is connected the wrong way round or the p.d. across the whole wire is less than the unknown p.d.

(ii) The balance point is found by repeating (i) for pairs of points that get progressively closer together, the protective resistor being shorted out near balance. The jockey should not be drawn along the wire or its uniformity will be lost.

(*iii*) The balance length is measured from the end A of the wire and should be reasonably long so that the percentage error in measuring it is small.

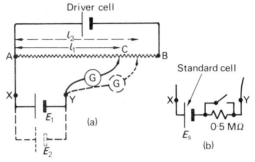

Fig. 3.34

(*c*) *Comparison of e.m.f.s of two cells.* The first cell, of e.m.f. E_1, is connected to XY and its balance length l_1 found, Fig. 3.34*a*. At balance no current is drawn from the cell and so

p.d. at terminals of cell = its e.m.f. = p.d. across AC

Hence

$$E_1 = \frac{l_1}{l} \cdot V$$ (see p. 89)

where l = AB and V = p.d. across AB due to driver cell.

Replacing the first cell by the second of e.m.f. E_2 and finding its balance length l_2 we have similarly

$$E_2 = \frac{l_2}{l} \cdot V$$

$$\therefore \quad \frac{E_1}{E_2} = \frac{l_1}{l_2}$$

If one of the cells is a Weston standard cell (p. 101) the e.m.f. of the other cell can be calculated accurately and, in effect, the potentiometer is calibrated. The Weston cell maintains a constant e.m.f. E_s of 1.0186 volts (at 20 °C) provided the current taken from it (at out of balance) is less than 10 μA. To ensure this a very high resistance (0.5 MΩ) is connected in series with it, Fig. 3.34*b*, and shorted out when the balance point is approached. Note that the high resistor does not affect the *position* of balance (since no current flows then) but only the precision with which it can be found.

(*d*) *Calibrating a voltmeter.* The potentiometer is first calibrated using a standard cell as explained in (*c*) so that the p.d. per cm of potentiometer wire

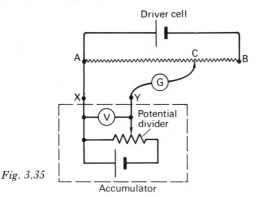

Fig. 3.35

is known. The circuit of Fig. 3.35 is suitable for calibrating a voltmeter in the range 0.2 to 2.0 volts. In the circuit below XY the variable resistor acts as a potential divider and enables various p.d.s to be applied to both the voltmeter (V) and the potentiometer. The p.d. can be calculated for each balance length.

Design a circuit to calibrate a voltmeter reading up to 10 volts using a 2 volt potentiometer.

In the circuit of Fig. 3.35 the balance lengths for p.d.s of less than 0.2 volt on a potentiometer wire 1–2 metres long would be too small for reasonable accuracy. The measurement of small p.d.s (and e.m.f.s) is achieved by reducing the p.d. across the whole potentiometer by a method similar to that now to be considered.

(*e*) *Measuring the e.m.f. of a thermocouple.* The e.m.f.s of thermocouples (p. 103) are of the order of a few millivolts and to ensure that an appreciable balance length is obtained when measuring them a high resistance R is joined in series with the potentiometer wire, Fig. 3.36. The value of R is chosen so that the p.d. across the whole wire AB is just greater than the maximum e.m.f. to be measured.

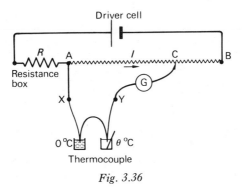

Fig. 3.36

Let V be the p.d. across R and AB in series and let R_{AB} be the resistance of AB then the current I through AB due to the driver cell is

$$I = \frac{V}{R + R_{AB}}$$

and

$$V_{AB} = IR_{AB} = \frac{V}{R + R_{AB}} R_{AB}$$

If $V = 2$ volts, $R_{AB} = 2$ ohms (as measured by a metre bridge) and $R = 1998$ ohms (a resistance box) then

$$V_{AB} = \frac{2 \times 2}{1998 + 2} = \frac{4}{2000} = \frac{2}{1000} \quad \text{volts}$$

$$= 2 \text{ mV}$$

If AB is 100 cm long, the p.d. across every cm is $2/100 = 0.02$ mV per cm and the e.m.f. of a thermocouple can then be found from the balance length so long as it does not exceed 2 mV. For example, the e.m.f. of a copper–iron thermocouple can be measured for different hot junction temperatures.

Where greater accuracy is required the potentiometer is calibrated using a standard cell and the circuit of Fig. 3.36 has to be modified as described in many practical books.

(*f*) *Calibrating an ammeter.* The potentiometer is first calibrated using a standard cell so that the p.d. per cm of the wire is known. The circuit is then connected as in Fig. 3.37*a* and the potentiometer used to measure the p.d. V

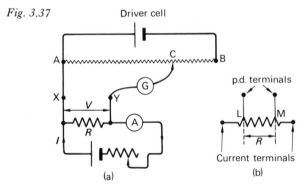

Fig. 3.37

(a)

(b)

across a suitable standard resistor R in series with the ammeter (A) to be calibrated. The current $I = V/R$; R is chosen to give a balance point near the end of the wire, i.e. V should be nearly 2 volts so that if $I = 0.5$ ampere then $R = 2/0.5 = 4$ ohms.

Unknown connection resistances at the terminals of two-terminal resistors are important in resistances of a few ohms and four-terminal types should be used if high accuracy is required in this type of measurement. The specified resistance exists between L and M in Fig. 3.37b; the resistance of the wires connecting L and M to the p.d. terminals does not affect things since at balance the current through them is zero.

(*g*) *Comparison of resistances.* The ratio of two resistances R_1 and R_2 can be found accurately by using a potentiometer to compare the p.d.s V_1 and V_2 across each when they are carrying the same current I. In the circuit of Fig. 3.38 X and Y are joined across R_1 and R_2 in turn and the corresponding balance lengths l_1 and l_2 measured. Then

$$\frac{l_1}{l_2} = \frac{V_1}{V_2} = \frac{IR_1}{IR_2} = \frac{R_1}{R_2}$$

To obtain balance lengths near the end of the wire V_1 and V_2 should approach 2 volts. If the value of I needed for this causes overheating of R_1 and R_2 then a smaller current must be used and a suitable resistor connected in series with the wire to make l_1 and l_2 large.

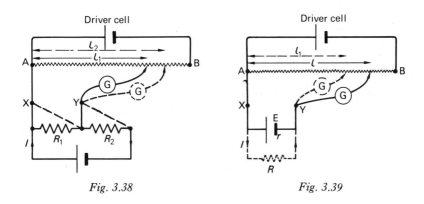

Fig. 3.38   Fig. 3.39

Using four-terminal-type resistors for R_1 and R_2 the potentiometer method is very suitable for resistances of less than 1 ohm since the resistance of connecting wires and terminal connections do not affect the result as they can in bridge methods.

(*h*) *Measuring internal resistance of a cell.* The balance length l is found first with the cell on open circuit, Fig. 3.39 (solid lines). The p.d. across XY therefore equals the p.d. at the terminals of the cell on open circuit, i.e. its e.m.f. E; therefore $E \propto l$. A known resistance R is then connected across the cell (dotted lines) and if l_1 is the new balance length, the p.d. across XY falls and equals the p.d. V across the cell when it maintains current through R; therefore $V \propto l_1$. Hence

$$\frac{E}{V} = \frac{l}{l_1}$$

If the current through R (at balance) is I and r is the internal resistance of the cell, Ohm's law applied first to the whole circuit and then to R alone gives

$$E = I(R + r) \quad \text{and} \quad V = IR$$

$$\therefore \quad \frac{E}{V} = \frac{R + r}{R} = \frac{l}{l_1}$$

$$\therefore \quad 1 + \frac{r}{R} = \frac{l}{l_1}$$

$$\frac{r}{R} = \frac{l}{l_1} - 1 = \frac{l - l_1}{l_1}$$

Hence r can be calculated.

Chemical effect

(a) *Electrolysis and the ionic theory.* Liquids which undergo chemical change when a current passes through them are called *electrolytes* and the process is known as *electrolysis*. Solutions in water of acids, bases and salts are electrolytes; liquids that conduct without suffering chemical decomposition are non-electrolytes and molten metals such as mercury are examples.

Conduction in an electrolyte is considered to be due to the movement of positive and negative ions. There is evidence from X-ray crystallography that in the solid state compounds such as sodium chloride consist of regular structures of positive and negative ions (p. 16) held together by electrostatic forces. We believe that when such substances are dissolved in water (and some other solvents) the interionic forces are weakened so much by the water that the ions can separate and move about easily in the solution. *Ionization* or *dissociation* is said to have occurred as a result of solution; the ionization of other salts, bases and acids is similarly explained. (*Note.* Whilst ions may exist in a solid, they are not free to move and so we do not consider the solid is ionized.)

In Fig. 3.40 when a p.d. is applied to the plates that dip into the electrolyte, i.e. the electrodes, an electric field is created which causes positive ions to

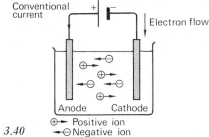

Fig. 3.40

⊕◄ Positive ion
◄⊖ Negative ion

move towards one electrode (the *cathode*), while negative ions are attracted to the other (the *anode*). The two streams of oppositely charged ions, drifting slowly in opposite directions (see p. 59), constitute the current in the electrolyte. At the electrodes, for conduction to continue, either (*i*) the ions must be discharged, i.e. give up their excess electrons if they are negative or accept electrons if they are positive, or (*ii*) fresh ions must be formed from the electrode and pass into solution. In any event the anode must gain electrons and the cathode lose them to maintain electron flow in the external circuit. After being discharged the ions usually come out of solution and are liberated as uncharged matter, being either deposited on the electrodes or released at them as bubbles of gas. (The electrodes must be made of metal or carbon. Why?)

Consider the electrolysis of copper sulphate solution with copper electrodes. The solution contains copper ions with a double positive charge (Cu^{2+}) and sulphate ions with a double negative charge ($SO_4{}^{2-}$). Under an applied p.d. the copper ions drift to the cathode where each receives two electrons ($2e$) and forms a copper atom that is deposited on the cathode.

$$Cu^{2+} + (2e \text{ from cathode}) \longrightarrow Cu$$

Sulphate ions collect round the anode and the most likely reaction to occur there, because it involves less energy than any other, is the formation of fresh copper ions by copper atoms of the anode going into solution. The anode is thus able to acquire electrons because every copper atom must lose two electrons to form a copper ion. Also, the fresh copper ions neutralize the negatively charged sulphate ions tending to gather round the anode.

$$Cu \longrightarrow Cu^{2+} + (2e \text{ to anode})$$

The net result is that copper is deposited on the cathode and goes into solution from the anode. In general, metals and hydrogen are liberated at the cathode and non-metals at the anode.

Electrolysis is used in many industrial processes. By allowing chemical reactions to occur at different places in the same solution, it keeps the products separate and makes feasible reactions that are otherwise impossible.

(b) *Faraday's laws of electrolysis.* Electrolysis was investigated quantitatively by Faraday. His results are summarized in two laws.

Law 1. The mass of any substance liberated in electrolysis is proportional to the quantity of electric charge passed.

If M is the mass liberated and Q is the charge passed then

$$M \propto Q$$

or

$$M = zQ$$

where z is a constant for a given substance, called its *electrochemical equivalent* (e.c.e.). If $Q = 1$ coulomb then $z = M$ and we can say that the *e.c.e. of a substance is the mass liberated in electrolysis by 1 coulomb*; it is expressed in grams per coulomb (g C^{-1}) or in kilograms per coulomb (kg C^{-1}). If Q is carried by a steady current I flowing for time t then $Q = It$ and we also have

$$M = zIt$$

Law 2. The number of moles of different substances liberated by the passage of the same charge are simply related to one another.

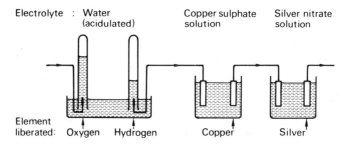

Fig. 3.41

In Fig. 3.41 the same charge passes through each electrolyte since they are in series. The number of moles liberated of oxygen, hydrogen, copper and silver are in the ratio $\frac{1}{2}:1:\frac{1}{2}:1$ respectively.

(c) *Charge on an ion.* It follows from the above that the charge which liberates one mole of hydrogen atoms (1.008 g) will also liberate half a mole of oxygen atoms (8.00 g) and so on.

Hydrogen is monovalent, oxygen is divalent. The valency of an atom or a radical is the number of hydrogen atoms it can combine with or replace.

The quantity of electric charge which liberates one mole of *any* monovalent ion is called the *Faraday constant* (*F*) and experiment gives its value as

$$F = 9.65 \times 10^4 \text{ coulombs per mole}$$

If *e* is the charge on say a hydrogen ion and N_A is the number of ions in 1 mole of hydrogen ions then

$$F = N_A e$$

since 1 mole of hydrogen ions is liberated by 9.65×10^4 coulombs (*F*). X-ray crystallography measurements give $N_A = 6.02 \times 10^{23}$ per mole and so

$$e = \frac{F}{N_A} = \frac{9.65 \times 10^4}{6.02 \times 10^{23}}$$

$$= 1.60 \times 10^{-19} \text{ coulomb}$$

This is the charge on a monovalent ion and is found to be the same as that on an electron. The above expression gives one of the most accurate ways of obtaining the electronic charge *e*.

In the case of copper, which is divalent in copper sulphate, half a mole is liberated by 9.65×10^4 coulombs. The atomic mass of copper is 63.6 and so 63.6 g of copper contain N_A atoms. Therefore 9.65×10^4 coulombs discharge $\frac{1}{2}N_A$ atoms of copper. But since this charge also liberates N_A atoms of a monovalent ion, we must conclude that each copper ion carries charge 2*e*. In general, the charge (in terms of *e*) on an ion equals the valency of the atom from which the ion is formed.

The charge represented by the Faraday is the number of electrons involved in the liberation of 1 mole of monovalent ions. A Faraday is therefore a 'mole of electrons'.

(*d*) *Theoretical derivation of Faraday's laws.* An ion of valency *v* has charge *ve*. If mass *M* of this ion is liberated in electrolysis by a total charge *Q* passing,

$$\text{number of ions discharged} = \frac{\text{total charge}}{\text{charge on one ion}} = \frac{Q}{ve}$$

But, $M = $ number of ions discharged $\times$ mass of an ion.

$$\therefore M = \frac{Q}{ve} \times \text{mass of an ion}$$

For a given ion, $M \propto Q$ (*Law 1*) since *v*, *e* and the mass of the ion are constant.

If *m* is the mass of a hydrogen ion then the mass of the ion of valency

v and atomic mass A_r which is liberated is $A_r m$ (nearly)[1]. Hence

$$\text{no. of moles of ion liberated} = n = \frac{M}{A_r}$$

$$\therefore M = A_r n = \frac{Q}{ve} \cdot A_r m$$

$$\therefore n = \frac{Qm}{e} \cdot \frac{1}{v}$$

Thus if Q is fixed, $n \propto 1/v$ since m and e are constant. In effect this is *Law 2*.

(*e*) *Ohm's law and electrolytes.* The variation of current with p.d. for an electrolyte may be investigated using the circuit of Fig. 3.42. The p.d. is

Fig. 3.42

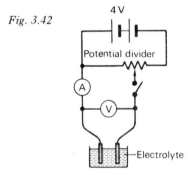

varied from 0–4 volts by the potential divider and measured with a high-resistance voltmeter.

With copper sulphate solution and copper electrodes, the graph of current against p.d. is a straight line through the origin, Fig. 3.43a, and Ohm's law is fairly well obeyed. The smallest p.d. causes current to flow and supports the ionic theory assumption that electrolytes, as soon as they dissolve, split into ions which are immediately available for conduction.

Fig. 3.43

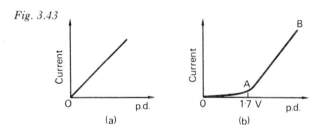

(a)   (b)

---

[1] Here we have based the definition of atomic mass on the obsolete scale which takes hydrogen as 1 instead of the carbon 12 scale (see p. 6). Strictly speaking m is 1 atomic mass unit but what we have said is true enough for our purposes.

Using water (acidulated) and platinum electrodes there is no appreciable current flow until the p.d. exceeds 1.7 volts. Thereafter increases of p.d. cause proportionate increases of current and hydrogen and oxygen are evolved at the cathode and anode respectively. The current–p.d. graph is shown in Fig. 3.43*b*. The virtual absence of current for p.d.s below 1.7 volts is attributed to the existence of a *back e.m.f.* of maximum value 1.7 volts which the applied p.d. must exceed before the electrolyte conducts. The back e.m.f. is due to polarization, i.e. the accumulation at the electrodes of products of electrolysis formed when the circuit is first made. In this case hydrogen at the cathode and oxygen at the anode effectively replace the platinum electrodes by gas electrodes and act as a chemical cell with an e.m.f. which opposes the applied p.d. (If the switch in Fig. 3.42 is opened the back e.m.f. is recorded on the voltmeter and falls rapidly.) AB in the graph of Fig. 3.43*b* is a straight line showing that if allowance is made for the back e.m.f., acidulated water obeys Ohm's law. The equation of AB is

$$V - E = IR$$

where R is the resistance of the electrolyte, I is the current when the applied p.d. is V and E is the back e.m.f.

(*f*) *Energy changes in electrolysis.* During electrolysis energy may be required (*i*) to drive ions through the electrolyte in the same way as it is required to drive electrons in a wire and (*ii*) to produce chemical changes at the electrodes. Re-arranging the above equation and multiplying each term by I

$$IV = I^2R + EI$$

IV is the electrical power supplied, I^2R is the heat produced per second due to (*i*) and EI is the power required for (*ii*).

In the electrolysis of copper sulphate solution $EI = 0$ since the net chemical change is zero, as much copper being deposited on the cathode as leaves the anode. With water, however, energy is needed to decompose it into hydrogen and oxygen and it would seem that a minimum p.d. is necessary to achieve this before electrolysis can start and current increase with p.d. We might expect this minimum p.d. to equal the back e.m.f. If we assume that the energy required to split water into hydrogen and oxygen is the same as the energy liberated when they combine to form water we can calculate the minimum p.d.

Experiment shows that when one mole of hydrogen atoms (1.008 g) and half a mole of oxygen atoms (8.00 g) combine, 1.47×10^5 joules of heat are released. These quantities of hydrogen and oxygen are produced in electrolysis by the passage of 9.65×10^4 coulombs. The energy required per coulomb, i.e. the p.d., is, therefore, $1.47 \times 10^5/9.65 \times 10^4$ joules per coulomb, i.e. 1.50 volts. The difference between this calculated value of the minimum p.d. to cause the electrolysis of water and the observed value of 1.70 volts is called the *overvoltage* and it has not been satisfactorily explained.

Electric cells

These are devices for converting chemical energy into electrical energy and consist of two different metals (or a metal and carbon) separated from each other by an electrolyte. Their e.m.f.s depend on the nature and concentration of the chemicals used, their size affects only the internal resistance.

Many different cells have been invented since the first was made by Volta at the end of the eighteenth century. Volta's consisted of plates of copper and zinc in dilute sulphuric acid. Those to be described are in use today; in all cases the chemical reactions are complex and will not be considered in detail.

(*a*) *Leclanché cell.* This consists of a positive pole of carbon, a negative pole of zinc, an electrolyte of ammonium chloride in water and a 'depolarizing' mixture of powdered manganese dioxide and carbon in a porous pot, Fig. 3.44. The chemical action produces hydrogen which tends to collect round

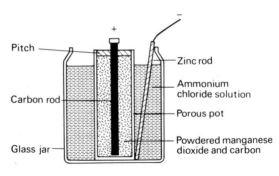

Fig. 3.44

the carbon rod, polarizing the cell and stopping its action. It is oxidized to water by the manganese dioxide, being aided by the powdered carbon which in effect extends the surface area of the positive pole and so reduces the depolarizing time. The porous pot keeps the depolarizing mixture packed round the carbon rod but allows ions to pass.

The cell has an e.m.f. of about 1.5 volts before it polarizes. It can only give small currents continuously and larger currents for short spells because of its slow depolarization. It is used in the laboratory for bridge circuits when a steady current is not required and its internal resistance of about 1 ohm limits the current to values not likely to damage resistance coils.

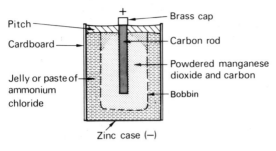

Pitch

Cardboard

Jelly or paste of ammonium chloride

Brass cap

Carbon rod

Powdered manganese dioxide and carbon

Bobbin

Zinc case (−)

Fig. 3.45

(*b*) *Dry cell*. This is a portable version of the Leclanché cell, much used in transistor radios and flashlamps, Fig. 3.45. A zinc case acts as the negative pole, the depolarizer forms a bobbin round the carbon rod and the electrolyte is in the form of a jelly or paste of ammonium chloride. It has the same e.m.f. as the 'wet' cell, depolarizes more quickly and is able to maintain a steady current for longer periods.

(*c*) *Weston standard cell*. A standard cell is used to provide an accurately known p.d. in potentiometer experiments (p. 90). In the Weston cell, Fig. 3.46, the positive pole is mercury, the negative an amalgam of cadmium

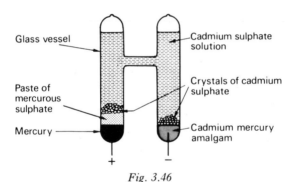

Glass vessel

Paste of mercurous sulphate

Mercury

Cadmium sulphate solution

Crystals of cadmium sulphate

Cadmium mercury amalgam

+   −

Fig. 3.46

in mercury, the electrolyte is a solution of cadmium sulphate (kept saturated by crystals) and the depolarizer is a paste of mercurous sulphate. The e.m.f. is 1.0186 volts at 20 °C and remains constant so long as the current does not exceed 10 μA. To ensure this a high resistor (about 0.5 MΩ) must be used in a series with it during the early stages of balancing a potentiometer.

(*d*) *Lead–acid accumulator*. When discharged this cell can be recharged by passing a current through it from another source in opposition to its e.m.f.; a 12 volt car battery consists of six such cells in series. The positive plate is of lead dioxide (brown), the negative of lead (grey) and the electrolyte is dilute sulphuric acid. The lead dioxide and lead, in the form of spongy

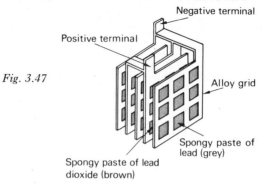

Fig. 3.47

pastes, are held in grids of lead–antimony alloy which are arranged alternately as in Fig. 3.47. During discharge the active material on both plates changes to lead sulphate, sulphuric acid is used up and water formed thus reducing the density of the acid.

The e.m.f. of a freshly charged accumulator is 2.2 volts and falls quickly to 2.0 volts where it remains until the accumulator is nearly discharged. The best indication of the state of charge of an accumulator is the relative density of the acid: fully charged it should be 1.25; when it falls to 1.18 recharging is necessary. If an accumulator is overdischarged, left discharged or idle for a long time, 'sulphation' occurs, i.e. the lead sulphate hardens on the plates and cannot be removed. The internal resistance of an accumulator is very small, less than 0.01 ohm. Consequently, for all ordinary currents the 'lost' volts are negligible but if the accumulator is short-circuited a current of several hundred amperes flows and can cause severe swelling or buckling of the plates.

The *capacity* of an accumulator is specified in *ampere-hours* (A h) for a particular discharge rate. Thus if the capacity is 30 A h at the 10 hour rate, a current of 3 A will be maintained for 10 hours. The capacity decreases as the discharge current increases and a current of 6 A would not be supplied for 5 hours. On the other hand we might expect to get 1 A for more than 30 hours.

A circuit for charging a battery of accumulators is shown in Fig. 3.48. The e.m.f. of the charging supply must be greater than that of the battery (about

Fig. 3.48

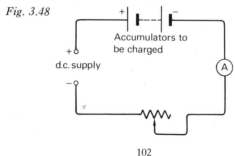

3 V per cell is usual) and its *positive* pole is joined to the *positive* of the battery. The rheostat allows adjustment of the charging current to the recommended value. Completion of charging is accompanied by 'gassing' at the electrodes, i.e. the evolution of hydrogen and oxygen due to the electrolysis of water. Before charging, the vent plugs are removed from the battery to allow these gases to escape and during charging naked lights must be kept well away.

Thermoelectric effect

If two different metals such as copper and iron are joined in a circuit and their junctions are kept at different temperatures, a small e.m.f. is produced and current flows, Fig. 3.49. The effect is known as the thermoelectric or *Seebeck effect* and the pair of junctions is called a *thermocouple*.

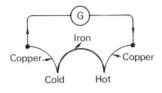

Fig. 3.49

The value of the thermo–e.m.f. depends on the metals used and the temperature difference between the junctions; the e.m.f.–temperature difference curve is always approximately a parabola. Fig. 3.50 shows the curves for (*a*) copper–iron and (*b*) iron–constantan which may be obtained

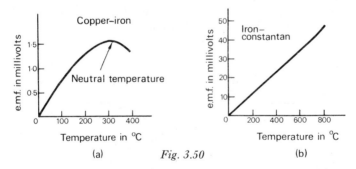

(a) Fig. 3.50 (b)

using a potentiometer as described previously (p. 91). The iron–constantan curve (although part of a parabola) is almost linear over a large range and produces about 10 times the e.m.f. of a copper–iron couple. The temperature of the hot junction at which the e.m.f. is a maximum is called the *neutral temperature*.

ELECTRICAL PROPERTIES

Thermocouples are used as thermometers particularly for measuring varying temperatures or the temperature at a point (p. 113).

The direct conversion of heat into electricity by metal thermocouples is a very inefficient process but better couples are now available based on semi-conductors such as iron disilicide. On account of their reliability, long life and cheapness, these are suitable as small power supply units in space satellites, weather buoys and weather ships. Radiation from a radioactive source (e.g. strontium 90) in the unit falls on the hot junction and produces the necessary temperature rise in it.

QUESTIONS

Current and charge

1. ¶If the heating element of an electric radiator takes a current of 4.0 A, what charge passes each point every minute? If the charge on an electron is 1.6×10^{-19} C how many electrons pass a given point in this time?

2. (a) If the density of copper is 9.0×10^3 kg m^{-3} and 63.5 kg of copper contains 6.0×10^{26} atoms, find the number of 'free' electrons per cubic metre of copper assuming that each copper atom has one 'free' electron.

   (b) How many 'free' electrons will there be in a 1.0 m length of copper wire of cross-section area 1.0×10^{-6} m^2 (i.e. 1.0 mm^2)?

   (c) Taking the charge on an electron as 1.6×10^{-19} C, what is the total charge of the 'free' electrons per metre of wire?

   (d) Assuming that the 'free' electrons are responsible for conduction, how long will the charge in (c) take to travel 1 m when a current of 2.0 A flows?

   (e) What is the drift velocity of the 'free' electrons?

3. Explain in terms of the motion of free electrons what happens when an electric current flows through a metallic conductor.

   A metal wire contains 5.0×10^{22} electrons per cm^3 and has a cross-sectional area of 1.0 mm^2. If the electrons move along the wire with a mean drift velocity of 1.0 mm s^{-1}, calculate the current in amperes in the wire if the electronic charge is 1.6×10^{-19} C.

   (O. and C. part qn.)

Potential difference : resistance : meters

4. (a) What is the p.d. between two points in a circuit if 200 joules of electrical energy are changed to other forms of energy when 25 coulombs of electric charge pass? If the charge flows in 10 seconds, what is the current?

   (b) What is the p.d. across an immersion heater which changes 3.6×10^3 joules of electrical energy to heat every second and takes a current of 15 amperes?

ELECTRICAL PROPERTIES

5. Three voltmeters $\widehat{V}$, $\widehat{V_1}$ and $\widehat{V_2}$ are connected as in Fig. 3.51.

(a) If $\widehat{V}$ reads 12 volts and $\widehat{V_1}$ reads 8.0 volts, what does $\widehat{V_2}$ read?

(b) If the ammeter $\widehat{A}$ reads 0.50 ampere, how much electrical energy is changed to heat and light by L_1 in 1 minute?

(c) Copy the diagram and mark with a + the positive terminals of the voltmeters and ammeter for correct connection.

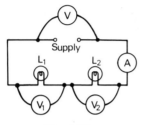

Fig. 3.51

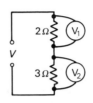

Fig. 3.52

6. A p.d. of V drives current through two resistors of 2 ohms and 3 ohms joined in series, Fig. 3.52.

(a) If voltmeter $\widehat{V_1}$ reads 4 volts, what is the current in the 2 ohm resistor?

(b) What is the current in the 3 ohm resistor?

(c) What does voltmeter $\widehat{V_2}$ read?

(d) What is the value of V?

(e) Find the value of the single equivalent resistor which, if it replaced the 2 ohm and 3 ohm resistors in series, would allow the same current to flow when joined to the same p.d. V.

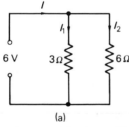

(a)

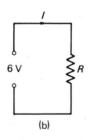

(b)

Fig. 3.53

7. Two resistors of 3 ohms and 6 ohms are connected in parallel across a p.d. of 6 volts, Fig. 3.53a.

(a) How are I, I_1 and I_2 related?

(b) What is the p.d. across each resistor?

(c) Find the values of I, I_1 and I_2 in amperes.

(d) Find the value of the single equivalent resistor R which, if it replaced the 3 ohm and 6 ohm resistors in parallel, would allow the value of I found in (c) to flow when the p.d. across it is 6 volts, Fig. 3.53b.

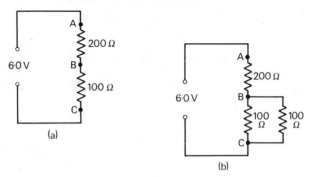

(a)

(b)

Fig. 3.54

8. (*a*) In the circuit of Fig. 3.54*a* what is the p.d. across (*i*) AB, (*ii*) BC?

(*b*) What do these p.d.s become when the circuit is altered as in Fig. 3.54*b*?

9. The circuit shown in Fig. 3.55 is used to provide a variable negative voltage, $-V$, with respect to earth, from a -200 V supply to a device drawing negligible current.

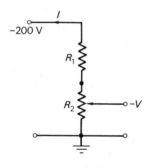

Fig. 3.55

The value of the fixed resistor, R_1, is known only to an accuracy (or tolerance) of $\pm 10\%$ and that of the variable resistor, R_2, to $\pm 20\%$. Choose values of R_1 and R_2 from the table shown, in order to satisfy the following conditions.

(*a*) The current in R_1 must be as low as possible but greater than 1 mA.

(*b*) It must be possible to vary the voltage $(-V)$ over a range of at least 0 to -20 V.

Resistor values

R_1 kilohms ($\pm 10\%$)	100	150	220	330	470	680
R_2 kilohms ($\pm 20\%$)	10	25	50			

(*J.M.B. Eng. Sc.*)

10. A resistor of 500 ohms and one of 2000 ohms are placed in series with a 60 volt supply. What will be the reading on a voltmeter of internal resistance 2000 ohms when placed across (*a*) the 500 ohms resistor, (*b*) the 2000 ohms resistor?

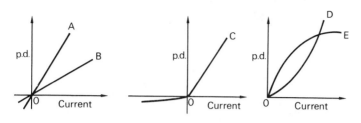

Fig. 3.56

11. Measurements of p.d. and current on five different 'devices' A, B, C, D, E gave the graphs in Fig. 3.56. Suggest, with reasons, what each might be.

12. If a moving-coil ammeter gives a full-scale deflection for a current of 15 mA and has a resistance of 5.0 ohms, how would you adapt it so that it could be used (*i*) as a voltmeter reading to 1.5 V, (*ii*) as an ammeter reading to 1.5 A? (*W. part qn.*)

Resistivity : temperature coefficient

13. Assuming that the resistivity of copper is half that of aluminium and that the density of copper is three times that of aluminium, find the ratio of the masses of copper and aluminium cables of equal resistance and length.

14. A wire has a resistance of 10.0 ohms at 20.0 °C and 13.1 ohms at 100 °C. Obtain a value for its temperature coefficient of resistance.

15. When the current passing through the Nichrome element of an electric fire is very small its resistance is found to be 50.9 Ω, room temperature being 20.0 °C. In use the current is 4.17 A on a 240 V supply. Calculate (*a*) the rate of energy conversion by the element, (*b*) the steady temperature reached by it. (The temperature coefficient of resistance of Nichrome may be taken to have the constant value 1.70×10^{-4} °C^{-1} over the temperature range involved.)

Electrical energy : e.m.f. : internal resistance

16. How much electrical energy does a battery of e.m.f. 12 volts supply when
  (*a*) a charge of 1 coulomb passes through it,
  (*b*) a charge of 3 coulombs passes through it,
  (*c*) a current of 4 amperes flows through it for 5 seconds?

17. Three accumulators each of e.m.f. 2 volts and internal resistance 0.01 ohm are joined in series and used as the supply for a circuit.
  (*a*) What is the total e.m.f. of the supply?
  (*b*) How much electrical energy per coulomb is supplied using (*i*) one accumulator, (*ii*) all three accumulators?
  (*c*) What is the total internal resistance of the supply?
  (*d*) What current would be driven by the supply through a resistance of 1.97 ohms?

18. (*a*) A flashlamp bulb is marked '2.5 V 0.30 A' and has to be operated from a dry battery of e.m.f. 3.0 V for the correct p.d. of 2.5 V to be produced across it. Why?

(*b*) How much heat and light energy is produced by a 100 watt electric lamp in 5 minutes?

(*c*) What is the resistance of a 240 V, 60 W bulb?

19. The p.d. across the terminals XY of a box is measured by a very-high-resistance voltmeter V. In arrangement A, Fig. 3.57, the voltmeter reads 105 volts. In arrangement

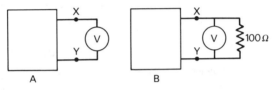

Fig. 3.57

B with the same box, the reading of the voltmeter is 100 volts. The inside of the box is not altered between the two arrangements.

Explain what you think may be in the box.

(You may, if you want to, draw in extra features on (copies of) the diagrams above to use in your explanation.) (*O. and C. Nuffield*)

20. What quantitative evidence could you bring forward in favour of the view that a cell may be looked upon as possessing a definite internal resistance? Describe an experiment you would perform to obtain such evidence.

A cell of e.m.f. 2.0 volts and internal resistance 1.0 ohm is connected in series with an ammeter of resistance 1.0 ohm and a variable resistor of R ohm. A voltmeter of resistance 1.0×10^2 ohm is connected across R. Find the value of R and the ammeter reading when the voltmeter reads 1.0 volt. Find also the power delivered to the external circuit.

(*S.*)

21. By considering a wire of radius r, length l and resistivity ρ, through which a current I flows show that

(*a*) the rate of production of heat by it is $I^2\rho l/\pi r^2$,

(*b*) the rate of loss of heat from its surface is $2\pi rlh$ where h is the heat lost per unit area of surface per second.

(*c*) the steady temperature it reaches is independent of its length and depends only on I.

Wheatstone bridge : potentiometer

22. Two resistance coils, P and Q, are placed in the gaps of a metre bridge. A balance point is found when the movable contact touches the bridge wire at a distance of 35.5 cm from the end joined to P. When the coil Q is shunted with a resistance of 10 ohms, the balance point is moved through a distance of 15.5 cm. Find the values of the resistances P and Q. (*W. part qn.*)

23. How would you investigate the way in which the current through a metal wire depends on the potential difference between its ends? What conditions should be fulfilled if Ohm's law is to hold?

Explain the theory of the Wheatstone bridge method of comparing resistances.

In an experiment with a simple metre bridge, the unknown X is kept in the left-hand gap and there is a fixed resistance in the right-hand gap. X is heated gradually, and when its temperature is 30 °C, the balance point on the bridge is found to be 51.5 cm from the left-hand end of the slide wire. When its temperature is 100 °C the balance point is 54.6 cm from that end. Find the temperature coefficient of resistance for the material of X, and calculate where the balance point would be if X were cooled to 0 °C.

(*O.*)

24. From the adoption of the fundamental units metre, kilogram, second, trace the steps necessary to define the volt and the ohm in terms of the ampere.

Discuss the suitability of (*a*) a moving-coil voltmeter, and (*b*) a slide-wire potentiometer for determining the potential differences in an experiment designed to verify Ohm's law.

Four resistors AB, BC, CD and DA of resistance 4.0 ohms, 8.0 ohms, 4.0 ohms and 8.0 ohms respectively are connected to form a closed loop, and a 6.0 volt battery of negligible resistance is connected between A and C. Calculate (*i*) the potential difference between B and D and (*ii*) the value of the additional resistance which must be connected between A and D so that no current flows through a galvanometer connected between B and D.

(*J.M.B.*)

25. Explain in detail how you would measure a small e.m.f. such as that of a thermocouple using a potentiometer method.

A 2 volt cell is connected in series with a resistance R ohms and a uniform wire AB of length 100 cm and resistance 4 ohms. One junction of a thermocouple is connected to A, and the other through a galvanometer to a tapping key. No current flows in the galvanometer when the key makes contact with the mid-point of the wire. If the e.m.f. of the couple was 4 millivolts what was the value of R? If the resistance of R is now increased by 4 ohms, by how much would the balance point change? (*S.*)

26. A two-metre potentiometer wire is used in an experiment to determine the internal resistance of a voltaic cell. The e.m.f. of the cell is balanced by the fall of potential along 90.6 cm of wire. When a standard resistor of 10.0 ohm is connected across the cell the balance length is found to be 75.5 cm. Draw a labelled circuit diagram and calculate, from first principles, the internal resistance of the cell.

How may the accuracy of this determination be improved? Assume that other electrical components are available if required. (*J.M.B.*)

Chemical effect

27. It is required to cover a plate of total surface area 50 cm^2 with a layer of copper 2.0 mm thick, using a steady current of 1.5 A. How long will this take? (Electrochemical equivalent of copper = 3.0×10^{-4} g C^{-1} and density of copper = 8.9 g cm^{-3}.)

28. The 1.8 kW starter motor of a car is operated by a 12 V battery. Find the current taken assuming complete efficiency. If the battery has a capacity of 50 A h, is used for an average of 10 s on each occasion and is not recharged in the intervals, how many times can the car be started before the battery is discharged?

4 Thermal properties

Temperature and thermometers

A knowledge of the thermal properties of materials is desirable when deciding, for example, what to use for making an electric storage heater or what to use as lagging in a refrigerator. Before studying some of these properties, certain basic ideas will first be considered.

(*a*) *Defining a temperature scale.* Temperature is sometimes called the degree of hotness and is a quantity which is such that when two bodies are placed in contact, heat flows from the body at the higher temperature to the one at the lower temperature. To measure temperatures, a *temperature scale* has to be established as follows.

(*i*) Some property of matter is selected whose value varies continuously with the degree of hotness. Suitable properties must be accurately measurable over a wide range of temperature with fairly simple apparatus and vary in a similar way to many other physical properties.

(*ii*) Two standard degrees of hotness are chosen—called the *fixed points* —and numbers assigned to them. On the Celsius method of numbering (until 1948 known as the centigrade method) the lower fixed point is the *ice point*, i.e. the temperature of pure ice in equilibrium with air-saturated water at standard atmospheric pressure[1] and is designated as 0 degrees Celsius (0 °C). The upper fixed point is the *steam point*, i.e. the temperature at which steam and pure boiling water are in equilibrium at standard atmospheric pressure and is taken as 100 °C.

(*iii*) The values X_{100} and X_0 of the temperature-measuring property are found at the steam and ice points respectively and $(X_{100} - X_0)$ gives the *fundamental interval* of the scale. If X_θ is the value of the property at some

---

[1] Standard atmospheric pressure is defined to be 1.013×10^5 Pa (1 Pa = 1 pascal = 1 N m^{-2}) and equals the pressure at the foot of a column of mercury 760 mm high of specified density and subject to a particular value of g.

other temperature θ which we wish to know then the value of θ in °C is given by the equation

$$\frac{\theta}{100} = \frac{X_\theta - X_0}{X_{100} - X_0}$$

This equation *defines* temperature θ in °C on the scale based on this particular temperature-measuring property. Note that it has been defined so that equal increases in the value of the property represent equal increases of temperature, i.e. the temperature scale is *defined* so that the property varies uniformly or linearly with temperature measured on its own scale.

Some thermometers using different temperature-measuring properties will now be considered briefly.

(b) *Mercury-in-glass thermometer*, Fig. 4.1. The change in length of a column of mercury in a glass capillary tube was one of the first thermometric

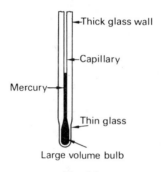

Fig. 4.1

properties to be chosen. If l_0 and l_{100} are the lengths of a mercury column at 0 °C and 100 °C respectively and if l_θ is the length at some other temperature θ then θ in °C is *defined* on the mercury-in-glass scale by the equation

$$\frac{\theta}{100} = \frac{l_\theta - l_0}{l_{100} - l_0}$$

For example, if a mercury thread has lengths 5.0 cm and 20 cm at the ice and steam points and is 8.0 cm long at another temperature θ then

$$\frac{\theta}{100} = \frac{8.0 - 5.0}{20 - 5.0} = \frac{3.0}{15}$$

$$\therefore \quad \theta = \frac{3.0}{15} \times 100 = 20 \text{ °C}$$

Inaccuracies arise in mercury thermometers from (*i*) non-uniformity of the bore of the capillary tube, (*ii*) the gradual change in the zero due to the bulb shrinking for a number of years after manufacture, and (*iii*) the mercury in the stem not being at the same temperature as that in the bulb.

Properties of this and other types of thermometer are summarized in Table 4.1, p. 114.

(*c*) *Constant-volume gas thermometer*. If the volume of a fixed mass of gas is kept constant, its pressure changes appreciably when the temperature changes. A temperature θ in °C is *defined* on the constant-volume gas scale by the equation

$$\frac{\theta}{100} = \frac{p_\theta - p_0}{p_{100} - p_0}$$

where p_0, p_θ and p_{100} are the pressures at the ice point, the required temperature θ and the steam point.

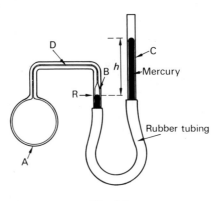

Fig. 4.2

A simple constant-volume thermometer is shown in Fig. 4.2. The gas (air in school models, hydrogen, helium or nitrogen in more accurate versions) is in bulb A which is at the temperature to be measured. As the temperature increases the gas expands pushing the mercury down in B and up in tube C. By raising C the mercury level in B is restored to the reference mark R and the volume of gas thus kept constant. The gas pressure is then $h + H$ where H is atmospheric pressure.

In accurate work corrections are made for (*i*) the gas in the 'dead-space' D not being at A's temperature, (*ii*) thermal expansion of A and (*iii*) capillary effects at the mercury surfaces (see p. 352).

(*d*) *Platinum resistance thermometer*. The electrical resistance of a pure platinum wire increases with temperature (by about 40% between the ice and steam points) and since resistance can be found very accurately it is a good property on which to base a temperature scale. A temperature θ in °C on the platinum resistance scale is *defined* by the equation

$$\frac{\theta}{100} = \frac{R_\theta - R_0}{R_{100} - R_0}$$

where R_0 and R_{100} are the resistances of the platinum wire at the ice and steam points respectively and R_θ is the resistance at the temperature required.

A platinum resistance thermometer, Fig. 4.3a, consists of a fine platinum wire wound on a strip of mica (an electrical insulator) and connected to thick

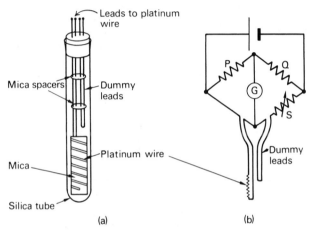

Fig. 4.3

copper leads. A pair of identical short-circuited dummy leads are enclosed in the same silica tube (which can withstand high temperatures) and compensate exactly for changes in resistance with temperature of the leads to the platinum wire. The thermometer is connected to a Wheatstone bridge circuit (p. 84) as in Fig. 4.3b and if $P = Q$, the resistance of the platinum wire equals that of S.

(e) Thermocouple thermometer. The thermoelectric effect was considered on p. 103 and is widely used to measure temperature. If great accuracy is not required, especially at high temperatures, the thermocouple can be connected across a galvanometer (rather than to a potentiometer). The meter may be marked to read temperatures directly if it is calibrated using the known melting points of metals.

It is not usual to define a thermoelectric scale of temperature but what would be the shape of a graph of thermocouple e.m.f. against temperature measured on such a scale?

THERMAL PROPERTIES

Table 4.1

Thermometer	Range in °C	Comments
Mercury-in-glass	− 39 to 500	Simple, cheap, portable, direct reading but not very accurate. Everyday use, clinical work and weather recording
Constant-volume gas	− 270 to 1500	Very wide range, very accurate, very sensitive but bulky, slow to respond and not direct reading. Used as a standard to calibrate other more practical types
Platinum resistance	− 200 to 1200	Wide range, very accurate but unsuitable for rapidly changing temperatures because of large heat capacity. Best for small steady temperature differences
Thermocouple	− 250 to 1500	Wide range, fairly accurate, robust and compact. Widely used in industry for rapidly changing temperatures and temperatures at a 'point'
Pyrometer	above 1000	The only thermometer for temperatures above 1500 °C

(*f*) *Disagreement between scales: thermodynamic scale.* Thermometers based on different properties give different values for the same temperature, except at the fixed points where they must agree by definition. All are correct according to their own scales and the discrepancy arises because, not unexpectedly, thermometric properties do not keep in step as the temperature changes. Thus when the length of the mercury column in a mercury-in-glass thermometer is, for example, mid-way between its 0 and 100 °C values (i.e. reading 50 °C) the resistance of a platinum resistance thermometer is not exactly mid-way between its 0 and 100 °C values.

The disagreement between scales, although small in the range 0 to 100 °C, is inconvenient. We could always state the temperature scale involved when giving a temperature, e.g. 50 °C on the mercury-in-glass scale, but a better procedure is to take one scale as a standard in which all temperatures are expressed, however they are measured. The one chosen is called the *absolute thermodynamic scale*. At present it is enough to say that it is the fundamental temperature scale in science and that the S I unit of temperature, the *kelvin* (denoted by K *not* °K) is defined in terms of it. The zero of this scale (0 K) is called *absolute zero* and it is thought that temperatures below this do not exist; certainly so far all attempts to reach it have been unsuccessful, although it has been approached very closely.

114

On the thermodynamic scale 0 °C = 273.15 K (273 K for most purposes) and 100 °C = 373.15 K, hence a temperature interval of one Celsius degree equals one kelvin. The symbol T is used for temperatures in kelvins and θ for those in degrees Celsius, the units in which most mercury thermometers are marked.

Heat and internal energy

Temperature is a useful idea when describing some aspects of the behaviour of matter in bulk. It is a quantity which is measurable in the laboratory as we have just seen and is capable of perception by the sense of touch. One of the aims of modern science is to relate macroscopic (i.e. large-scale) properties such as temperature, to the masses, speeds, energies, etc., of the constituent atoms and molecules. That is, to explain the macroscopic in terms of the microscopic.

The kinetic theory regards the atoms of a solid as vibrating to and fro about their equilibrium positions, alternately attracting and repelling one another. Their energy, called *internal energy*, is considered to be partly kinetic and partly potential. The kinetic component is due to the vibratory motion of the atoms and according to the theory depends on the temperature; the potential component is stored in the interatomic bonds that are continuously stretched and compressed as the atoms vibrate and it depends on the forces between the atoms and their separation. In a solid both forms of energy are present in roughly equal amounts and there is continual interchange between them. In a gas, where the intermolecular forces are weak, the internal energy is almost entirely kinetic. The kinetic theory thus links temperature with the kinetic energy of atoms and molecules.

Heat, in science, is defined as *the energy which is transferred from a body at a higher temperature to one at a lower temperature by conduction, convection or radiation.* Like other forms of energy it is measured in joules. When a transfer of heat occurs the internal energy of the body receiving the heat increases and if the kinetic component increases, the temperature of the body rises. Heat was previously regarded as a fluid called 'caloric', which all bodies were supposed to contain. It was measured in calories—a unit now going out of use—one-thousand of which equal the dietician's Calorie.

The internal energy of a body can also be increased by doing work, i.e. by a force undergoing a displacement in its own (or a parallel) direction. Thus the temperature of the air in a bicycle pump rises when it is compressed, i.e. it becomes hotter. Work done by the compressing force has become internal energy of the air and its temperature rises, as it would by heat transfer. It is therefore impossible to tell whether the temperature rise of a given sample of hot air is due to compression (i.e. work done) or to heat flow from a hotter body.

The expression 'heat in a body', although often used in everyday life, is thus misleading, for it may be that the body has become hot yet no heat flow has occurred. We should talk about the 'internal energy' of the body. It is sometimes said that 'the quantity of heat contained *in* a cup of boiling water is greater than that *in* a spark of white-hot metal'. What is really meant is that the boiling water has more *internal energy* and more heat *can be obtained* from it than from the spark.

The internal energy of a body may be changed in two ways: by doing *work* or by transferring *heat*. Work and heat are both concerned with energy *in the process of transfer* and when the transfer is over, neither term is relevant. Work is energy being transferred by a force moving its point of application (p. 115), and the force may arise from a mechanical, gravitational, electrical or magnetic source; heat flow arises from a temperature difference.

In a wire carrying a current, electrical energy is changed to internal energy (i.e. more vigorous vibration of the atoms of the wire) and a temperature rise occurs. Subsequently this energy may be given out by the wire to the surroundings as heat. We sum up the whole process by saying that an electric current has a 'heating effect'.

Specific heat capacity

(a) *Definition of c*. Materials differ from one another in the quantity of heat needed to produce a certain rise of temperature in a given mass. The *specific heat capacity c* enables comparisons to be made. Thus if a quantity of heat δQ raises the temperature of a mass m of a material by $\delta\theta$ then c is defined by the equation

$$c = \frac{\delta Q}{m \, \delta\theta}$$

In words, we can say that c is *the quantity of heat required to produce unit rise of temperature in unit mass*. (In modern terminology the word 'specific' before a quantity means per unit mass.)

The unit of c is joule per kilogram degree Celsius ($J \ kg^{-1} \ ^\circ C^{-1}$) or joule per kilogram kelvin ($J \ kg^{-1} \ K^{-1}$), since in the above expression δQ is in joule, m in kilogram and $\delta\theta$ in $^\circ C$ or K. Sometimes it is more convenient to consider mass in grams when the units are $J \ g^{-1} \ ^\circ C^{-1}$ or $J \ g^{-1} \ K^{-1}$.

In the expression for c, as the temperature rise $\delta\theta$ tends to zero, c approaches the specific heat capacity at a particular temperature and experiment shows that its value for a given material is not constant but varies slightly with temperature. Mean values are thus obtained over a temperature range and to be strictly accurate this range should be stated. For ordinary purposes, however, it is often assumed constant.

The approximate specific heat capacity of water at room temperature is

4.2×10^3 J kg^{-1} °C^{-1} (or 4.2 J g^{-1} °C^{-1}) and is large compared with the values for most substances. It means that 4.2×10^3 J of heat will raise the temperature of 1 kg of water by 1 °C, from say 15 °C to 16 °C. At temperatures approaching absolute zero (0 K) all values of c tend to zero. Values for some other materials at ordinary temperatures are shown in Table 4.2.

Table 4.2
Mean specific heat capacities in J kg^{-1} °C^{-1} *(or* J kg^{-1} K^{-1})

aluminium	9.1×10^2	ice	2.1×10^3
brass	3.8×10^2	rubber	1.7×10^3
copper	3.9×10^2	wood	1.7×10^3
glass (ordinary)	6.7×10^2	alcohol	2.5×10^3
iron	4.7×10^2	glycerine	2.5×10^3
mercury	1.4×10^2	paraffin oil	2.1×10^3
lead	1.3×10^2	turpentine	1.8×10^3

High specific heat capacity is desirable in a material if only a small temperature rise is required for a given heat input. This accounts for the efficiency as a coolant of water in a car radiator and of hydrogen gas in enclosed electric generators (the latter also because of its comparatively good thermal conductivity). Severe demands are often made on the thermal properties of materials used in space travel. In the heat shield of a space craft, Fig. 4.4a,

Fig. 4.4a

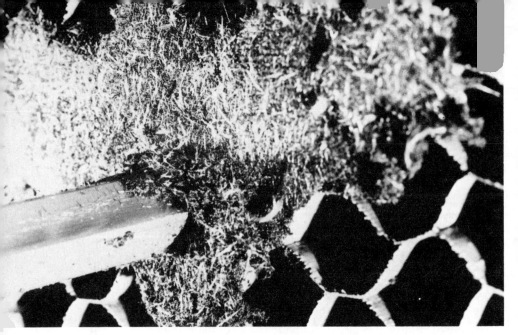

Fig. 4.4b

fibre-glass honeycomb is first bonded to the surface of the vehicle and then each cell is packed with silica-reinforced plastic, Fig. 4.4*b*. On re-entering the earth's atmosphere the shield, which may reach a temperature of

Fig. 4.4c

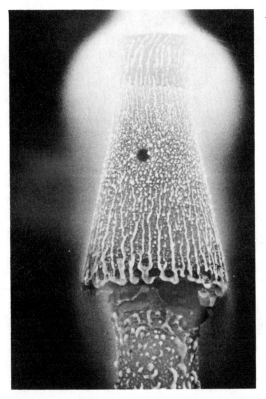

5000 °C, melts and burns off. Inside the craft the temperature is no more than about 25 °C. The laboratory photograph, Fig. 4.4c, shows the effect of a hot blast at nearly 11 000 °C. What kind of thermal properties should the heat shield have?

(b) *Molar heat capacity.* If heat capacities are referred to 1 mole (pp. 6–7) of the material instead of to unit mass, the quantity obtained by multiplying the specific heat capacity by the atomic or molecular mass (pp. 6–7) is called the *atomic* or *molar heat capacity.* It is very nearly 25 J mol^{-1} °C^{-1} for many solids. This fact is known as Dulong and Petit's law. Since 1 mole of any substance contains the same number of atoms or molecules, the heat required per atom or molecule to raise the temperature of many solids by a given amount is about the same. The implication is that the heat capacity of a solid depends on the *number* of atoms or molecules present, not on their mass and is further evidence for the atomic theory of matter.

(c) *Useful equation.* The equation defining specific heat capacity may be written

$$Q = mc(\theta_2 - \theta_1)$$

This expression is useful in heat calculations and gives the quantity of heat Q taken in by a body of mass m and mean specific heat capacity c when its temperature rises from θ_1 to θ_2. It also gives the heat lost by the body when its temperature falls from θ_2 to θ_1. In words, we can say

$$\begin{array}{c} heat\ given\ out \\ (or\ taken\ in) \end{array} = mass \times \begin{array}{c} specific\ heat \\ capacity \end{array} \times \begin{array}{c} temperature \\ change \end{array}$$

Thus if the temperature of a body of mass 0.5 kg and specific heat capacity 400 J kg^{-1} °C^{-1} rises from 15 °C to 20 °C the heat taken in is

$$\begin{aligned} Q &= (0.5\ \text{kg}) \times (400\ \text{J kg}^{-1}\ °\text{C}^{-1}) \times (5\ °\text{C}) \\ &= 0.5 \times 400 \times 5 \quad \text{kg} \times \text{J kg}^{-1}\ °\text{C}^{-1} \times °\text{C} \\ &= 1000\ \text{J} \end{aligned}$$

(d) *Heat capacity.* The *heat capacity* or *thermal capacity of a body* is a term in common use and is defined as *the quantity of heat needed to produce unit rise of temperature in the body.* It is measured in joules per degree Celsius (J °C^{-1}) and from the definition of specific heat capacity it follows that

$$heat\ capacity = mass \times specific\ heat\ capacity$$

Thus the heat capacity of a copper vessel of mass 0.1 kg and specific heat capacity 390 J kg^{-1} °C^{-1} is 39 J °C^{-1}.

Measuring specific heat capacities

1. Electrical method

(a) *Solids.* The method is suitable for metals such as copper and aluminium that are good thermal conductors. A cylindrical block of the material is used having holes for an electric heater (12 V, 2–4 A) and a thermometer, Fig. 4.5. The mass m of the block is found and its initial temperature θ_1 recorded. The block is lagged with expanded polystyrene and a suitable

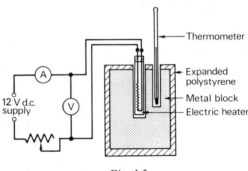

Fig. 4.5

steady current switched on as a stop clock is started. The voltmeter and ammeter readings V and I are noted. When the temperature has risen by about 10 °C, the current is stopped and the time t taken for which it passed. The highest reading θ_2 on the thermometer is noted.

Assuming that no energy loss occurs we have,

electrical energy supplied by heater = heat received by block

$$ItV = mc(\theta_2 - \theta_1)$$

where c is the specific heat capacity of the metal. Hence

$$c = \frac{ItV}{m(\theta_2 - \theta_1)}$$

Notes. (i) If I is in amperes, t in seconds, V in volts, m in g, θ_1 and θ_2 in °C then c is in J g^{-1} °C^{-1}.

(ii) The small amount of heat received by the thermometer and heater has been neglected.

(b) *Liquids.* The apparatus is shown in Fig. 4.6, a calorimeter being a vessel in which heat measurements are made. The procedure is similar to that for solids except that the liquid is stirred continuously during the heating. If m is the mass of liquid, c its specific heat capacity, m_c the mass of the

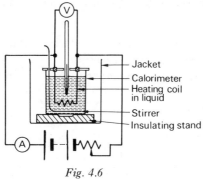

Fig. 4.6

calorimeter and stirrer, c_c the specific heat capacity of the material of the calorimeter and stirrer, and if θ_1, θ_2, I, V and t have their previous meanings then assuming

$$\frac{\text{energy supplied}}{\text{by heater}} = \frac{\text{energy received}}{\text{by liquid}} + \frac{\text{energy received by}}{\text{calorimeter and stirrer}}$$

we have

$$ItV = mc(\theta_2 - \theta_1) + m_c c_c(\theta_2 - \theta_1)$$
$$= (mc + m_c c_c)(\theta_2 - \theta_1)$$

whence c can be found if c_c is known.

2. *Method of mixtures*

(a) *Solids.* The solid is weighed to find its mass m, heated in boiling water at temperature θ_3 for 10 minutes, Fig. 4.7a, and then quickly transferred to

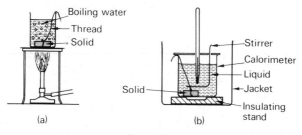

Fig. 4.7

a calorimeter of mass m_c containing a mass of water m_w at temperature θ_1, Fig. 4.7b. The water is stirred and the highest reading θ_2 on the thermometer noted.

Assuming no heat loss from the calorimeter when the hot solid is dropped into it, we have

heat given out	heat received by	heat received by
by solid cooling =	water warming +	calorimeter warming
from θ_3 to θ_2	from θ_1 to θ_2	from θ_1 to θ_2

If c is the specific heat capacity of the solid, c_w that of water and c_c that of the calorimeter, then

$$mc(\theta_3 - \theta_2) = m_w c_w(\theta_2 - \theta_1) + m_c c_c(\theta_2 - \theta_1)$$

$$= (m_w c_w + m_c c_c)(\theta_2 - \theta_1)$$

$$\therefore \quad c = \frac{(m_w c_w + m_c c_c)(\theta_2 - \theta_1)}{m(\theta_3 - \theta_2)}$$

Hence c can be found knowing c_w and c_c.

(b) *Liquids.* In this case a hot solid of known specific heat capacity is dropped into the liquid whose specific heat capacity is required, otherwise the procedure and calculation are the same as for a solid.

3. Continuous flow method

The method was devised in 1899 by Callendar and Barnes for measuring the specific heat capacity of water. A simple form of the apparatus is shown in Fig. 4.8. It consists of a wire carrying a steady electric current which heats

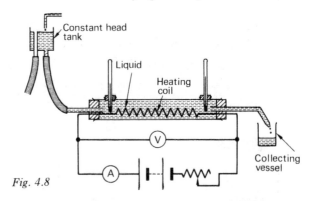

Fig. 4.8

a liquid flowing at a constant rate through a glass tube from a constant head tank to a collecting vessel. Two thermometers measure the entrance and exit temperatures of the liquid.

After a time the thermometer readings become steady and none of the electrical energy supplied by the current warms the apparatus (since it is at a constant temperature); it is either used to heat the liquid or is lost to the surroundings.

THERMAL PROPERTIES

Let the current in the wire be I_1 and the p.d. across it be V_1, then if θ_1 and θ_2 are the thermometer readings and m_1 is the mass of liquid collected in time t, we have

$$I_1 t V_1 = m_1 c(\theta_2 - \theta_1) + h$$

where c is the specific heat capacity of the liquid and h is the heat lost in time t to the surroundings.

The rate of flow is changed and the p.d. and current adjusted to V_2 and I_2 so that the entrance and exit temperatures are still θ_1 and θ_2. This ensures that h is the same as in the first case since the average temperature excess of the apparatus over the surroundings is unaltered. If m_2 is now the mass collected in the same time t then

$$I_2 t V_2 = m_2 c(\theta_2 - \theta_1) + h$$

Subtracting the two equations

$$(I_1 V_1 - I_2 V_2)t = c(m_1 - m_2)(\theta_2 - \theta_1)$$

$$\therefore \quad c = \frac{(I_1 V_1 - I_2 V_2)t}{(m_1 - m_2)(\theta_2 - \theta_1)}$$

The advantages of this method are (i) the heat capacity of the apparatus does not need to be known and (ii) consideration of heat loss is unnecessary. The chief difficulty is ensuring the liquid is mixed sufficiently to keep θ_2 constant.

In their more elaborate form of the apparatus Callendar and Barnes surrounded the glass tube by a vacuum jacket to make h very small, the currents and p.d.s were measured accurately by potentiometer methods and the temperatures were taken with platinum resistance thermometers (p. 112).

4. *Mechanical method*

The first part of the experiment described on p. 72 to calibrate a voltmeter can also be used to find the specific heat capacity of a solid in the form of the material of the drum. If M is the mass of the drum, c its specific heat capacity and θ the temperature *rise* produced by n revolutions with a mass m attached we can say,

mechanical energy supplied $= mg\pi \, dn$

and heat needed for a temperature rise $\theta = Mc\theta$

Assuming all the mechanical energy appears as heat,

therefore $$Mc\theta = mg\pi \, dn$$

$$\therefore \quad c = \frac{mg\pi \, dn}{M\theta}$$

123

Heat loss and cooling corrections

In experiments with calorimeters certain precautions can be taken to mini-mize heat losses. These include (*i*) polishing the calorimeter to reduce radiation loss, (*ii*) surrounding it by an outer container or a jacket of a poor heat conductor to reduce convection and conduction loss and (*iii*) support-ing it on an insulating stand or supports to minimize conduction.

When the losses, despite all precautions, are not small, or where great accuracy is required, an estimate can be made of the temperature that would have been reached, i.e. a 'cooling correction' is made which when added to the observed maximum temperature gives the estimated maximum tem-perature had no heat been lost. Alternatively the need to make a cooling correction can be eliminated, as in the continuous flow method described above, or in other ways, one of which is explained under the second pro-cedure.

1. Graphical method. As well as being suitable for electrical heating ex-periments, this method is convenient when finding the specific heat capacity of a bad thermal conductor (e.g. glass or rubber) by the method of mixtures. In the latter case the hot solid is slow to transfer heat to the calorimeter and water and some time elapses before the mixture reaches its maximum tem-perature. During this time appreciable cooling occurs even if the calorimeter is lagged.

To make the cooling correction, the temperature is taken at half-minute intervals starting *just before* the hot solid is added to the calorimeter and ending when the temperature has fallen by at least 1 °C from its observed maximum value. A graph of temperature against time is plotted. In that

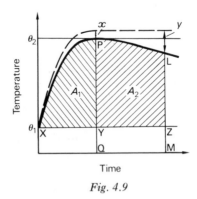

Fig. 4.9

shown in Fig. 4.9, θ_1 is the initial temperature of the calorimeter and contents (i.e. room temperature) and θ_2 is the observed maximum temperature. The dotted line shows how the temperature might have risen if no heat were lost.

The cooling correction required is x. To obtain it, PQ is drawn through the top of the curve parallel to the temperature axis and similarly LM further along the curve so that y is 1 °C. XYZ is then drawn through θ_1, parallel to the time axis. The areas A_1 and A_2 are found by counting the squares on the graph paper and it can be shown that the cooling correction is given by

$$x = \frac{A_1}{A_2} \times y$$

where $y = 1$ °C for convenience. The estimated maximum temperature is then $\theta_2 + x$.

This method is based on the assumption that the rate of loss of heat is directly proportional to the temperature difference between the body (e.g. calorimeter) and its surroundings. This is true for heat loss by (a) conduction (see p. 137), (b) convection so long as it is forced (i.e. a draught) or if natural, provided the temperature difference is small (see below) and (c) radiation if the temperature excess is small.

In electrical heating experiments, temperature–time readings are taken during and immediately after the heating, and the cooling correction obtained from a graph as explained above.

2. *Initial cooling method.* If the calorimeter and its contents are cooled to about 5 °C below room temperature and then heated steadily during the experiment to about 5 °C above, the heat gained from the surroundings during the first half of the time will be nearly equal to that lost to the surroundings during the second half. No cooling correction is then necessary. The method is suitable when finding the specific heat capacity of a liquid by electrical heating.

Cooling laws and temperature fall

(a) *Five-fourths power law.* For cooling in still air by natural convection the five-fourths power law holds. It states

$$rate\ of\ loss\ of\ heat \propto (\theta - \theta_0)^{5/4}$$

where θ is the temperature of the body in surroundings at temperature θ_0. If the temperature excess $(\theta - \theta_0)$ is small, the relation becomes approximately linear.

(b) *Newton's law of cooling.* Under conditions of forced convection, i.e. in a steady draught, Newton's law applies. It states

$$rate\ of\ loss\ of\ heat \propto (\theta - \theta_0)$$

and is true for quite large temperature excesses.

(c) *Rate of fall of temperature.* As well as the temperature excess, the rate of loss of heat from a body depends on the area and nature of its surface (i.e. whether it is dull or shiny). Hence for a body having a uniform temperature θ and a surface area A we can say, if Newton's law holds

$$rate\ of\ loss\ of\ heat\ =\ kA(\theta\ -\ \theta_0)$$

where k is a constant depending on the nature of the surface.

If the temperature θ of the body falls we can also write

$$rate\ of\ loss\ of\ heat\ =\ mc\ \times\ rate\ of\ fall\ of\ temperature$$

where m is the mass of the body and c is its specific heat capacity. Hence

$$rate\ of\ fall\ of\ temperature\ =\ kA(\theta\ -\ \theta_0)/(mc)$$

The mass of a body is proportional to its volume and so the rate of fall of temperature of a body is proportional to the ratio of its surface area to its volume, i.e. is inversely proportional to a linear dimension. A small body, therefore, cools faster than a large one (its temperature falls faster), as everyday experience confirms. In calorimeter experiments the use of large apparatus, etc., minimizes the effect of errors due to heat loss.

Latent heat

The heat which a body absorbs in melting, evaporating or sublimating and gives out in freezing or condensing is called *latent* (hidden) *heat* because it does not produce a change of temperature in the body—it causes a change of state or phase. Thus when water is boiling its temperature remains steady at 100 °C (at s.t.p.) even although heat, called latent heat of vaporization, is being supplied to it. Similarly the temperature of liquid naphthalene stays at 80 °C whilst it is freezing; there is no fall of temperature until all the liquid has solidified but heat, called latent heat of fusion, is still being given out by the liquid.

The kinetic theory sees the supply of latent heat to a melting solid as enabling the molecules to overcome sufficiently the forces between them for the regular crystalline structure of the solid to be broken down. The molecules then have the greater degree of freedom and disorder that characterize the liquid state. Thus, whilst heat which increases the kinetic energy component of molecular internal energy causes a temperature rise, the supply of latent heat is regarded as increasing the potential energy component since it allows the molecules to move both closer together and farther apart.

When vaporization of a liquid occurs a large amount of energy is needed to separate the molecules and allow them to move around independently as gas molecules. In addition some energy is required to enable the vapour to expand against the atmospheric pressure. The energy for both these operations is supplied by the latent heat of vaporization and like latent heat of fusion, we regard it as increasing the potential energy of the molecules.

(a) *Specific latent heat of fusion.* This is defined as *the quantity of heat required to change unit mass of a substance from solid to liquid without change of temperature.* It is denoted by the symbol l and is measured in $J\ kg^{-1}$ or $J\ g^{-1}$.

The specific latent heat of fusion of ice can be determined by the method of mixtures. A calorimeter of mass m_c is two-thirds filled with a mass m_w of water warmed to about 5 °C above room temperature. The temperature θ_1 of the water is noted, then a sufficient number of small pieces of ice, carefully dried in blotting paper, are added one at a time and the mixture stirred, until the temperature is about 5 °C below room temperature. The lowest temperature θ_2 is noted. The calorimeter and contents are then weighed to find the mass of ice m added.

The heat given out by the calorimeter and warm water in cooling from θ_1 to θ_2 does two things. First it supplies the latent heat needed to melt the ice at 0 °C to water at 0 °C and second it provides the heat to raise the now melted ice from 0 °C to the final temperature of the mixture θ_2. Hence

heat given out	heat used to	heat used to warm
by calorimeter	= melt ice at 0 °C +	melted ice from
and water cooling		0 °C to θ_2

If c_c and c_w are the specific heat capacities of the calorimeter and water respectively and l is the specific latent heat of fusion of ice then

$$m_c c_c(\theta_1 - \theta_2) + m_w c_w(\theta_1 - \theta_2) = ml + mc_w(\theta_2 - 0)$$

Hence

$$(m_c c_c + m_w c_w)(\theta_1 - \theta_2) = m(l + c_w \theta_2)$$

$$\therefore \quad l + c_w \theta_2 = \frac{(m_c c_c + m_w c_w)(\theta_1 - \theta_2)}{m}$$

and

$$l = \frac{(m_c c_c + m_w c_w)(\theta_1 - \theta_2)}{m} - c_w \theta_2$$

For ice the accepted value of l is 334 $J\ g^{-1}$. No cooling correction is necessary (see 'Initial cooling method' on p. 125) but the temperature of the mixture must not be taken more than 5 °C below room temperature otherwise water vapour in the air may condense to form dew on the calorimeter and give up latent heat to it.

(*b*) *Specific latent heat of vaporization.* This is *the quantity of heat required to change unit mass of a substance from liquid to vapour without change of temperature.* It is also denoted by *l* and measured in $J\ kg^{-1}$ or $J\ g^{-1}$.

A value can be found for *l* by a continuous flow-type method using the apparatus of Fig. 4.10. The liquid is heated electrically by a coil carrying a

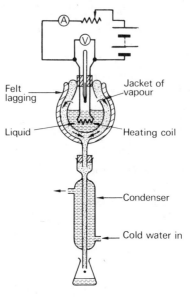

Fig. 4.10

steady current *I* and having a p.d. *V* across it. Vapour passes down the inner tube of a condenser where it is changed back to liquid by cold water flowing through the outer tube.

After the liquid has been boiling for some time it becomes surrounded by a 'jacket' of vapour at its boiling point and a steady state is reached when the rate of vaporization equals the rate of condensation. All the electrical energy supplied is then used to supply latent heat to the liquid (and none to raise its temperature) and to make good any heat loss from the 'jacket'. If a mass *m* of liquid is now collected in time *t* from the condenser, we have

$$ItV = ml + h$$

where *l* is the specific latent heat of vaporization of the liquid and *h* is the heat lost from the 'jacket' in time *t*. The 'jacket' of vapour makes *h* small and if it is neglected *l* can be found. Alternatively it may be eliminated as in the Callendar and Barnes experiment (p. 122) by a second determination with a different power input.

The specific latent heat of vaporization of water is $2260\ J\ g^{-1}$.

Heat calculations

1. *A piece of copper of mass 100 g is heated to 100 °C and then transferred to a well-lagged copper can of mass 50.0 g containing 200 g of water at 10.0 °C. Neglecting heat loss, calculate the final steady temperature of the water after it has been well stirred. Take the specific heat capacities of copper and water as 4.00×10^2 J kg^{-1} °C^{-1} and 4.20×10^3 J kg^{-1} °C^{-1} respectively.*

Let the final steady temperature $= \theta$

∴ Fall in temperature of piece of copper $= (100 - \theta)$ °C

Rise in temperature of can and water $= (\theta - 10)$ °C

Expressing masses in kg,

heat given out by copper $= 0.1 \times 400 \times (100 - \theta)$ J

heat received by copper can $= 0.05 \times 400 \times (\theta - 10)$ J

heat received by water $= 0.2 \times 4200 \times (\theta - 10)$ J

heat given out $=$ heat received

Therefore
$$40(100 - \theta) = 20(\theta - 10) + 840(\theta - 10)$$
$$= (20 + 840)(\theta - 10)$$
$$4000 - 40\theta = 860\theta - 8600$$
$$900\theta = 12\,600$$

Hence
$$\theta = 14.0 \text{ °C}$$

2. *The rate of flow of liquid through a continuous flow calorimeter is 15 g s^{-1} and the electric heating element dissipates 200 W, a steady difference of temperature of 3.0 °C being maintained. To maintain the same temperatures, 80 W is necessary when the flow is reduced to 5.0 g s^{-1}. Assuming the temperature of the surroundings to be the same in the two cases, calculate the specific heat capacity of the liquid and the rate at which heat is lost to the surroundings.*

In the steady state the apparatus absorbs no heat, hence

$$\frac{\text{electrical energy}}{\text{supplied per second}} = \frac{\text{heat received by}}{\text{water per second}} + \frac{\text{heat lost to surroundings}}{\text{per second}}$$

Since 1 watt $=$ 1 joule per second, then in the first case the electrical energy supplied per second $=$ 200 J s^{-1}. Also, heat received by water per second $= m \times c \times (\theta_1 - \theta_2)$ where $m = 15$ g, $c =$ specific heat capacity of liquid in J g^{-1} °C^{-1} and $(\theta_1 - \theta_2) = 3$ °C. If h is the heat lost to the surroundings per second then,

$$200 = 15 \times c \times 3 + h \qquad (1)$$

In the second case,

$$80 = 5 \times c \times 3 + h \qquad (2)$$

Subtracting (2) from (1),

$$200 - 80 = 45c - 15c$$
$$\therefore \quad 120 = 30c$$
$$\therefore \quad c = 120/30 = 4.0 \text{ J g}^{-1} \text{ }^{\circ}\text{C}^{-1}$$

Substituting for c in (1),

$$200 = 45 \times 4 + h$$
$$\therefore \quad h = 200 - 180 = 20 \text{ J s}^{-1}$$

3. When a current of 2.0 A is passed through a coil of constant resistance 15 Ω immersed in 500 g of water at 0 °C in a vacuum flask, the temperature of the water rises to 8 °C in 5 minutes. If instead the flask originally contained 250 g of ice and 250 g of water, what current must be passed through the coil if the mixture is to be heated to the same temperature in the same time? (Specific heat capacity of water = 4.2 J g^{-1} °C^{-1}; specific latent heat of fusion of ice = 3.3 × 10^2 J g^{-1}.)

Assuming no heat is lost from the vacuum flask then in time t,

$$\frac{\text{electrical energy}}{\text{supplied}} = \frac{\text{heat received}}{\text{by water}} + \frac{\text{heat received by}}{\text{vacuum flask}}$$

But, $\qquad$ electrical energy supplied $= I^2 Rt$

where $\qquad I = 2 \text{ A}, \quad R = 15 \text{ }\Omega \quad$ and $\quad t = 5 \times 60 = 300 \text{ s}$

Also, $\qquad$ heat received by water $= mc(\theta_1 - \theta_2)$

where $m = 500 \text{ g}, \quad c = 4.2 \text{ J g}^{-1} \text{ }^{\circ}\text{C}^{-1} \quad$ and $\quad (\theta_1 - \theta_2) = 8 \text{ }^{\circ}\text{C}$

If h is the heat received by the vacuum flask in time t,

$$4 \times 15 \times 300 = 500 \times 4.2 \times 8 + h$$
$$\therefore \quad 18\ 000 = 16\ 800 + h$$
$$\therefore \quad h = 1200 \text{ J}$$

In the second part, in the same time t,

$$\begin{array}{llll} \text{electrical} & \text{heat to melt} & \text{heat to warm} & \text{heat received} \\ \text{energy} & = 250 \text{ g ice} & + 500 \text{ g water} & + \text{ by vacuum} \\ \text{supplied} & \text{at 0 °C} & \text{from 0 °C to 8 °C} & \text{flask (i.e. } h) \end{array}$$

$$\therefore \quad I^2 \times 15 \times 300 = 250 \times 330 + 500 \times 4.2 \times 8 + 1200$$

where I is the current required to warm the ice and water to 8 °C in 300 s.

$$\therefore \quad I^2 \times 4500 = 82\,500 + 16\,800 + 1200 = 100\,500$$

$$\therefore \quad I^2 = \frac{100\,500}{4500} = 22.3$$

$$\therefore \quad I = 4.7 \text{ A}$$

Expansion of solids

(a) *Linear expansion.* The change of length which occurs with temperature change in a solid has to be allowed for in the design of many devices. The variation is described by the *linear expansivity* α. Thus if a solid of length l increases in length by δl due to a temperature rise $\delta\theta$, α for the material is defined by the equation

$$\alpha = \frac{\delta l}{l} \cdot \frac{1}{\delta\theta}$$

In words, α is *the fractional increase of length* (i.e. $\delta l/l$) *per unit rise of temperature.* The unit of α is $°C^{-1}$ (or K^{-1}) since δl and l have the same units (metres) and so $\delta l/l$ is a ratio.

As the temperature rise $\delta\theta$ tends to zero, α approaches the linear expansivity at a particular temperature and experiment shows that its value for a given material is not constant but varies slightly with temperature. Mean values are therefore obtained over a temperature range and in accurate work this range is stated. For ordinary purposes α can be assumed constant in the range 0 to 100 °C. The mean value for copper (at room temperatures) is $1.7 \times 10^{-5}\ °C^{-1}$; a copper rod 1 metre long therefore increases in length by 1.7×10^{-5} metre for every 1 °C temperature rise.

A useful expression is obtained if we consider a solid of original length l_0 which increases to l_θ for a temperature *rise* of θ. Replacing δl by $l_\theta - l_0$ and $\delta\theta$ by θ in the expression for α we get

$$\alpha = \frac{l_\theta - l_0}{l_0 \theta}$$

Rearranging $\qquad\qquad l_\theta - l_0 = l_0 \alpha \theta$

$$\therefore \quad l_\theta = l_0(1 + \alpha\theta)$$

Note that l_0 is the *original* length and, since values of α are very small, it need not be the length at 0 °C. (Unlike R_0 in the temperature coefficient of resistance formula $R_\theta = R_0(1 + \alpha\theta)$ on p. 77, which is generally taken to be the resistance at 0 °C.) For example, if the temperature of a 2-metre-long copper rod rises from 15 °C to 25 °C then $l_0 = 2$ m, $\theta = (25 - 15) = 10$ °C, $\alpha = 1.7 \times 10^{-5}$ °C^{-1} and

$$l_\theta - l_0 = l_0 \alpha \theta$$

$$= (2 \text{ m}) \times (1.7 \times 10^{-5} \text{ °C}^{-1}) \times (10 \text{ °C})$$

$$= 2 \times 1.7 \times 10^{-5} \times 10 \qquad \text{m} \times \text{°C}^{-1} \times \text{°C}$$

$$= 3.4 \times 10^{-4} \text{ m}$$

$$= 0.34 \text{ mm}$$

Thermal expansion of a solid can be explained on the atomic scale with the help of Fig. 4.11a which shows that the repelling forces between atoms

Fig. 4.11

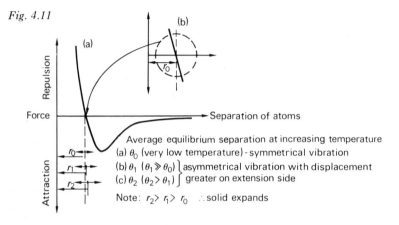

Average equilibrium separation at increasing temperature
(a) θ_0 (very low temperature) - symmetrical vibration
(b) θ_1 $(\theta_1 \gg \theta_0)$ } asymmetrical vibration with displacement
(c) θ_2 $(\theta_2 > \theta_1)$ } greater on extension side

Note: $r_2 > r_1 > r_0$ ∴ solid expands

increases more rapidly than the attractive forces as their separation varies. A rise of temperature increases the amplitude of vibration of the atoms about their equilibrium position but this will be greater on the extension side of that position, for the reason just given, and the average separation of the atoms increases. At low temperatures the amplitudes of oscillation are small and symmetrical, Fig. 4.11b, about the equilibrium position and so we would not expect expansion with increase of temperature—a fact confirmed by experiment. A further conclusion from this argument is that linear expansivities should increase with rising temperature, as they do.

(b) *Area expansion.* The change of area of a surface with temperature change is described by the *area* or *superficial expansivity* β. Thus if an area A increases by δA due to a temperature rise $\delta \theta$ then β is given by

$$\beta = \frac{\delta A}{A} \cdot \frac{1}{\delta \theta}$$

In words, β is *the fractional increase of area* (i.e. $\delta A/A$) *per unit rise of temperature.*

The variation of area with temperature is given by an equation similar to that for linear expansion, thus

$$A_\theta = A_0(1 + \beta\theta)$$

where A_0 and A_θ are the original and new areas respectively and θ is the temperature rise.

Fig. 4.12

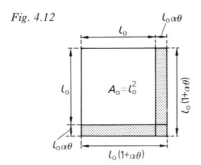

It can be shown that for a given material $\beta \simeq 2\alpha$. Consider a square plate of side l_0, Fig. 4.12. We have

$$A_0 = l_0{}^2$$

A temperature rise of θ causes the length of each side to become $l_0(1 + \alpha\theta)$ if the material is isotropic, i.e. has the same properties in all directions. Hence the new area A_θ is

$$A_\theta = l_0{}^2(1 + \alpha\theta)^2$$
$$= l_0{}^2(1 + 2\alpha\theta + \alpha^2\theta^2)$$

Now $\alpha^2\theta^2$ is very small compared with $2\alpha\theta$ and since $A_0 = l_0{}^2$ we have

$$A_\theta \simeq A_0(1 + 2\alpha\theta)$$

Comparing this with $A_\theta = A_0(1 + \beta\theta)$ it follows that

$$\beta \simeq 2\alpha$$

(c) *Volume expansion.* Changes of volume of a material with temperature are expressed by the *cubic expansivity* γ. Thus if a volume V increases by δV for a temperature rise $\delta\theta$ then γ is given by

$$\gamma = \frac{\delta V}{V} \cdot \frac{1}{\delta\theta}$$

In words, γ is *the fractional increase of volume* (i.e. $\delta V/V$) *per unit rise of temperature*.

The equation $V_\theta = V_0(1 + \gamma\theta)$ is also useful and it can be shown that for a given material $\gamma \simeq 3\alpha$. The proof involves calculating the volume change of a cube in terms of the linear expansion of each side—in a similar manner to that adopted for areas. Cubic and area expansivities for solids are not given in tables of physical constants since they are readily calculated from linear expansivities. The comment on the constancy of α (p. 131) also applies to β and γ.

A hollow body such as a bottle expands as if it were solid throughout, otherwise it would not retain the same shape when heated.

Thermal stress

Forces are created in a structure when thermal expansion or contraction is resisted. An idea of the size of such forces can be obtained by considering a metal rod of initial length l_0, cross-section area A and linear expansivity α, supported between two fixed end plates, Fig. 4.13. If the temperature of the

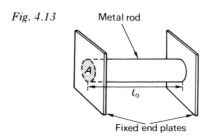

Fig. 4.13   Metal rod

l_0

Fixed end plates

rod is raised by θ °C, it tries to expand but is prevented by the plates and a compressive stress arises in it.

Removing one of the plates would allow the rod to expand freely and its new length l_0 would be $l_0 + l_0\alpha\theta$. We can therefore look upon the plate, when fixed in position, as exerting a force F on the rod and reducing its length from l_θ to l_0. Then (from p. 30),

$$\text{compressive strain} = \frac{\text{change in length}}{\text{original length}}$$

$$= \frac{l_\theta - l_0}{l_\theta} = \frac{l_0\alpha\theta}{l_\theta}$$

Also,

$$\text{compressive stress} = \frac{F}{A}$$

If Young's modulus for the material of the rod is E (the compressive modulus is the same as the tensile one for small compressions) then

$$E = \frac{\text{stress}}{\text{strain}} = \frac{Fl_\theta}{Al_0\alpha\theta}$$

The difference between l_θ and l_0 would be small compared with either and so to a good approximation $l_\theta = l_0$. Therefore

$$E = \frac{F}{A\alpha\theta} \quad \text{and} \quad F = EA\alpha\theta$$

For a steel girder with $E = 2.0 \times 10^{11}\,\text{Pa}\,(\text{N m}^{-2})$, $A = 100\,\text{cm}^2 = 10^{-2}\,\text{m}^2$, $\alpha = 1.2 \times 10^{-5}\,°\text{C}^{-1}$ and a temperature rise $\theta = 20\,°\text{C}$,

$$F = (2 \times 10^{11}\,\text{N m}^{-2}) \times (10^{-2}\,\text{m}^2) \times (1.2 \times 10^{-5}\,°\text{C}^{-1}) \times (20\,°\text{C})$$

$$= 2 \times 10^{11} \times 10^{-2} \times 1.2 \times 10^{-5} \times 20 \quad \text{N m}^{-2} \times \text{m}^2 \times °\text{C}^{-1} \times °\text{C}$$

$$= 4.8 \times 10^5\,\text{N}$$

A sizeable force. The original length of the rod does not affect the force but long rods tend to buckle at lower compressive stresses than short ones.

Thermal stress is put to good use in the technique of shrink fitting in which, for example, a large gear wheel is fitted on to a shaft of the same material. The diameter of the central hole in the wheel is smaller at room temperature than the outside diameter of the shaft. If the shaft is cooled with solid carbon dioxide ('dry ice') at $-78\,°\text{C}$ it can be fitted into the wheel. At room temperature the shaft is under compression and the wheel under tension and a tight-fitting joint results.

Expansion of liquids

(a) *Real and apparent expansion.* Only the cubic expansivity has meaning in the case of a liquid since its shape depends on the containing vessel. The cubic expansivities of liquids are generally greater than those for solids and like the latter they vary with temperature. Mean values are therefore obtained for the temperature range considered.

The expansion of a liquid is complicated by the fact that the vessel expands as well and makes the expansion of the liquid appear less than it really is. Consequently two expansivities are defined.

The apparent cubic expansivity γ_{app} of a liquid in a vessel of a particular material is the apparent fractional increase of volume per unit rise of temperature.

The real or absolute cubic expansivity γ_{real} of a liquid is the actual fractional increase of volume per unit rise of temperature and is always greater than γ_{app}.

It can be shown that

$$\gamma_{real} \simeq \gamma_{app} + \gamma_{material\ of\ vessel}$$

This relationship enables the apparent expansion of a liquid relative to its container to be calculated. For example, if the temperature of 100 cm^3 of mercury in a glass vessel is raised from 10 °C to 100 °C and γ_{real} for mercury $= 1.82 \times 10^{-4}$ °C^{-1} and $\alpha_{glass} = 8.00 \times 10^{-6}$ °C^{-1} then since

$$\gamma_{glass} = 3\alpha_{glass} = 2.40 \times 10^{-5}\ °\text{C}^{-1}$$

we have
$$\gamma_{app} \simeq \gamma_{real} - \gamma_{glass}$$
$$\simeq 1.82 \times 10^{-4} - 2.40 \times 10^{-5}$$
$$\simeq 1.58 \times 10^{-4}\ °\text{C}^{-1}$$

Hence

$$\text{apparent expansion of mercury} = V_0\gamma_{app}\theta$$
$$= 100 \times 1.58 \times 10^{-4} \times (100 - 10)$$
$$= 1.42\ \text{cm}^3$$

(*b*) *Variation of density with temperature.* It is sometimes more useful to know how the density rather than the volume of a liquid changes with temperature. Consider a fixed mass m of liquid of real cubic expansivity γ which occupies volume V_0 and has density ρ_0 at a certain temperature. Let the temperature *rise* by θ causing the volume to increase to V_θ and the density to decrease to ρ_θ then as for a solid,

$$V_\theta = V_0(1 + \gamma\theta)$$

$$\therefore \quad \frac{V_\theta}{V_0} = 1 + \gamma\theta$$

But
$$m = V_0\rho_0 = V_\theta\rho_\theta$$

$$\therefore \quad \frac{V_\theta}{V_0} = \frac{\rho_0}{\rho_\theta}$$

Thus
$$\frac{\rho_0}{\rho_\theta} = 1 + \gamma\theta$$

or
$$\rho_0 = \rho_\theta(1 + \gamma\theta)$$

(c) *Anomalous expansion of water.* At 4 °C the density of water is a maximum; from 4 °C to 0 °C it expands and the expansion is said to be anomalous, i.e. abnormal. It is this abnormal behaviour of water which results in convection currents ceasing when all the water in a pond has reached 4 °C, assuming, of course, a surface air temperature of that value or below. The expansion between 4 °C and 0 °C is explained on the assumption that at 4 °C the expansion due to the dissociation of complex molecules such as H_4O_2 and H_6O_3, already present in the water, more than cancels out the contraction due to the fall of temperature.

Water is also unusual because it expands on freezing, every 100 cm³ of water becoming 109 cm³ of ice. This accounts for the bursting of water pipes in very cold weather.

Thermal conductivity

Heat transfer, i.e. the passage of energy from a body at a higher temperature to one at a lower temperature, occurs by the three processes of conduction, convection and radiation although evaporation and condensation may often play an important part. In some cases the aim of the heat engineer is to encourage heat flow (as in a boiler) while in others it is to minimize it (e.g. lagging a house). Here we shall discuss *conduction*, i.e. the transfer of energy due to the temperature difference between neighbouring parts of the same body.

(a) *Definition of k.* Consider a *thin* slab of material of thickness δx and uniform cross-section area A between whose faces a *small* temperature difference $\delta\theta$ is maintained, Fig. 4.14. If a quantity of heat δQ passes

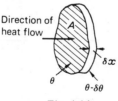

Direction of heat flow

θ

$\theta - \delta\theta$

δx

A

Fig. 4.14

through the slab by conduction in time δt, the *thermal conductivity k* of the material is defined by the equation

$$\frac{\delta Q}{\delta t} = -kA\frac{\delta\theta}{\delta x}$$

The negative sign indicates that heat flows towards the lower temperature, i.e. as x increases θ decreases thus making $\delta\theta/\delta x$—called the *temperature gradient*—negative. Inserting the negative sign ensures that $\delta Q/\delta t$ and k will be positive. In words, we may define k as *the rate of flow of heat through a material per unit area, per unit temperature gradient.*

The unit of k is $\text{J s}^{-1}\,\text{m}^{-1}\,{}^{\circ}\text{C}^{-1}$ or $\text{W m}^{-1}\,{}^{\circ}\text{C}^{-1}$ as can be seen by inserting units for the various quantities in the previous expression and remembering that 1 watt is 1 joule per second. Its value for copper is $390\ \text{W m}^{-1}\,{}^{\circ}\text{C}^{-1}$ and for asbestos $0.08\ \text{W m}^{-1}\,{}^{\circ}\text{C}^{-1}$ (both at room temperature).

When $\delta\theta \to 0$, k approaches the thermal conductivity at a particular temperature and experiment shows that its value for a given material varies slightly with temperature. (In the limiting case when $\delta x \to 0$, a cross-section is then being considered and the equation defining k can be more precisely written in calculus notation as $dQ/dt = -kA(d\theta/dx)$.) If measurements are not made over too great a temperature range, a constant mean value for k is usually assumed.

(*b*) *Temperature gradients.* When heat has been passing along a conductor for some time from a source of fixed high temperature, a steady state may be reached with the temperature at each point of the conductor becoming constant.

In the *unlagged* bar of Fig. 4.15*a* the quantity of heat passing in a given time through successive cross-sections decreases due to heat loss from the sides. The lines of heat flow are divergent and the temperature falls faster near the hotter end. For steady state conditions a graph of temperature

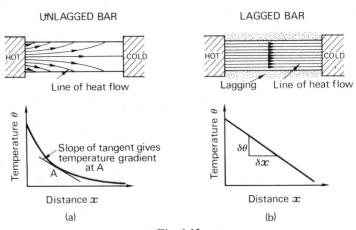

Fig. 4.15

θ against distance x from the hot end is as shown. The temperature gradient at any point is given by the slope of the tangent at that point (in calculus notation by $d\theta/dx$).

In a *lagged* bar whose sides are well wrapped with a good insulator, Fig. 4.15b, heat loss from the sides is negligible and the rate of flow of heat is the same all along the bar. The lines of heat flow are parallel and, in the steady state, the temperature falls at a constant rate as shown. The temperature gradient in this case is the slope of the graph, i.e. $\delta\theta/\delta x$.

There is a useful expression applicable to many simple problems in which the *lines of heat flow are parallel*; they are in a lagged bar and in a plate of large cross-section area. Why the latter? Consider a conductor of length x, cross-section area A and thermal conductivity k whose opposite ends are maintained at temperatures θ_2 and θ_1 ($\theta_2 > \theta_1$). From what has been said in the previous paragraph it follows that the quantity of heat Q passing any point in time t when the *steady state* has been reached is given by

$$\frac{Q}{t} = kA\left(\frac{\theta_2 - \theta_1}{x}\right)$$

We will now use this expression, which is sometimes called *Fourier's law*.

(c) *Composite slab problem.* Suppose we wish to find the rate of flow of heat through a plaster ceiling which measures 5 m × 3 m × 15 mm (*i*) without and (*ii*) with a 45 mm thick layer of insulating fibre-glass if the inside and outside surfaces are at the surrounding air temperatures of 15 °C and 5 °C respectively. ($k_{\text{plaster}} = 0.60 \text{ W m}^{-1} \,°\text{C}^{-1}$ and $k_{\text{fibre-glass}} = 0.040 \text{ W m}^{-1} \,°\text{C}^{-1}$.)

Assuming steady states are reached and lines of heat flow are parallel we can use rate of flow of heat $= Q/t = kA(\theta_2 - \theta_1)/x$. We have $A = 5 \times 3 = 15 \text{ m}^2$.

(*i*) Without fibre-glass, Fig. 4.16a, $x = 15 \text{ mm} = 0.015 \text{ m}$,

$$\frac{Q}{t} = \frac{(0.60 \text{ W m}^{-1} \,°\text{C}^{-1}) \times (15 \text{ m}^2) \times (10 \,°\text{C})}{(0.015 \text{ m})}$$

$$= \frac{0.60 \times 15 \times 10}{0.015} \quad \frac{\text{W m}^{-1} \,°\text{C}^{-1} \times \text{m}^2 \times °\text{C}}{\text{m}}$$

$$= 6.0 \times 10^3 \text{ W}$$

(In practice it will be very much less than this, see later.)

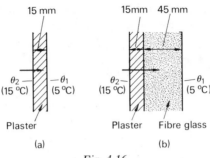

Fig. 4.16

(*ii*) With fibre-glass, Fig. 4.16*b*. Let the temperature of the plaster–fibre-glass boundary be θ. *The rate of flow of heat is the same through both materials.*

$$\therefore \quad \frac{Q}{t} = \frac{0.60 \times 15 \times (15 - \theta)}{0.015}$$

$$= \frac{0.04 \times 15 \times (\theta - 5)}{0.045}$$

Solving for θ we get

$$\theta = 14.8 \text{ °C}$$

$$\therefore \quad \frac{Q}{t} = \frac{0.60 \times 15 \times (15 - 14.8)}{0.015}$$

$$= 1.2 \times 10^2 \text{ W}$$

The previous example shows that when heat flows through a composite slab the temperature fall per mm is greater across the poorer conductor. This is of practical importance where a good conductor is in contact with a bad conductor such as air (or a liquid); the latter in fact controls the rate of conduction of heat through the good conductor. If it is assumed that the surface of a good conductor is at the same temperature as that of the surrounding air, heat flows will be obtained that are of the order of one hundred times too large. In Fig. 4.17 most of the temperature drop occurs in the layer

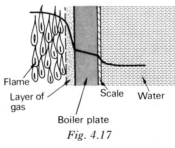

Fig. 4.17

of gas between the flame and the boiler plate and in any scale deposited on the plate by the water.

(*d*) *Mechanisms of thermal conduction.* In solids (and liquids) two processes seem to be involved. The first concerns atoms and the second 'free' electrons.

Atoms at a higher temperature vibrate more vigorously about their equilibrium positions in the lattice than their colder neighbours. But because they are coupled to them by interatomic bonds they pass on some of their vibratory energy and cause them to vibrate more energetically as well. These in turn affect other atoms and thermal conduction occurs. However, the process is generally *slow* because atoms, compared with electrons, are massive and the increases in vibratory motion are therefore fairly small. Consequently materials such as electrical insulators, in which this is the main conduction mechanism, are usually poor thermal conductors.

This effect is often regarded as resembling the passage of elastic waves through the material and because light waves are considered to have a dual nature, sometimes behaving as particles called photons, so too are these elastic waves considered to have particle-like forms called *phonons*. On this view, thermal conduction is said to be due to phonons having collisions with, and transferring energy to, atoms in the lattice.

The second process concerns materials with a supply of 'free' electrons. In these the electrons share in any gain of energy due to temperature rise of the material and their velocities increase much more than those of the atoms in the lattice since they are considerably lighter. They are thus able to move over larger distances and pass on energy *quickly* to cooler parts. Materials such as electrical conductors in which this mechanism predominates are therefore good thermal conductors; 'free' electrons are largely responsible for both properties.

Lest it be thought that only metals are good thermal conductors it should be stated that phonons can be a very effective means of heat transfer, especially at low temperatures. Thus at about -180 °C synthetic sapphire (Al_2O_3) is a better conductor than copper.

Methods of measuring k

Since we use the expression $Q/t = kA(\theta_2 - \theta_1)/x$ we must ensure that (*i*) the specimen is in a steady state and (*ii*) the lines of heat flow are parallel. By measuring the rate of flow of heat Q/t through the specimen, the temperature gradient $(\theta_2 - \theta_1)/x$ and the cross-section area A, k can be calculated.

THERMAL PROPERTIES

(a) *Good conductors.* The problem here is to obtain a measurable temperature gradient and is solved by using a *bar* of the material which is long compared with its diameter. Large x and small A then make $(\theta_2 - \theta_1)$ sufficiently big whilst still giving a satisfactory heat flow rate.

The apparatus due to Searle is shown in Fig. 4.18. The bar under study is heavily lagged and heated at one end by steam (or by an electrical heating coil in some arrangements). A spiral copper tube soldered to the bar carries

Fig. 4.18

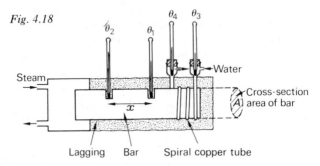

Steam

Water

Cross-section area of bar

x

Lagging   Bar   Spiral copper tube

a steady flow of water which is warmed at the cold end by the heat conducted down the bar. Two thermometers record the entrance and exit temperatures θ_3 and θ_4 of the water. Which will be the greater? The thermometers giving the temperatures θ_2 and θ_1 at a known separation x on the bar are inserted in holes containing mercury. Why?

When all four thermometer readings are constant they are noted and the mass m of water flowing through the copper tube in time t is found with a measuring cylinder and stop-watch. If Q is the quantity of heat flowing down the bar also in time t and A is the cross-section area of the bar then

$$\frac{Q}{t} = kA\left(\frac{\theta_2 - \theta_1}{x}\right)$$

This heat is taken from the bar by the mass m of cooling water of specific heat capacity c and so

$$\frac{Q}{t} = mc(\theta_4 - \theta_3)$$

Hence

$$kA\left(\frac{\theta_2 - \theta_1}{x}\right) = mc(\theta_4 - \theta_3)$$

from which k can be found.

In this method once a steady state has been reached the rate of flow of heat and the temperature gradient are the same for *any* section of the bar, since it is lagged. It is then possible to measure each at different parts of the bar.

142

For accurate work thermocouples bound to the bar replace the mercury thermometers.

(b) *Poor conductors.* In this case even a thin specimen gives a measurable temperature gradient; the difficulty is getting an adequate rate of heat flow. It is overcome by using a thin disc of the material (i.e. x small and A large); this shape also helps to reduce heat loss from the sides of the specimen thereby giving parallel lines of heat flow.

A simple form of the apparatus adapted from one due to Lees, is shown in Fig. 4.19a. A disc D of the material under test rests on a thick brass slab B containing a thermometer and is heated from above by a steam-chest C whose thick base also carries a thermometer.

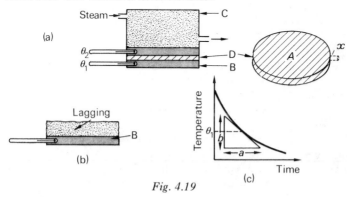

Fig. 4.19

In the first part of the experiment steam is passed until temperatures θ_2 and θ_1 are steady. We can then say that the heat passing per second through D, to B from C, equals the heat lost per second by B at temperature θ_1 to the surroundings. If this is Q then

$$Q = kA\left(\frac{\theta_2 - \theta_1}{x}\right)$$

A, x, θ_2 and θ_1 can all be measured.

In the second part D is removed and B heated directly from C until its temperature is about 5 °C above what it was in the first part, i.e. about 5 °C above θ_1. C is now removed and a thick felt pad placed on B, Fig. 4.19b. Temperature–time readings are taken and a cooling curve plotted. The rate of fall of temperature of B (in °C per second) at θ_1 equals the slope (b/a) of the tangent to the curve at θ_1, Fig. 4.19c. If we assume that the heat lost per second by B at θ_1, cooling under more or less the same conditions as in the first part (since the felt minimizes heat loss from the top surface) is Q then

$$Q = mc\left(\frac{b}{a}\right)$$

143

where m is the mass of B and c its specific heat capacity. Hence k can be found from

$$kA \left(\frac{\theta_2 - \theta_1}{x} \right) = mc \left(\frac{b}{a} \right)$$

Note that whilst B's temperature is steady in the first part when it is supplied with heat from C, in the second part it is falling since it is drawing on its own internal energy.

QUESTIONS

Temperature: thermometers

1. Explain what is meant by a *scale of temperature* and how a temperature is defined in terms of a specified property.

When a particular temperature is measured on scales based on different properties it has a different numerical value on each scale except at certain points. Explain why this is so and state (*a*) at what points the values agree and (*b*) what scale of temperature is used as a standard.

Explain the principles of two different types of thermometer one of which is suitable for measuring a rapidly varying temperature and the other for measuring a steady temperature whose value is required to a high degree of accuracy. Give reasons for your choice of thermometer in each case. Experimental details are not required. (*L.*)

2. (*a*) What is the value of the temperature θ in $^\circ$C on the scale of a platinum resistance thermometer if $R_0 = 2.000$ ohms, $R_{100} = 2.760$ ohms and $R_\theta = 2.480$ ohms?

(*b*) The resistance of a wire at a temperature θ $^\circ$C measured on a standard scale is given by

$$R_\theta = R_0(1 + A\theta + 10^{-3}A\theta^2)$$

where A is a constant. When the thermometer is at a temperature of 50.0 $^\circ$C on the standard scale, what will be the temperature indicated on the resistance scale?

(*J.M.B. part* $\cdot$*qn.*)

Specific heat capacity: latent heat

3. A current of 2.50 A passing through a heating coil immersed in 180 g of paraffin (specific heat capacity 2.00 J g^{-1} K^{-1}) contained in a 100 g calorimeter (specific heat capacity 0.400 J g^{-1} K^{-1}) raises the temperature from 5 $^\circ$C below room temperature to 5 $^\circ$C above room temperature in 100 s. What should be the reading of a voltmeter connected across the heating coil? (*S.*)

4. When water was passed through a continuous flow calorimeter the rise in temperature was from 16.0 to 20.0 $^\circ$C, the mass of water flowing was 100 g in 1 minute, the potential difference across the heating coil was 20.0 V and the current was 1.50 A.

THERMAL PROPERTIES

Another liquid at 16.0 °C was then passed through the calorimeter and to get the same change in temperature the potential difference was changed to 13.0 V, the current to 1.20 A and the rate of flow to 120 g in 1 minute. Calculate the specific heat capacity of the liquid if the specific heat capacity of water is assumed to be 4.20 J g^{-1} $°C^{-1}$.

State two advantages of the continuous flow method of calorimetry. (*J.M.B.*)

5. Give a labelled diagram of a continuous-flow apparatus which could be used to determine the specific heat capacity of water. The diagram should include the electrical circuit, but a description of the apparatus is *not* required.

In such an experiment, the following readings were taken:

Current in heating coil	2.0 A	1.5 A
Potential difference across coil	6.0 V	4.5 V
Mass of water collected	42.3 g	70.2 g
Time of flow	60 s	180 s
Inlet temperature	38.0 °C	38.0 °C
Outlet temperature	42.0 °C	42.0 °C

Explain how each reading would be taken, and use the figures to obtain a value for the specific heat capacity of water in J g^{-1} $°C^{-1}$. If each temperature reading was subject to an uncertainty of ± 0.1 °C, find the resulting percentage uncertainty in the specific heat capacity due to this cause alone. (*C. part qn.*)

6. Describe how you would determine experimentally the specific heat capacity of *either* copper *or* water by a direct mechanical method.

(You may assume, as necessary, that the values of other specific heats are known.)

A metal disc of radius 0.050 m and thickness 0.10 m is turned at 10 revolutions per second against a friction band, the tensions on the two sides of the band being 45 N and 5.0 N respectively. The density of the metal is 8.0 × 10^3 kg m^{-3} and its specific heat capacity 4.0 × 10^2 J kg^{-1} $°C^{-1}$. Find the rate at which the temperature of the disc rises initially.

For what speed of revolution would a disc of the same dimensions, but consisting of metal of density 2.7 × 10^3 kg m^{-3} and specific heat capacity 8.0 × 10^2 J kg^{-1} $°C^{-1}$, give the same initial rate of temperature rise when rotated against the friction band under the same tensions as before? (*O.*)

7. Define the terms *specific heat capacity, specific latent heat of vaporization.*

Describe how the specific latent heat of vaporization of a liquid such as alcohol may be determined by an electrical method explaining how a correction for heat losses may be made in the experiment.

A well-lagged copper calorimeter of mass 100 g contains 200 g of water and 50.0 g of ice all at 0 °C. Steam at 100 °C containing condensed water at the same temperature is passed into the mixture until the temperature of the calorimeter and its contents is 30.0 °C. If the increase in mass of the calorimeter and its contents is 25.0 g calculate the percentage of condensed water in the wet steam. (Assume that specific latent heat of vaporization of water at 100 °C = 2.26 × 10^3 J g^{-1}; specific latent heat of fusion of ice at 0 °C = 3.34 × 10^2 J g^{-1}; specific heat capacity of copper = 0.400 J g^{-1} $°C^{-1}$; mean specific heat capacity of water = 4.18 J g^{-1} $°C^{-1}$ (*L.*)

THERMAL PROPERTIES

Expansion

8. The steel cylinder of a car engine has an aluminium alloy piston. At 15 °C the internal diameter of the cylinder is exactly 8.0 cm and there is an all-round clearance between the piston and the cylinder wall of 0.050 mm. At what temperature will they fit perfectly? (Linear expansivities of steel and the aluminium alloy are 1.2×10^{-5} and 1.6×10^{-5} °C^{-1} respectively.)

9. The height of the mercury column in a barometer is 76.46 cm as read at 15 °C by a brass scale which was calibrated at 0 °C. Calculate the error caused by the expansion of the scale and hence find the true height of the column. (Linear expansivity of brass is 1.900×10^{-5} °C^{-1}.)

10. Calculate the minimum tension with which platinum wire of diameter 0.10 mm must be mounted between two points in a stout invar frame if the wire is to remain taut when the temperature rises 100 °C. Platinum has linear expansivity 9.0×10^{-6} °C^{-1} and Young's modulus 1.7×10^{11} Pa. The thermal expansion of invar may be neglected. (O. and C. part qn.)

11. (a) Distinguish between the *real* and *apparent* cubic expansivities.

A glass vessel contains some tungsten and is then filled with mercury to a certain mark. It is found that the mercury level remains at this mark despite changes of temperature. What is the ratio of volumes of the mercury and tungsten? (Linear expansivities of tungsten and glass are 4.4×10^{-6} and 8.0×10^{-6} °C^{-1} respectively, and real cubic expansivity of mercury is 1.8×10^{-4} °C^{-1}.) (L. part qn.)

(b) The density of a certain oil at 15 °C is 1.03 g cm^{-3} and its cubic expansivity is 8.50×10^{-4} °C^{-1}; the density of water at 4 °C is 1.00 g cm^{-3} and its mean cubic expansivity over the range concerned is 2.10×10^{-4} °C^{-1}. Find the temperature at which drops of the oil will just float in water. (O. part qn.)

Thermal conductivity

12. A cubical container full of hot water at a temperature of 90 °C is completely lagged with an insulating material of thermal conductivity 6.4×10^{-4} W cm^{-1} °C^{-1} (6.4×10^{-2} W m^{-1} °C^{-1}). The edges of the container are 1.0 m long and the thickness of the lagging is 1.0 cm. Estimate the rate of flow of heat through the lagging if the external temperature of the lagging is 40 °C. Mention any assumptions you make in deriving your result.

Discuss qualitatively how your result will be affected if the thickness of the lagging is increased considerably assuming that the temperature of the surrounding air is 18 °C. (J.M.B.)

13. Explain what is meant by *temperature gradient*.

An ideally lagged compound bar 25 cm long consists of a copper bar 15 cm long joined to an aluminium bar 10 cm long and of equal cross-sectional area. The free end of the copper is maintained at 100 °C and the free end of the aluminium at 0 °C. Calculate the temperature gradient in each bar when steady state conditions have been reached.

Thermal conductivity of copper $= 3.9$ W cm^{-1} °C^{-1}.
Thermal conductivity of aluminium $= 2.1$ W cm^{-1} °C^{-1} (J.M.B.)

14. Define thermal conductivity and explain how you would measure its value for a good conductor. Explain why the method which you describe is not suitable for a poor conductor.

The walls of a container used for keeping objects cool consist of two thicknesses of wood 0.50 cm thick separated by a space 1.0 cm wide packed with a poorly conducting material. Calculate the rate of flow of heat per unit area into the container if the temperature difference between the internal and external surfaces is 20 °C. (Thermal conductivity of wood = 2.4×10^{-3} W cm^{-1} °C^{-1}, of the poorly conducting material = 2.4×10^{-4} W cm^{-1} °C^{-1}.) (A.E.B.)

15. The ends of a straight uniform metal rod are maintained at temperatures of 100 °C and 20 °C, the room temperature being below 20 °C. Draw sketch-graphs of the variation of the temperature of the rod along its length when the surface of the rod is (a) lagged, (b) coated with soot, (c) polished. Give a qualitative explanation of the form of the graphs.

A liquid in a glass vessel of wall area 595 cm^2 and thickness 2.0 mm is agitated by a stirrer driven at a uniform rate by an electric motor rated at 100 W. The efficiency of conversion of electrical to mechanical energy in the motor is 75%. The temperature of the outer surface of the glass is maintained at 15.0 °C. Estimate the equilibrium temperature of the liquid, stating any assumptions you make.

Thermal conductivity of glass = 0.840 W m^{-1} °C^{-1}. (C.)

5 Optical properties

Introduction

The scientific study of light and optical materials is involved in the making of spectacles, cameras, projectors, binoculars, microscopes and telescopes. Whilst the most important of all optical materials are the various kinds of glass, others such as plastics, Polaroid, synthetic and natural crystals have useful applications.

In this chapter we shall consider the behaviour of certain optical components and instruments. Light will be treated as a form of energy which travels in straight lines called *rays*, a collection of rays being termed a *beam*. The ray treatment of light is known as geometrical optics and is developed from

  (i) rectilinear propagation, i.e. straight-line travel
 (ii) the laws of reflection
(iii) the laws of refraction.

When light comes to be regarded as waves it will be seen that shadows cast by objects are not as sharp as rectilinear propagation suggests. However, for the present it will be sufficiently accurate to assume that light does travel in straight lines so long as we exclude very small objects and apertures (those with diameters less than about 10^{-2} mm).

Reflection at plane surfaces

(a) *Laws of reflection.* When light falls on a surface it is partly reflected, partly transmitted and partly absorbed. Considering the part reflected, experiments with rays of light and mirrors show that two laws hold.

1. *The angle of reflection equals the angle of incidence,* i.e. $i_1 = i_2$ in Fig. 5.1.

2. *The reflected ray is in the same plane as the incident ray and the normal to the mirror at the point of incidence,* i.e. the reflected ray, is not turned to either side of the normal as seen from the incident ray.

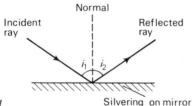

Fig. 5.1   Silvering on mirror

Note that the angles of incidence and reflection are measured to the normal to the surface and not with the surface itself.

(b) *Regular and diffuse reflection.* A mirror in the form of a highly polished metal surface or a piece of glass with a deposit of silver on its back surface reflects a high percentage of the light falling on it. If a parallel beam of light falls on a plane (i.e. flat) mirror in the direction IO, it is reflected as a parallel beam in the direction OR and *regular reflection* is said to occur, Fig. 5.2a.

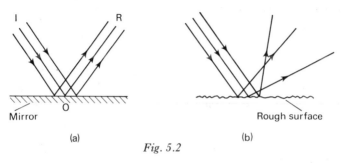

(a)   (b)

Fig. 5.2

Most objects, however, reflect light diffusely and the rays in an incident parallel beam are reflected in many directions as in Fig. 5.2b. *Diffuse reflection* is due to the surface of the object not being perfectly smooth like a mirror and although at each point on the surface the laws of reflection are observed, the angle of incidence and therefore the angle of reflection varies from point to point.

(c) *Rotation of a mirror.* When a mirror is rotated through a certain angle, the reflected ray turns through *twice* that angle. This is a useful fact which can be proved by considering the plane mirror of Fig. 5.3a. When it is in position MM, the ray IO, incident at angle i, is reflected along OR so that $\angle RON = i$.

If the mirror is now rotated through an angle θ to position M'M' and the *direction of the incident ray kept constant*, Fig. Fig. 5.3b, the angle of incidence

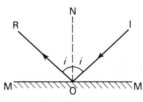

Fig. 5.3a

149

$\angle$ ION' becomes $(i + \theta)$ since the angle between the first and second positions of the normals, i.e. $\angle$ NON', is also θ. Let OR' be the new direction of the reflected ray then $\angle$ R'ON' $= (i + \theta)$. The reflected ray is thus turned through $\angle$ ROR' and

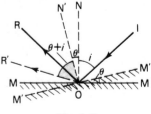

Fig. 5.3b

$$\angle ROR' = \angle R'ON - \angle RON$$
$$= \angle R'ON' + \angle N'ON - \angle RON$$
$$= (\theta + i) + \theta - i$$
$$= 2\theta$$
$$= \text{twice the angle of rotation of the mirror}$$

(*d*) *Optical lever and light-beam galvanometers.* Some sensitive galvanometers use a beam of light in conjunction with a small mirror as a pointer. The arrangement is called an *optical lever*, it uses the 'rotation of a mirror' principle and increases the ability of the meter to detect small currents, i.e. makes it more sensitive.

A tiny mirror M is attached to the part (e.g. a coil of wire) of the meter which rotates when a current flows in it, Fig. 5.4. A beam of light from a fixed lamp falls on M and is reflected onto a scale S. For a given current, the

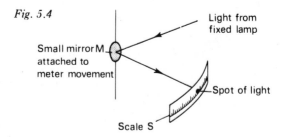

Fig. 5.4

Light from fixed lamp

Small mirror M attached to meter movement

Spot of light

Scale S

longer the pointer (i.e. the reflected beam) the greater will be the deflection observed on S. Besides being almost weightless the arrangement has the additional advantage of doubling the rotation of the moving part since the angle the reflected beam turns through is twice the angle of rotation of the mirror, the direction of the incident ray remaining fixed.

Images in plane mirrors

(*a*) *Point object.* The way in which the image of a point object is seen in a plane mirror is shown in Fig. 5.5. Rays from the object at O are reflected according to the laws of reflection so that they *appear* to come from a point I

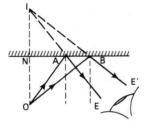

Fig. 5.5

behind the mirror and this is where the observer imagines the image to be. The image at I is called an unreal or *virtual* image because the rays of light do not actually pass through it, they only seem to come from it. It would not be obtained on a screen placed at I as would a *real* image which is one where rays really do meet. (The image produced on a cinema screen by a projector is a real image.) Rays OA and AE are real rays, IA is a virtual ray that appears to have travelled a certain path but has not; it gives rise to a virtual image.

Everyday observation suggests and experiment shows that *the image in a plane mirror is as far behind the mirror as the object is in front and that the line joining the object to the image is perpendicular to the mirror*, i.e. in Fig. 5.5 ON = NI and OI is at right angles to the mirror. However, you should be able to show, using the first law of reflection and congruent triangles, that this is so and also that *all* rays from O, after reflection, appear to intersect at I. A perfect image is thus obtained, i.e. all rays from the point object pass through *one* point on the image; a plane mirror is one of the few optical devices achieving this.

It is possible for a plane mirror to give a real image. Thus in Fig. 5.6a a convergent beam is reflected so that the reflected rays actually pass through a point I in front of the mirror. There is a real image at I which can be picked up on a screen. At the point O, towards which the incident beam was converging before it was intercepted by the mirror, there is considered to be a *virtual object*. Later we will find it useful on occasion to treat a convergent beam in this way. Comparing Figs. 5.6a and b we see that in the first, a *convergent beam* regarded as a virtual point object, gives a real point image

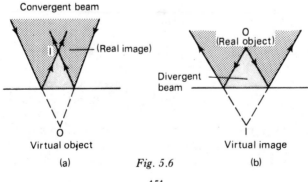

Fig. 5.6

(a) (b)

151

whilst in the second a *divergent beam* from a real point object gives a virtual point image. In both cases object and image are equidistant from the mirror.

(*b*) *Extended object.* Each point on an extended (finite-sized) object produces a corresponding point image. In Fig. 5.7 the image of a point A on the object is at A′, the two points being equidistant from the mirror. The image

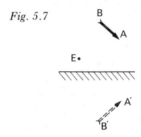

Fig. 5.7

of point B is at B′. If an eye at E views the object directly it sees A on the right of B, but if it observes the image in the mirror A′ is on the left. The right-hand side of the object thus becomes the left-hand side of the image and vice versa. The image is said to be *laterally inverted*, i.e. the wrong way round, as you can check by looking at yourself in a mirror.

(*c*) *No parallax method of locating images.* Suppose the object is a small pin O placed in front of a plane mirror M. To find the position of its virtual image I, a large locating pin P is placed behind M, Fig. 5.8*a*, and moved towards or away from M until P and the image of O always appear to move

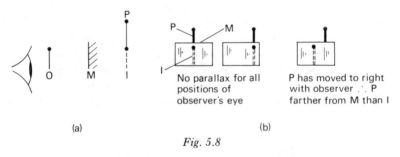

Fig. 5.8

together when the observer moves his head from side to side. P and I are then coincident and P is at the position of the image of O. When P and I do not coincide there is relative movement, called *parallax*, between them when the observer's head is moved sideways. The location of I can be achieved more quickly by remembering that if P is farther from M than I, then P appears to move in the *same* direction as the observer, Fig. 5.8*b*.

The no parallax method is used to find real as well as virtual images as we shall later when curved mirrors and lenses are considered.

(d) *Inclined mirrors.* Two mirrors M_1 and M_2 at $90°$ form *three* images of an object P placed between them, Fig. 5.9a. I_1 is formed by a single reflection at M_1, I_2 by a single reflection at M_2 and I_3 by reflections at M_1 and M_2. The line joining each image and its object is perpendicularly bisected by the mirror involved (we can think of I_3 as being either the image of I_1 acting as an object for mirror M_2 extended to the left or the image of I_2 as an object for mirror M_1 extended downwards) and so it follows that OP $=$ OI$_1$ $=$ OI$_2$ $=$ OI$_3$. Justify this. Hence P, I_1, I_2 and I_3 lie on a circle centre O.

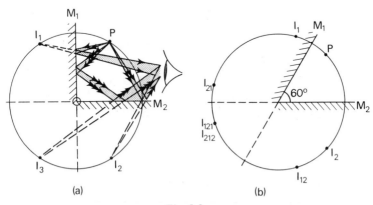

(a)   (b)

Fig. 5.9

Two mirrors at $60°$ to each other form *five* images, Fig. 5.9b. As the angle between the mirrors decreases, the number of images increases and in general for an angle θ (which is such that $360/\theta$ is an integer) it can be shown that $[(360/\theta) - 1]$ images are formed. When the mirrors are parallel $\theta = 0°$ and in theory an infinite number of images should be obtained, all lying on a straight line passing through the object and perpendicular to the mirrors, Fig. 5.10. In practice some light is lost at each reflection and a limited number only is seen. If the distances of P from M_1 and M_2 are a and b respectively, prove that the separation of the images is successively $2a$, $2b$, $2a$, $2b$, etc.

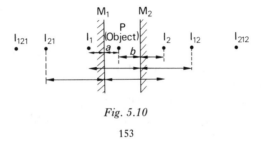

Fig. 5.10

153

Curved mirrors

Curved mirrors are used as car driving-mirrors and as reflectors in car head-lamps, searchlights and flashlamps. They are an essential component of the largest telescopes. We shall consider mainly spherical mirrors, i.e. those which are part of a spherical surface.

(a) *Terms and definitions.* There are two types of spherical mirror, con-cave and convex, Figs. 5.11a and b. In a concave mirror the centre C of the

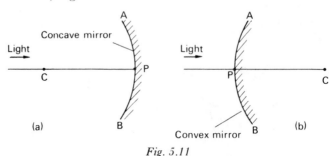

Fig. 5.11

sphere of which the mirror is a part is in front of the reflecting surface, in a convex mirror it is behind. C is the *centre of curvature* of the mirror and P, the centre of the mirror surface, is called the *pole*. The line CP produced is the *principal axis*. AB is the *aperture* of the mirror.

Observation shows that a *narrow* beam of rays, parallel and near to the principal axis, is reflected from a concave mirror so that all rays converge to a point F on the principal axis, Fig. 5.12. F is called the *principal focus* of the mirror and it is a *real* focus since light actually passes through it. Concave mirrors are also known as converging mirrors because of their action on a parallel beam of light.

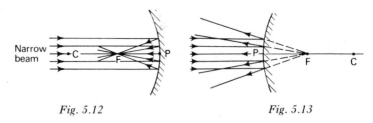

Fig. 5.12   Fig. 5.13

A narrow beam of rays, parallel and near to the principal axis, falling on a convex mirror is reflected to form a divergent beam which *appears* to come from a point F behind the mirror, Fig. 5.13. A convex mirror thus has a *virtual* principal focus; it is also called a diverging mirror.

Rays which are close to the principal axis and make small angles with it, i.e. are nearly parallel to the axis, are called *paraxial* rays. Our treatment of spherical mirrors will be restricted at present to such rays, which, in effect, means we shall consider only mirrors of small aperture. In diagrams, however, they will be made larger for clarity.

As well as bringing to a point focus F, paraxial rays parallel to the principal axis, spherical mirrors form a point image of *all* paraxial rays from a point object (as we will see shortly, p. 159).

(*b*) *Relation between f and r.* The distance PC from the pole to the centre of curvature of a spherical mirror is called its *radius of curvature* (*r*); the distance PF from the pole to the principal focus is its *focal length* (*f*). A simple relation exists between *f* and *r*.

In Fig. 5.14 a ray AM, parallel to the principal axis of a concave mirror of small aperture, is reflected through the principal focus F. If C is the centre

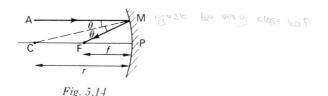

Fig. 5.14

of curvature, CM is the normal to the mirror at M because the radius of a spherical surface is perpendicular to the surface. Hence by the first law of reflection

$$\angle AMC = \angle CMF = \theta$$

But $\qquad\qquad \angle AMC = \angle MCF \quad$ (alternate angles)

$$\therefore \quad \angle CMF = \angle MCF$$

$\triangle FCM$ is thus isosceles and FC = FM. The rays are paraxial and so M is very close to P; therefore to a good approximation FM = FP.

$$\therefore \quad FC = FP \quad or \quad FP = \tfrac{1}{2}CP$$

That is $\qquad\qquad\qquad f = r/2$

Thus the *focal length of a spherical mirror is approximately half the radius of curvature.* Check this relation for a convex mirror using Fig. 5.15.

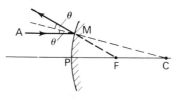

Fig. 5.15

OPTICAL PROPERTIES

Images in spherical mirrors

In general the *position* of the image formed by a spherical mirror and its *nature* (i.e. whether it is real or virtual, inverted or upright, magnified or diminished) depend on the distance of the object from the mirror. Information about the image in any case can be obtained either by drawing a ray diagram or by calculation using formulae.

(*a*) *Ray diagrams.* We shall assume that small objects on the principal axes of mirrors of small aperture are being considered so that all rays are paraxial. Point images will therefore be formed of points on the object.

To construct the image, *two* of the following three rays are drawn from the *top of the object.*

(*i*) A ray parallel to the principal axis which after reflection actually passes through the principal focus or appears to diverge from it.

(*ii*) A ray through the centre of curvature which strikes the mirror normally and is reflected back along the same path.

(*iii*) A ray through the principal focus which is reflected parallel to the principal axis, i.e. a ray travelling the reverse path to that in (*i*).

Fig. 5.16

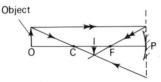

(i) *Object* beyond C.
Image between C and F, real, inverted, diminished.

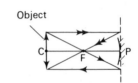

(ii) *Object* at C.
Image at C, real, inverted, same size.

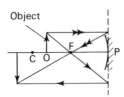

(iii) *Object* between C and F.
Image beyond C, real, inverted, magnified.

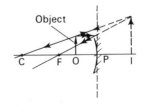

(iv) *Object* between F and P.
Image behind mirror, virtual, upright, magnified.

Notes : 1. In (i) and (iii) O and I are interchangeable; such positions of object and image are called *conjugate points.*

2. C is a *self-conjugate point*—as (ii) shows, object and image are coincident at C.

3. If the object is at infinity (i.e. a long way off), a real image is formed at F. Conversely an object at F gives a real image at infinity.

4. In all cases the foot of the object is on the principal axis and its image also lies on this line.

Since we are considering paraxial rays the mirror must be represented by a *straight line* in accurate diagrams. It should also be appreciated that the rays drawn are *constructional* rays and are not necessarily those by which the image is seen.

The diagrams for a concave mirror are shown in Fig. 5.16 and for a convex mirror in Fig. 5.17. In the latter case no matter where the object is, the image

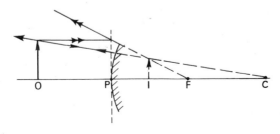

Fig. 5.17

is always virtual, upright and diminished. Fig. 5.18 shows that a convex mirror gives a wider *field of view* than a plane mirror which explains its use as a driving mirror and on the stairs of double-decker buses. It does make

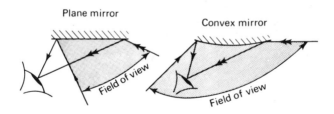

Fig. 5.18

the estimation of distances more difficult, however, because there is only small movement of the image for large movement of the object.

(*b*) *The mirror formula.* In Figs. 5.19*a* and *b* a ray OM from a point object O on the principal axis is reflected at M so that the angles θ, made by the incident and reflected rays with the normal CM, are equal. A ray OP strikes the mirror normally and is reflected back along PO. The intersection I of the reflected rays MI and PO in (*a*) gives a *real* point image of O, and of MI and PO both produced backwards in (*b*) gives a *virtual* point image of O.

Let angles α, β and γ be as shown. In $\triangle$CMO, since the exterior angle of a triangle equals the sum of the interior opposite angles,

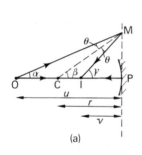 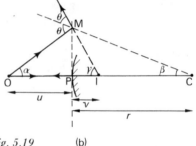

(a)

Fig. 5.19

(b)

| *Concave* | *Convex* |

$$\beta = \alpha + \theta \qquad \therefore \quad \theta = \beta - \alpha$$

In $\triangle$CMI
$$\gamma = \beta + \theta$$
$$\therefore \quad \theta = \gamma - \beta$$
$$\therefore \quad \beta - \alpha = \gamma - \beta$$
$$\therefore \quad 2\beta = \gamma + \alpha \qquad (1)$$

$$\theta = \alpha + \beta$$
In $\triangle$CMI
$$\gamma = \theta + \beta$$
$$\therefore \quad \theta = \gamma - \beta$$
$$\therefore \quad \alpha + \beta = \gamma - \beta$$
$$\therefore \quad 2\beta = \gamma - \alpha \qquad (1)'$$

If the mirror is of small aperture, the rays are *paraxial*, M will be close to P and α, β and γ are small. Then

$$\alpha \text{ (in radians)} \simeq \tan \alpha \qquad (1 \text{ radian} = 57.3°)$$

$$\therefore \quad \alpha \simeq \frac{MP}{OP} \qquad \text{where OP is the object distance}$$

Similarly
$$\beta \simeq \frac{MP}{CP} \qquad \text{where CP is the radius of curvature of the mirror}$$

$$\gamma \simeq \frac{MP}{IP} \qquad \text{where IP is the image distance}$$

Substituting in (1)

$$\frac{2MP}{CP} = \frac{MP}{IP} + \frac{MP}{OP}$$

$$\therefore \quad \frac{2}{CP} = \frac{1}{IP} + \frac{1}{OP} \qquad (2)$$

Substituting in (1)'

$$\frac{2MP}{CP} = \frac{MP}{IP} - \frac{MP}{OP}$$

$$\therefore \quad \frac{2}{CP} = \frac{1}{IP} - \frac{1}{OP} \qquad (2)'$$

If we now introduce a *sign convention* so that distances are given a positive or a negative sign, the same equation is obtained for both concave and convex mirrors irrespective of whether objects and images are real or virtual. We shall adopt the 'real is positive' rule which states:

A real object or image distance is positive.
A virtual object or image distance is negative.

158

The focal length of a concave mirror is thus positive (since its principal focus is real) and of a convex mirror negative. The radius of curvature takes the same sign as the focal length.

If we now let u, v and r stand for the *numerical values* and *signs* of the object and image distances and radius of curvature respectively then for both cases we get the same algebraic relationship

$$\frac{1}{v} + \frac{1}{u} = \frac{2}{r}$$

Also, since $r = 2f$, we have

$$\frac{1}{v} + \frac{1}{u} = \frac{1}{f}$$

Notes. (*i*) The formula is independent of the angle the incident ray makes with the axis, therefore *all* paraxial rays from point object O must, after reflection, pass through I to give a point image.

(*ii*) When numerical values for u, v, r or f are substituted in the formula, the appropriate sign *must* also be included; the sign (as well as the value) of the distance to be found comes out in the answer and so even if the sign is known from other information it must *not* be inserted in the equation.

(*iii*) Only two cases have been considered but it can be shown that the same formula holds for others, e.g. a concave mirror forming a virtual image of a real object, a convex mirror giving a real image of a virtual object (i.e. of converging light).

(*c*) *Magnification.* The lateral, transverse or linear magnification m (abbreviated to magnification) produced by a mirror is defined by

$$m = \frac{\text{height of image}}{\text{height of object}}$$

In Fig. 5.20, II' is the real image formed by a concave mirror of a finite object OO'. A paraxial ray from the top O' of the object, after reflection at

Fig. 5.20

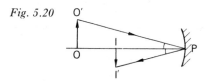

say P, passes through the top I' of the image. Since the principal axis is the normal to the mirror at P, $\angle$O'PO = $\angle$I'PI, by the first law of reflection. Triangles O'PO and I'PI are therefore similar and so

$$\frac{\text{height of image}}{\text{height of object}} = \frac{\text{I'I}}{\text{O'O}} = \frac{\text{IP}}{\text{OP}}$$

$$\therefore \quad m = \frac{v}{u}$$

For example, if the image distance is twice the object distance, the image is twice the height of the object.

Notes. (*i*) No signs need be inserted in this formula for m, i.e. it is a numerical and not an algebraic formula.

(*ii*) The same result can be derived for other cases.

Mirror calculations

1. An object is placed 15 cm from (a) a concave mirror, (b) a convex mirror, of radius of curvature 20 cm. Calculate the image position and magnification in each case.

(*a*) *Concave mirror*

The object is real, therefore $u = +15$ cm.
Since the mirror is concave $r = +20$ cm, therefore $f = +10$ cm.
Substituting values and signs in $1/v + 1/u = 1/f$

$$\frac{1}{v} + \frac{1}{(+15)} = \frac{1}{(+10)}$$

$$\therefore \quad \frac{1}{v} = \frac{1}{10} - \frac{1}{15} = \frac{1}{30}$$

$$\therefore \quad v = +30 \text{ cm}$$

The image is real since v is positive and it is 30 cm in front of the mirror. Also,

$$\text{magnification } m = \frac{v}{u} \quad \text{(numerically)}$$

$$= \frac{30}{15} = 2.0$$

The image is twice as high as the object (see Fig. 5.16*iii*).

(*b*) *Convex mirror*

We have $u = +15$ cm but $r = -20$ cm and $f = -10$ cm since the mirror is convex.
Substituting as before in $1/v + 1/u = 1/f$

$$\frac{1}{v} + \frac{1}{(+15)} = \frac{1}{(-10)}$$

$$\therefore \quad \frac{1}{v} = -\frac{1}{10} - \frac{1}{15} = -\frac{5}{30}$$

$$\therefore \quad v = -\frac{30}{5} = -6.0 \text{ cm}$$

The image is virtual since v is negative and it is 6.0 cm behind the mirror. Also

$$m = \frac{v}{u} \quad \text{(numerically)}$$

$$= \frac{6}{15} = \frac{2}{5}$$

The image is two-fifths as high as the object (see Fig. 5.17).

2. When an object is placed 20 cm from a concave mirror, a real image magnified three times is formed. Find (a) the focal length of the mirror, (b) where the object must be placed to give a virtual image three times the height of the object.

(a) The object is real, therefore $u = +20$ cm.
   Also,

$$m = 3 = \frac{v}{u} = \frac{v}{20} \quad \text{(numerically)}$$

The image is real

$$\therefore \quad v = +3 \times 20 = +60 \text{ cm}$$

Substituting in $1/v + 1/u = 1/f$

$$\frac{1}{(+60)} + \frac{1}{(+20)} = \frac{1}{f}$$

$$\therefore \quad \frac{1}{f} = \frac{4}{60}$$

$$\therefore \quad f = +15 \text{ cm}$$

(b) Let the *numerical* value of the object distance $= x$.
   Therefore, $u = +x$ since the object is real and $v = -3x$ since the image is virtual; $f = +15$ cm.

Using the mirror formula

$$\frac{1}{(+x)} + \frac{1}{(-3x)} = \frac{1}{(+15)}$$

$$\therefore \quad \frac{3}{3x} - \frac{1}{3x} = \frac{1}{15}$$

$$\therefore \quad \frac{2}{3x} = \frac{1}{15}$$

$$\therefore \quad x = 10 \text{ cm}$$

The object should be 10 cm in front of the mirror.

Note. By letting x be the numerical value of u we are able to substitute for u since we know its sign—a useful dodge.

Methods of measuring f for spherical mirrors

(a) Concave mirror

Rough method. The image formed by the mirror of a distant (several metres away) window is focused sharply on a screen, Fig. 5.21a. The distance between the mirror and the screen is f since rays of light from a point on such an object are approximately parallel, Fig. 5.21b.

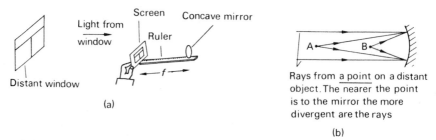

(a)

Rays from a point on a distant object. The nearer the point is to the mirror the more divergent are the rays

(b)

Fig. 5.21

Self-conjugate point method. The position of an object is adjusted until it coincides in position with its own image. In this position it is at the centre of curvature and distant r, i.e. $2f$, from the mirror. A point at which an object and its image coincide is said to be 'self-conjugate'.

The object can be a pin moved up and down above the mirror until there is no parallax between it and its real, inverted image, Fig. 5.22a. If an illuminated object is used, it is moved to and from the mirror until a clear image is obtained on a screen beside the object, Fig. 5.22b.

The pin/no parallax method generally gives more accurate results.

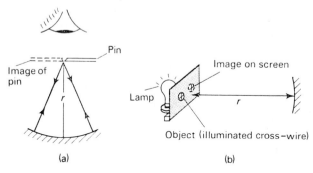

Fig. 5.22

Mirror formula method. Several values of the image distance v corresponding to different values of the object distance u are found using either the pin/no parallax method or an illuminated object and screen. For each pair of values, f is calculated from $1/f = 1/v + 1/u$ and the average taken.

A better plan is to plot a graph of $1/v$ against $1/u$ and draw the best straight line AB through the points, Fig. 5.23. Each pair of values of u and v gives

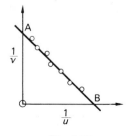

Fig. 5.23

two points on the graph because u and v are interchangeable; they are called 'conjugate points'. The intercepts OA and OB on the axes are both equal to $1/f$. Prove this.

(b) *Convex mirror*

Auxiliary converging lens method. A convex mirror normally forms a virtual image of a real object. Such an image cannot be located by a screen and is not easy to find by a pin/no parallax method. With the help of a converging lens, however, a real image can be obtained.

163

In Fig. 5.24 the lens L forms a real image at C of an object O when the convex mirror is absent. This image is located and the distance LC noted. The mirror is then placed between L and C and moved until O coincides in position with its own image.

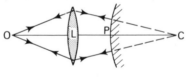

Fig. 5.24

The light from L is then falling normally on the mirror and is retracing its path to form a real inverted image at O. If produced, the rays from L must pass through the centre of curvature of the mirror since they are normal to the mirror. C, the position of the image of O formed by L alone, must therefore also be the centre of curvature of the mirror and so $PC = r$. Distance LP is measured and then $r = 2f = PC = LC - LP$.

Refraction at plane surfaces

(*a*) *Laws of refraction*. When light passes from one medium, say air, to another, say glass, Fig. 5.25, part is reflected back into the first medium and the rest passes into the second medium with its direction of travel changed.

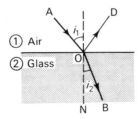

Fig. 5.25

The light is said to be bent or *refracted* on entering the second medium and the angle of refraction is the angle made by the refracted ray OB with the normal ON. There are two laws of refraction.

1. *For two particular media, the ratio of the sine of the angle of incidence to the sine of the angle of refraction is constant*, i.e. $\sin i_1/\sin i_2 =$ a constant in Fig. 5.25. (This is known as Snell's law after its discoverer.)

2. *The refracted ray is in the same plane as the incident ray and the normal to the mirror at the point of incidence but on the opposite side of the normal from the incident ray.*

The constant ratio $\sin i_1 / \sin i_2$ is called the *refractive index* for light passing from the first to the second medium. If the media containing the incident and refracted rays are denoted by ① and ② respectively, the refractive index is written as $_1n_2$. That is,

$$_1n_2 = \frac{\sin i_1}{\sin i_2}$$

The ratio depends on the colour of the light and is usually stated for yellow light. If medium ① is a vacuum (or in practice, air) we refer to the *absolute* refractive index of medium ② and denote it by $_vn_2$ or $_an_2$ or simply by n_2. The absolute refractive index of water is 1.33, of crown glass about 1.5 and of air at normal pressure about 1.0003—which is 1 near enough, the same as for a vacuum.

The greater the refractive index (absolute) of a medium the greater is the change in direction suffered by a ray of light when it passes from air to the medium. Refraction is therefore greater from air to crown glass than from air to water. In both cases the refracted ray is bent *towards* the normal, i.e. towards ON in Fig. 5.26a and the light is travelling into an 'optically denser' medium. A ray travelling from glass or water to air is bent *away from* the normal, Fig. 5.26b.

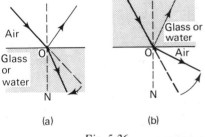

(a)   (b)

Fig. 5.26

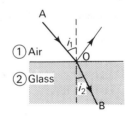

Fig. 5.27

Refraction can be attributed to the fact that light has different speeds in different media.

(b) *Refractive index relationships*

(i) $_1n_2 = 1/{_2n_1}$. Consider a ray AO travelling from air (medium ①) to glass (medium ②) and refracted along OB as in Fig. 5.27. We have

$$_1n_2 = \frac{\sin i_1}{\sin i_2}$$

Since light rays are reversible, a ray travelling along BO in glass (medium ②) will be refracted along OA in air (medium ①), hence

$$_2n_1 = \frac{\sin i_2}{\sin i_1}$$

$$\therefore \quad _1n_2 = \frac{1}{_2n_1}$$

For example, if the refractive index from air to water $(_an_w)$ is 4/3 then that from water to air $(_wn_a)$ is 3/4.

(ii) $_1n_3 = {}_1n_2 \times {}_2n_3$. Suppose a ray AB travels from air (medium ①), to glass (medium ②), to water (medium ③), to air (medium ①), as in Fig. 5.28. Experiment shows that *if the boundaries of the media are parallel*, the

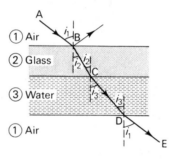

Fig. 5.28

emergent ray DE, although laterally displaced, is parallel to the incident ray AB. The incident and emergent angles are thus equal and are denoted by i_1. We then have

$$_1n_2 = \frac{\sin i_1}{\sin i_2} \qquad _2n_3 = \frac{\sin i_2}{\sin i_3} \qquad _3n_1 = \frac{\sin i_3}{\sin i_1}$$

$$\therefore \quad _1n_2 \times {}_2n_3 \times {}_3n_1 = \frac{\sin i_1}{\sin i_2} \times \frac{\sin i_2}{\sin i_3} \times \frac{\sin i_3}{\sin i_1} = 1$$

$$\therefore \quad _1n_2 \times {}_2n_3 = \frac{1}{_3n_1}$$

From (i),
$$_1n_3 = \frac{1}{_3n_1}$$

$$\therefore \quad _1n_3 = {}_1n_2 \times {}_2n_3$$

166

The order of the subscripts aids memorization of the relation. For example, if we wish to know the refractive index for water to glass ($_wn_g$) and we know air to water ($_an_w = \frac{4}{3}$) and air to glass ($_an_g = \frac{3}{2}$) then

$$_wn_g = _wn_a \times _an_g = \frac{1}{_an_w} \times _an_g$$

$$= \tfrac{3}{4} \times \tfrac{3}{2} = \tfrac{9}{8}$$

Will a ray of light be bent towards or away from the normal on travelling from water to glass?

(iii) $n_1 \sin i_1 = n_2 \sin i_2$. This is a more symmetrical form of Snell's law, useful in calculations, which will now be derived.

From Fig. 5.27 we can say

$$_1n_2 = \frac{\sin i_1}{\sin i_2}$$

From (ii) above,

$$_1n_2 = _1n_a \times _an_2 \quad (\text{a} = \text{air or a vacuum})$$

$$= \frac{_an_2}{_an_1} = \frac{n_2}{n_1}$$

where n_1 and n_2 are absolute refractive indices of media ① and ②respectively.

$$\therefore \quad \frac{n_2}{n_1} = \frac{\sin i_1}{\sin i_2}$$

$$\therefore \quad n_1 \sin i_1 = n_2 \sin i_2$$

For example, if a ray of light is incident on a water-glass boundary at $30°$ then $i_1 = i_w = 30°$ and if $n_1 = n_w = \frac{4}{3}$ and $n_2 = n_g = \frac{3}{2}$, the angle of refraction $i_2 = i_g$ is given by

$$n_w \sin i_w = n_g \sin i_g$$

i.e. $$\tfrac{4}{3} \sin 30 = \tfrac{3}{2} \sin i_g$$

$$\therefore \quad \sin i_g = \tfrac{4}{3} \times \tfrac{1}{2} \times \tfrac{2}{3} \quad (\sin 30 = \tfrac{1}{2})$$

$$= \tfrac{4}{9}$$

$$\therefore \quad i_g = 26°$$

(c) *Real and apparent depth.* Because of refraction the apparent depth of a pool of clear water, when viewed from above the surface, is less than its real depth; also, an object under water is not where it seems to be to an outside observer.

In Fig. 5.29 rays from a point O under water are bent away from the normal

Fig. 5.29

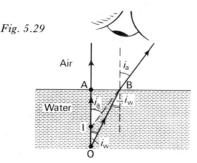

at the water–air boundary and appear to come from I, the image of O. For refraction at B from water to air,

$$n_w \sin i_w = n_a \sin i_a = \sin i_a \qquad (n_a = 1)$$

$$\therefore \quad n_w = \frac{\sin i_a}{\sin i_w}$$

$$= \frac{AB/IB}{AB/OB} = \frac{OB}{IB}$$

If the observer is directly above O, i_w and i_a are small, rays OB and IB are close to OA, thus making OB $\simeq$ OA and IB $\simeq$ IA. Hence

$$n_w = \frac{OA}{IA} \quad \text{(approximately)}$$

$$= \frac{\text{real depth}}{\text{apparent depth}}$$

Taking $n_w = \frac{4}{3}$, what will be the apparent depth of a pond actually 2 metres deep?

The distance OI is called the *displacement d* of the object and if t is the real depth then

$$d = OA - IA = t - \frac{t}{n_w}$$

$$= t\left(1 - \frac{1}{n_w}\right)$$

The same expression gives the displacement of an object which is some distance in air below a parallel-sided block of material, as can be seen from

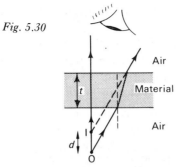

Fig. 5.30

Fig. 5.30. When viewed through several media whose boundaries are parallel, the total displacement is the sum of the displacements that would be produced by each medium if the others were absent.

A pool of water appears even shallower when viewed obliquely rather than from vertically above. As the observer moves, the image of a point O traces out a curve, called a *caustic*, whose apex is at I_1, Fig. 5.31.

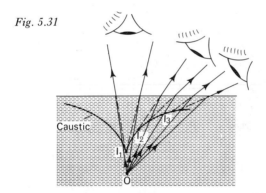

Fig. 5.31

(*d*) *Multiple images in mirrors.* Several images are seen when an object is viewed obliquely in a thick glass mirror with silvering on the back surface. In Fig. 5.32, I_1 is a faint image of object O formed by the weak reflected ray

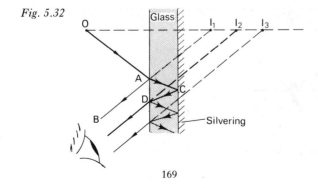

Fig. 5.32

AB from the front surface of the mirror. I_2, the main reflection, is bright and is due to the refracted ray AC being reflected at the back (silvered) surface and again refracted at the front surface. I_3 and other weaker images are formed as shown. The net effect of these multiple reflections and refractions is to reduce the sharpness of the primary image I_2. Front-silvered mirrors eliminate the secondary images but are liable to be scratched and to tarnish.

(*e*) *Mirages.* These are often seen as small, distant pools of water on a hot tar-macadam road, particularly when vision is very oblique as it is for anyone in a car. They are caused by refraction in the atmosphere.

The air near a road heated by the sun is hot; higher up the air is cool and its density greater. Consequently rays from the sky travelling towards the road are gradually refracted away from the normal as they pass from denser to less dense air. Upward bending of the light occurs, Fig. 5.33, and the blue light from the sky then seems to an observer to have been reflected from the road and gives the appearance of puddles.

Mirages are sometimes seen in the desert as distant, shimmering lakes.

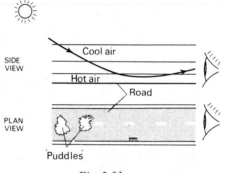

Fig. 5.33

Total internal reflection

(*a*) *Critical angle.* For small angles of incidence a ray of light travelling from one medium to another of smaller refractive index, say from glass to air, is refracted away from the normal, Fig. 5.34*a*; a weak internally reflected ray is also formed. Increasing the angle of incidence increases the angle of

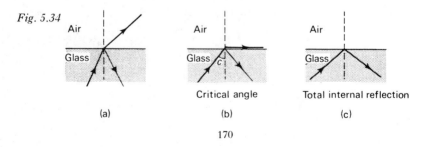

Fig. 5.34

(a)   (b)   Critical angle   (c)   Total internal reflection refraction and at a certain angle of incidence c, called the *critical angle*, the refracted ray just emerges along the surface of the glass and the angle of refraction is 90°, Fig. 5.34b. At this stage the internally reflected ray is still weak but just as c is exceeded it suddenly becomes bright and the refracted ray disappears, Fig. 5.34c. *Total internal reflection* is now said to be occurring since all the incident light is reflected inside the optically denser medium.

Applying Snell's law in the form $n_1 \sin i_1 = n_2 \sin i_2$ to the critical ray at a glass–air boundary, we have

$$n_1 = \text{refractive index of glass} = n_g$$

$$i_1 = \text{critical angle for glass} = c$$

$$n_2 = \text{refractive index of air} = 1$$

$$i_2 = \text{angle of refraction in air} = 90°$$

$$\therefore \quad n_g \sin c = 1 \sin 90 = 1 \quad (\sin 90 = 1)$$

$$\therefore \quad n_g = \frac{1}{\sin c}$$

Taking $n_g = \frac{3}{2}$ (crown glass), $\sin c = \frac{2}{3}$ and so $c \simeq 42°$. Thus if the incident angle in the crown glass exceeds 42°, total internal reflection occurs. Can it occur when a ray of light in glass ($n_g = \frac{3}{2}$ say) is incident on a boundary with water ($n_w = \frac{4}{3}$)?

(*b*) *Totally reflecting prisms.* The disadvantages of plane mirrors (p. 169), silvered on either the back or front surface, can be overcome by using right-angled isosceles prisms (angles 90°, 45°, 45°) as reflectors.

The critical angle of crown glass is about 42° and a ray OA incident normally on face PQ of such a prism, Fig. 5.35a, suffers total internal reflection at face PR since the angle of incidence in the optically denser medium is 45°. A bright ray AB emerges at right angles to face QR since the angle of reflection at QR is also 45°. The prism thus reflects the ray through 90°.

Light can be reflected through 180° and an erect image obtained of an inverted one (as in prism binoculars, p. 213) if the prism is arranged as in Fig. 5.35b.

Fig. 5.35

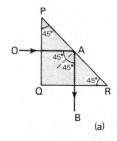

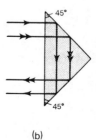

(a)   (b)

(c) *Fibre optics*. Light can be confined within a bent glass rod by total internal reflection and so 'piped' along a twisted path, as in Fig. 5.36. The beam is reflected from side-to-side practically without loss (except for that due to absorption in the glass) and emerges only at the end of the rod where

Fig. 5.36

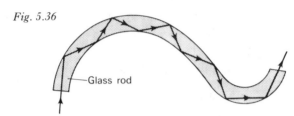

Glass rod

it strikes the surface almost normally, i.e. at an angle less than the critical angle. A single, very thin, solid glass-fibre behaves in the same way and if several thousand are taped together a flexible light pipe is obtained that can be used, as it has been in medicine and engineering, to illuminate some otherwise inaccessible spot. One difficulty which arises in a bundle of fibres is leakage of light at places of contact between the fibres. This can be reduced by coating each fibre with glass of lower refractive index than its own, thereby encouraging total internal reflection.

If it is desired to transport an *image* and not simply to transport *light*, the fibres must occupy the same positions in the bundle relative to each other. Such bundles are more difficult to make and cost more. Fig. 5.37a shows

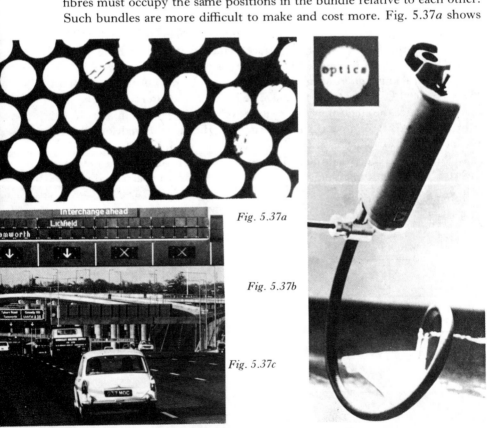

Fig. 5.37a

Fig. 5.37b

Fig. 5.37c

part of the end-section of a bundle of fibres and Fig. 5.37*b* is a fibre-optics viewing instrument with bent light pipe and an inset of the word 'optics' as seen by a camera above the eyepiece at the top. Figure 5.37*c* shows a motorway sign illuminated by light pipes. Fibre optics could become an important research technique in medicine and engineering.

Refraction through prisms

A prism has two plane surfaces inclined to each other as are LMQP and LNRP in Fig. 5.38. Angle MLN is called the *refracting angle* of the prism,

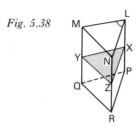

Fig. 5.38

LP is the *refracting edge* and any plane such as XYZ which is perpendicular to LP is a *principal plane*.

The importance of the prism really depends on the fact that the angle of deviation suffered by light at the first refracting surface, say LMQP, is not cancelled out by the deviation at the second surface LNRP (as it is in a parallel-sided block), but is added to it. This is why it can be used in a spectrometer, an instrument for analysing light into its component colours. In what follows, expressions for the angle of deviation will be obtained and subsequently used.

(*a*) *General formulae.* In Fig. 5.39, EFGH is a ray lying in a principal plane XYZ of a prism of refracting angle A and passing from air, through the prism and back to air again. KF and KG are normals at the points of incidence and emergence of the ray.

For refraction at XY

$$\text{angle of deviation} = \text{angle JFG} = i_1 - r_1$$

For refraction at XZ

$$\text{angle of deviation} = \text{angle JGF} = i_2 - r_2$$

Since both deviations are in the same direction, the total deviation D is given by

$$\text{angle TJH} = \text{angle JFG} + \text{angle JGF}$$

i.e.
$$D = (i_1 - r_1) + (i_2 - r_2) \tag{1}$$

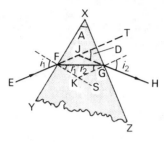

Fig. 5.39

Another expression arises from the geometry of Fig. 5.39. In quadrilateral XFKG

$$\text{angle XFK} + \text{angle XGK} = 180°$$

$$\therefore \quad A + \text{angle FKG} = 180°$$

But since FKS is a straight line

$$\text{angle GKS} + \text{angle FKG} = 180°$$

$$\therefore \quad \text{angle GKS} = A$$

In triangle KFG, angle GKS is an exterior angle

$$\therefore \quad \text{angle GKS} = r_1 + r_2$$

$$\therefore \quad A = r_1 + r_2 \tag{2}$$

Equations (1) and (2) are true for any prism. (The position and shape of the third side of the prism does not affect the refraction under consideration and so is shown as an irregular line in Fig. 5.39.)

(b) *Minimum deviation.* It is found that the angle of deviation D varies with the angle of incidence i_1 of the ray incident on the first refracting face of the prism. The variation is shown in Fig. 5.40a and for one angle of

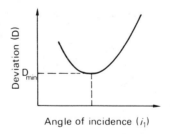

Angle of incidence (i_1)

Fig. 5.40a

174

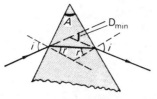

Fig. 5.40b

incidence it has a minimum value D_{min}. At this value the *ray passes symmetrically through the prism* (a fact that can be proved theoretically as well as be shown experimentally), i.e. the angle of emergence of the ray from the second face equals the angle of incidence of the ray on the first face: $i_2 = i_1 = i$, Fig. 5.40b. It therefore follows that $r_1 = r_2 = r$. Hence from equation (1) of the previous section the angle of minimum deviation D_{min} is given by

$$D_{min} = (i - r) + (i - r) = 2(i - r) \tag{3}$$

Also, from equation (2)

$$A = r + r = 2r$$

$$\therefore \quad r = \frac{A}{2}$$

Substituting for r in (3)

$$D_{min} = 2i - A$$

$$\therefore \quad i = \frac{A + D_{min}}{2}$$

If n is the refractive index of the material of the prism then

$$n = \frac{\sin i}{\sin r}$$

$$= \frac{\sin [(A + D_{min})/2]}{\sin (A/2)}$$

Thus if $A = 60°$ and $D_{min} = 40°$, then $(A + D_{min})/2 = 50°$ and so $n = \sin 50/\sin 30 = 1.5$.

Two points for you to consider. First, no values of D are shown on the graph of Fig. 5.40a for small values of i (less than about 30° for a crown glass prism of refracting angle 60° and $n = 1.5$). Why? Second, the above formula for minimum deviation only holds for a prism of angle A less than twice the critical angle. Why?

175

(c) *Small-angle prism.* The expression for the deviation in this case will be used later for developing lens theory.

Consider a ray falling almost normally in air on a prism of small angle A (less than about $6°$ or 0.1 radian) so that angle i_1 in Fig. 5.41 is small. Now

Fig. 5.41

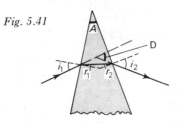

(angles exaggerated for clarity)

$n = \sin i_1/\sin r_1$ where n is the refractive index of the material of the prism, therefore r_1 will also be small. Hence, since the sine of a small angle (like the tangent) is nearly equal to the angle in radians, we have

$$i_1 = nr_1$$

Also, $A = r_1 + r_2$ (see p. 174), and so if A and r_1 are small, r_2 and i_2 will also be small. From $n = \sin i_2/\sin r_2$ we can say

$$i_2 = nr_2$$

The deviation D of a ray passing through any prism is given by (p. 174)

$$D = (i_1 - r_1) + (i_2 - r_2)$$

Substituting for i_1 and i_2

$$D = nr_1 - r_1 + nr_2 - r_2$$
$$= n(r_1 + r_2) - (r_1 + r_2)$$
$$= (n - 1)(r_1 + r_2)$$

But $A = r_1 + r_2$

$$\therefore \quad D = (n - 1)A$$

This expression shows that for a given angle A *all* rays entering a *small-angle* prism at *small angles of incidence* suffer the *same* deviation.

(d) *Dispersion.* Newton found that when a beam of white light (e.g. sunlight) passes through a prism it is spread out by the prism into a band of all the colours of the rainbow from red to violet. The band of colours is called a *spectrum* and the separation of the colours by the prism is known as *dispersion.* He concluded that white light is a mixture of light of various colours and identified red, orange, yellow, green, blue, indigo, violet.

Red is deviated least by the prism and violet most as shown by the exaggerated diagram of Fig. 5.42*a*. The refractive index of the material of the prism for violet light is thus greater than for red light since the angle of incidence in the air is the same for red and violet rays.

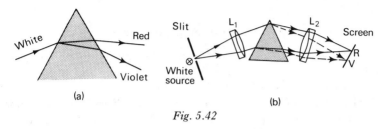

(a)

(b)

Fig. 5.42

A method of producing a *pure* spectrum, i.e. one in which the different colours do not overlap (as they do when a prism is used on its own), is shown in Fig. 5.42*b*. A diverging beam of white light, emerging from a very narrow slit, is made parallel by lens L_1 and then dispersed by the prism into a number of different coloured parallel beams, each travelling in a slightly different direction. Lens L_2 brings each colour to a separate focus on a screen. The spectrum is a series of monochromatic images of the slit and the narrower this is the purer the spectrum. (L_1 and L_2 are achromatic doublets, see p. 196.)

Methods of measuring *n*

(*a*) *Real and apparent depth method* (solids and liquids). A travelling microscope is focused on a pencil dot O on a sheet of white paper lying on the bench and the reading on the microscope scale noted, Fig. 5.43. Let it be *x*.

Fig. 5.43

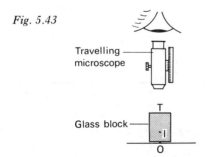

Travelling microscope

Glass block

If the refractive index of glass is required, a block of the material is placed over the dot and the microscope refocused on the image I of O as seen through the block. Let the reading be *y*. Finally the microscope is focused on the top T of the block, made visible by a sprinkling of lycopodium powder.

Suppose the reading is now z, then

$$\text{real depth of O} = \text{OT} = z - x$$

$$\text{apparent depth of O} = \text{IT} = z - y$$

$$\therefore \quad n = \frac{\text{real depth}}{\text{apparent depth}} = \frac{z - x}{z - y}$$

The refractive index of a liquid in a beaker can be found by a similar procedure.

The method satisfies the assumption made in deducing the expression for n (p. 168) because the microscope collects only rays very close to the normal OT; accuracy of $\pm 1\%$ is possible if the microscope has a small depth of focus.

(*b*) *Minimum deviation method* (solids and liquids). A solid prism of the material is placed on the table of a spectrometer, A and D_{min} measured as described on p. 216 and n calculated from $n = \sin\left[(A + D_{min})/2\right]/\sin(A/2)$. The method is suitable for liquids if a hollow prism with perfectly parallel, thin walls is used. Accuracy of $\pm 0.1\%$ is possible.

(*c*) *Concave mirror method* (liquids). The centre of curvature C of the mirror is first found by moving an object pin up and down above the mirror until it coincides in position with its image (Method 2, p. 162). Some liquid is then poured into the mirror and the object pin moved until point O is found where it again coincides with its image. In Fig. 5.44 ray ONB must be retracing its own path after striking the mirror normally at B and if BN is produced it will pass through C.

Fig. 5.44

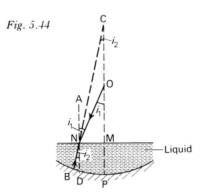

For the refraction at N

i_1 = angle of incidence = $\angle$ONA = $\angle$NOM   (alt. angles)

i_2 = angle of refraction = $\angle$BND = $\angle$NCM   (corr. angles)

178

The refractive index n of the liquid is given by

$$n = \frac{\sin i_1}{\sin i_2} = \frac{\sin \text{NOM}}{\sin \text{NCM}} = \frac{\text{NM/NO}}{\text{NM/NC}} = \frac{\text{NC}}{\text{NO}}$$

If ray ONB is close to the principal axis CP of the mirror then to a good approximation NC = MC and NO = MO

$$\therefore \quad n = \frac{\text{MC}}{\text{MO}}$$

Both distances can be measured and n found. The method is useful when only a small quantity of liquid is available.

Thin lenses

Lenses are of two basic types, *convex* which are thicker in the middle than at the edges and *concave* for which the reverse holds. Fig. 5.45 shows examples of both types, bounded by spherical or plane surfaces.

Fig. 5.45

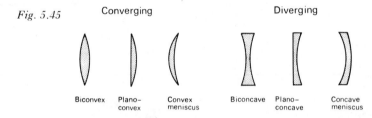

Converging   Diverging

Biconvex   Plano–convex   Convex meniscus   Biconcave   Plano–concave   Concave meniscus

The *principal axis* of a spherical lens is the line joining the centres of curvature of its two surfaces, Fig. 5.46. For the present our treatment will be confined to *paraxial* rays, i.e. rays close to the axis and making very small angles with it. In effect this means we shall only consider lenses of small aperture but in diagrams both angles and lenses will be made larger for clarity. The case of wide angle beams will be considered briefly later (p. 194).

The *principal focus* F of a thin lens is the point on the principal axis towards which all paraxial rays, parallel to the principal axis, converge in the case of a convex lens or from which they appear to diverge in the case of a concave lens, after refraction, Fig. 5.46a and b. Since light can fall on either

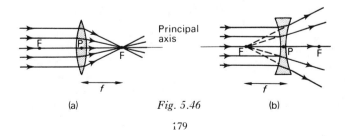

Principal axis

(a)   *Fig. 5.46*   (b)

surface, a lens has two principal foci, one on each side, and these are equidistant from its centre P (if the lens is thin and has the same medium on both sides, e.g. air). The distance FP is the *focal length f* of the lens. A convex lens is a *converging* lens[1] and has real foci. A concave lens is a *diverging* lens and has virtual foci.

A parallel beam at a small angle to the axis of a lens is refracted to converge to, or to appear to diverge from, a point in the plane containing F, perpendicular to the axis and known as the *focal plane*, Fig. 5.47a and b.

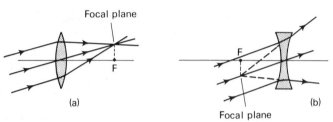

Fig. 5.47

As we shall see shortly the important property of a lens is that it focuses *all* paraxial rays from a point object (and not just parallel, paraxial rays) to form a point image.

Images formed by thin lenses

Information as to the position and nature of the image in any case can be obtained either from a ray diagram or by calculation.

(*a*) *Ray diagrams.* To construct the image of a small object perpendicular to the axis of a lens, *two* of the following three rays are drawn from the top of the object.

(*i*) A ray parallel to the principal axis which after refraction passes through the principal focus or appears to diverge from it.

(*ii*) A ray through the centre of the lens (called the *optical centre*) which continues straight on undeviated (it is only slightly displaced laterally because the middle of the lens acts like a thin parallel-sided block), Fig. 5.48.

Fig. 5.48

[1] This is only true if the convex lens has a greater refractive index than the surrounding medium. In water a biconvex air lens diverges light.

(*iii*) A ray through the principal focus which is refracted parallel to the principal axis, i.e. a ray travelling the reverse path to that in (*i*).

The diagrams for a converging lens are shown in Fig. 5.49*a* to *e* and for a diverging lens in Fig. 5.49*f*. The latter, like a convex mirror, always forms a virtual, upright and diminished image whatever the object position. Note that a thin lens is represented by a straight line at which all the refraction is considered to occur; in practice it is usually refracted both on entering and leaving the lens.

It must also be emphasized that the lines drawn are *constructional* ones; two narrow cones of rays that actually enter the eye of an observer from the top and bottom of an object are shown shaded in Fig. 5.49*e*. They are obtained by working back from the eye.

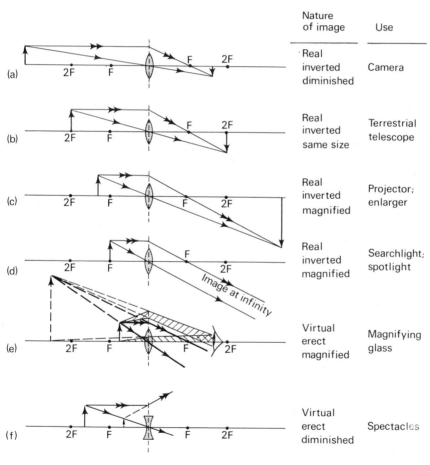

	Nature of image	Use
(a)	Real inverted diminished	Camera
(b)	Real inverted same size	Terrestrial telescope
(c)	Real inverted magnified	Projector; enlarger
(d)	Real inverted magnified	Searchlight; spotlight
(e)	Virtual erect magnified	Magnifying glass
(f)	Virtual erect diminished	Spectacles

Fig. 5.49

181

OPTICAL PROPERTIES

(b) *Simple formula for a thin lens*. We can regard a thin lens as made up of a large number of small-angle prisms whose angles increase from zero at the middle of the lens to a small value at its edge. Consider one such prism at distance h from the optical centre P of a lens, Fig. 5.50. If a paraxial ray parallel to the axis is incident on this prism it suffers small deviation D (since the prism is small-angled) and is refracted through the principal focus F. Hence, since the tangent of a small angle equals the small angle in radians,

$$D = \frac{h}{FP} \tag{1}$$

Converging   *Diverging*

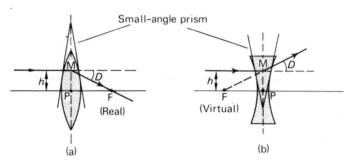

Fig. 5.50

Now consider a point object O on the axis which gives rise to a point image I, Fig. 5.51. If a paraxial ray from O is incident on the small-angle prism at distance h from the axis, it must also suffer deviation D (since all rays entering a small-angle prism at small angles of incidence suffer the same deviation (p.176).

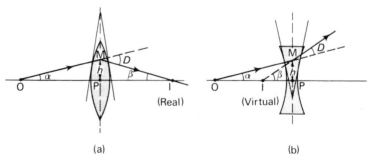

Fig. 5.51

Let angles α and β be as shown. In triangle IOM, since the exterior angle of a triangle equals the sum of the interior opposite angles,

$$D = \alpha + \beta \qquad\qquad D = \beta - \alpha$$

$$= \frac{h}{OP} + \frac{h}{IP} \qquad\qquad = \frac{h}{IP} - \frac{h}{OP}$$

Therefore from (1)

$$\frac{h}{FP} = \frac{h}{OP} + \frac{h}{IP} \quad (2) \qquad\qquad \frac{h}{FP} = \frac{h}{IP} - \frac{h}{OP} \quad (2)'$$

If we introduce the 'real is positive' sign convention given on p. 158, the *focal length of a converging lens is positive* and of *a diverging lens negative*. If u, v and f stand for the *numerical values* and *signs* of the object and image distances and focal length respectively then for *both* cases we get the *algebraic* relationship

$$\frac{1}{v} + \frac{1}{u} = \frac{1}{f}$$

Notes. (*i*) The formula is independent of the angle the incident ray makes with the axis, therefore *all* paraxial rays from point object O must, after refraction, pass through I to give a point image, i.e. the small angle prisms to which the lens is equivalent deviate the various rays from O by varying amounts depending on the angle of the prism but always so that they all pass through I.

(*ii*) When numerical values for u, v and f are inserted in the formula, the appropriate sign *must* also be included.

(*c*) *Magnification.* The lateral, transverse or linear magnification m (abbreviated to magnification) produced by a lens is defined by

$$m = \frac{\text{height of image}}{\text{height of object}}$$

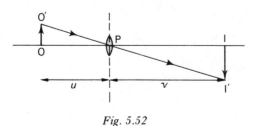

Fig. 5.52

In Fig. 5.52, II' is the real image formed by a converging lens of a finite object OO'. Triangles O'PO and I'PI are similar, therefore

$$\frac{\text{height of image}}{\text{height of object}} = \frac{\text{I'I}}{\text{O'O}} = \frac{\text{IP}}{\text{OP}}$$

$$\therefore \quad m = \frac{v}{u}$$

This is a numerical formula and no signs need be inserted.

Lens calculations

1. An object is placed 20 cm from (a) a converging lens, (b) a diverging lens, of focal length 15 cm. Calculate the image position and magnification in each case.

(a) *Converging lens*

The object is real, therefore $u = +20$ cm.
Since the lens converges, $f = +15$ cm.
Substituting values and signs in $1/v + 1/u = 1/f$

$$\frac{1}{v} + \frac{1}{(+20)} = \frac{1}{(+15)}$$

$$\therefore \quad \frac{1}{v} = \frac{1}{15} - \frac{1}{20} = \frac{4}{60} - \frac{3}{60} = \frac{1}{60}$$

$$\therefore \quad v = +60 \text{ cm}$$

The image is real since v is positive and it is 60 cm from the lens. Also,

$$\text{magnification } m = \frac{v}{u} \quad \text{(numerically)}$$

$$= \frac{60}{20} = 3.0$$

The image is three times as high as the object (see Fig. 5.49c).

(b) *Diverging lens*

We have $u = +20$ cm and $f = -15$ cm.
Substituting as before in $1/v + 1/u = 1/f$

$$\frac{1}{v} + \frac{1}{(+20)} = \frac{1}{(-15)}$$

$$\therefore \quad \frac{1}{v} = -\frac{1}{15} - \frac{1}{20} = -\frac{7}{60}$$

$$\therefore \quad v = -\frac{60}{7} = -8.6 \text{ cm}$$

The image is virtual since v is negative and it is 8.6 cm from the lens. Also,

$$m = \frac{v}{u} \quad \text{(numerically)}$$

$$= \frac{60/7}{20} = \frac{3}{7}$$

The image is three-sevenths as high as the object (see Fig. 5.49f).

2. *An object is placed 6.0 cm from a thin converging lens A of focal length 5.0 cm. Another thin converging lens B of focal length 15 cm is placed coaxially with A and 20 cm from it on the side away from the object. Find the position, nature and magnification of the final image.*

For lens A,

$$u = +6.0 \text{ cm}, \qquad f = +5.0 \text{ cm}$$

Substituting in $1/v + 1/u = 1/f$

$$\frac{1}{v} + \frac{1}{(+6)} = \frac{1}{(+5)}$$

$$\therefore \quad \frac{1}{v} = \frac{1}{5} - \frac{1}{6} = \frac{6}{30} - \frac{5}{30} = \frac{1}{30}$$

$$\therefore \quad v = +30 \text{ cm}$$

Image I_1 in Fig. 5.53 is real, therefore *converging* light falls on lens B and I_1 acts as a virtual object for B. Applying $1/v + 1/u = 1/f$ to B we have,

$$u = -(AI_1 - AB) = -(30 - 20) = -10 \text{ cm}, \qquad f = +15 \text{ cm}$$

Fig. 5.53

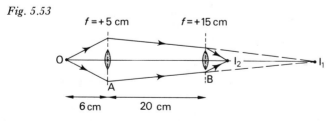

185

$$\therefore \quad \frac{1}{v} + \frac{1}{(-10)} = \frac{1}{(+15)}$$

$$\therefore \quad \frac{1}{v} = \frac{1}{15} + \frac{1}{10} = \frac{2}{30} + \frac{3}{30} = \frac{5}{30}$$

$$\therefore \quad v = +6.0 \text{ cm}$$

Image I_2 is real and is formed 6.0 cm beyond B.

$$\text{Magnification by A} = m_1 = \frac{v}{u} = \frac{30}{6} = 5.0$$

$$\text{Magnification by B} = m_2 = \frac{v}{u} = \frac{6}{10} = \frac{3}{5}$$

$$\text{Total magnification } m = m_1 \times m_2 = 5 \times \frac{3}{5}$$

$$= 3.0$$

The final image is three times the size of the object.
(*Note*: m_1 and m_2 are multiplied and *not* added. Why?)

Other thin lens formulae

(*a*) *Full formula for a thin lens.* We require to find a relationship, some-times called the 'lens-maker's formula', between the focal length of a thin lens, the radii of curvature of its surfaces and the refractive index of the lens material. It will be assumed that (*i*) the lens can be replaced by a system of small-angle prisms and (*ii*) all rays falling on the lens are paraxial, i.e. the lens has a small aperture and all objects are near the axis.

Consider the prism of small angle A which is formed by the tangents XL and XM to the lens surfaces at L and M, Fig. 5.54. XL and XM are per-pendicular to the radii of curvature C_1L and C_2M respectively, C_1 and C_2

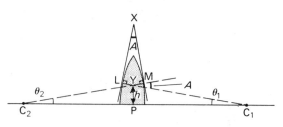

Fig. 5.54

being the centres of curvature of the surfaces. Therefore angle LXM (i.e. A) between the tangents equals angle MYC_1 between the radii

$$\therefore \quad \angle MYC_1 = A = \theta_1 + \theta_2$$

(ext. angle of triangle YC_2C_1 = sum of int. opp. angles)

But since θ_1 and θ_2 are small we can say $\theta_1 = \tan \theta_1$ and $\theta_2 = \tan \theta_2$

$$\therefore \quad A = \frac{h}{C_1P} + \frac{h}{C_2P} \qquad (h = YP)$$

The deviation D produced in *any* ray incident at a small angle on a prism of small angle A and refractive index n is (p.176)

$$D = (n - 1)A$$

$$= (n - 1)\left(\frac{h}{C_1P} + \frac{h}{C_2P}\right)$$

If we now consider a ray parallel to the axis and at height h above it, it suffers the same deviation D as any other paraxial ray and since it is refracted through the principal focus F, $D = h/FP$ (from equation (1), p. 182). Hence

$$\frac{h}{FP} = (n - 1)\left(\frac{h}{C_1P} + \frac{h}{C_2P}\right)$$

$$\therefore \quad \frac{1}{FP} = (n - 1)\left(\frac{1}{C_1P} + \frac{1}{C_2P}\right)$$

Introducing a sign convention for distances converts this numerical relationship to an algebraic one applicable to all lenses and cases. Thus if f, r_1 and r_2 stand for the numerical values and *signs* of the focal length and radii of curvature respectively of the lens then we have

$$\frac{1}{f} = (n - 1)\left(\frac{1}{r_1} + \frac{1}{r_2}\right)$$

In the 'real is positive' convention the rule for the sign of a radius of curvature is—*a surface convex to the less dense medium has a positive radius while a surface concave to the less dense medium has a negative radius.* A positive surface thus converges light, a negative one diverges it.

A numerical example may help. For the convex meniscus lens of Fig. 5.55a we have $n = 1.5$, $r_1 = +10$ cm (since it is convex to the air on its left), $r_2 = -15$ cm (since it is concave to the air on its right).

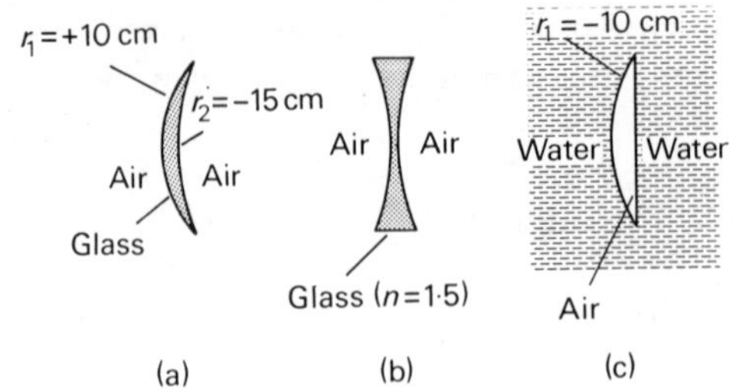

$r_1 = +10$ cm

$r_2 = -15$ cm

Air | Air

Air || Air

Glass

Glass $(n=1.5)$

$r_1 = -10$ cm

Water | Water

Air

(a) (b) (c)

Fig. 5.55

$$\therefore \quad \frac{1}{f} = (1.5 - 1)\left(\frac{1}{(+10)} + \frac{1}{(-15)}\right) = \frac{1}{2}\left(\frac{3}{30} - \frac{2}{30}\right)$$

$$= \frac{1}{60}$$

$$\therefore \quad f = +60 \text{ cm}$$

Now calculate f for the biconcave lens in Fig. 5.55b whose radii of curvature are each 20 cm. (*Ans.* -20 cm.)

A more general form of the formula is

$$\frac{1}{f} = \left(\frac{n_2}{n_1} \sim 1\right)\left(\frac{1}{r_1} + \frac{1}{r_2}\right)$$

where n_2 is the refractive index of the lens material and n_1 that of the surrounding medium. $(n_2/n_1 \sim 1)$ *is always taken to be positive* since refractive indices do not have signs. ($\sim$ means the 'difference between'.)

For a plano-convex *air* lens of radius 10 cm, in *water* of refractive index $\frac{4}{3}$ we have $n_2 = 1$, $n_1 = \frac{4}{3}$, $r_1 = -10$ cm (since it is concave to the air), $r_2 = \infty$, Fig. 5.55c.

$$\therefore \quad \frac{1}{f} = \left(\frac{1}{4/3} \sim 1\right)\left(\frac{1}{(-10)} + \frac{1}{\infty}\right) = \frac{1}{4} \times \left(\frac{-1}{10} + 0\right)$$

$$\therefore \quad f = -40 \text{ cm} \quad \text{(a diverging lens)}$$

(*b*) *Focal length of two thin lenses in contact.* Combinations of lenses in contact are used in many optical instruments to improve their performance. In Fig. 5.56, A and B are two thin lenses in contact, of focal lengths, f_1 and f_2. Paraxial rays from point object O on the principal axis are refracted through A and would, in the absence of B, give a real image of O at I'. Hence for A, $u = +x$ and $v = +y$. From the simple formula for a thin lens

$$\frac{1}{(+x)} + \frac{1}{(+y)} = \frac{1}{f_1}$$

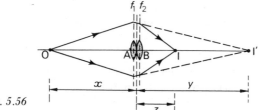

Fig. 5.56

For B, I′ acts as a *virtual object* (i.e. converging light falls on B from A) giving a real image of O at I and so for B, $u = -y$ and $v = +z$

$$\therefore \quad \frac{1}{(-y)} + \frac{1}{(+z)} = \frac{1}{f_2}$$

Adding

$$\frac{1}{(+x)} + \frac{1}{(+z)} = \frac{1}{f_1} + \frac{1}{f_2}$$

Considering the combination, I is the real image formed of O by both lenses, therefore $u = +x$ and $v = +z$

$$\therefore \quad \frac{1}{(+x)} + \frac{1}{(+z)} = \frac{1}{f}$$

where f is the combined focal length, i.e. the focal length of the single lens that would be exactly equivalent to the two in contact

$$\therefore \quad \frac{1}{f} = \frac{1}{f_1} + \frac{1}{f_2}$$

For example if a converging lens of 5.0 cm focal length is in contact with a diverging lens of 10 cm focal length, then $f_1 = +5.0$ cm, $f_2 = -10$ cm and the combined focal length f is given by

$$\frac{1}{f} = \frac{1}{(+5)} + \frac{1}{(-10)} = \frac{1}{5} - \frac{1}{10} = + \frac{1}{10}$$

$$\therefore \quad f = +10 \text{ cm} \quad \text{(a converging combination)}$$

(c) *Power of a lens.* The shorter the focal length of a lens, the more does it converge or diverge light. The *power* F of a lens is defined as the reciprocal of its focal length f in metres.

$$F = \frac{1}{f}$$

The unit of power is now the *radian per metre* (rad m^{-1}) since $f = h/D$ (p. 182), where the distance h is in metres and the angle of deviation D is in radians. (The former unit was the dioptre, 1 dioptre = 1 radian per metre.) The power of a lens of focal length (i) 1 metre is 1 rad m^{-1}, (ii) 25 cm (0.25 m) is 4.0 rad m^{-1}. The sign of F is the same as f, i.e. positive for a converging lens and negative for a diverging one.

Opticians obtain the power of a lens using a 'lens measurer', Fig. 5.57. This has three legs, the centre one being spring-loaded and connected to a pointer moving over a scale. By measuring the surface curvature, the power

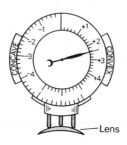

Fig. 5.57

may be obtained quickly and accurately of any lens made of material of a certain refractive index. Lenses of materials of other refractive indices are catered for by using a scale of refractive indices along with the instrument.

Whilst f is useful for constructing ray diagrams, it is more convenient to use F when calculating the combined effect of several optical parts. Thus the combined power F of three thin lenses of powers F_1, F_2 and F_3 in contact is

$$F = F_1 + F_2 + F_3$$

Methods of measuring f for lenses

(a) Converging lens

Rough method. The image formed by the lens of a *distant* window is focused sharply on a screen. The distance between the lens and the screen is f. Why?

Plane mirror method. Using the arrangement shown in Fig. 5.58, a pin or illuminated object is adjusted until it coincides in position with its image, located by no parallax or by a screen. The rays from the object must emerge from the lens and fall on the plane mirror normally to retrace their path. The object is therefore at the principal focus.

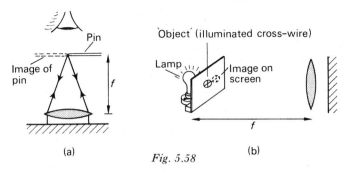

(a)

Fig. 5.58

(b)

Lens formula method. Several values of the image distance v, corresponding to different values of the object distance u are found using either the pin no parallax method or an illuminated object and screen. For each pair of values, f is calculated from $1/f = 1/v + 1/u$ and the average taken.

A better plan is to plot a graph of $1/v$ against $1/u$, draw the best straight line AB through the points, Fig. 5.59. The intercepts OA and OB on the

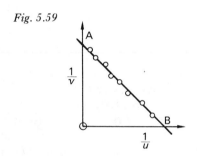

Fig. 5.59

axes are both equal to $1/f$ since when $1/u = 0$, $1/v = OA = 1/f$, from the simple lens equation.

Two-position method. The image I of an illuminated object O is obtained on a screen, Fig. 5.60. With the object and screen in the same position, the lens is moved from A to B where another sharp image is obtained. (The image is magnified for position A of the lens and diminished for B. Why?) The distances between O and I, d, and between A and B, l, are measured then f calculated from

$$f = \frac{d^2 - l^2}{4d}$$

By changing d, a set of values of d and l can be found to give an average value of f.

To derive the above expression for f we use the fact that O and I are inter-

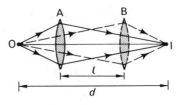

Fig. 5.60

changeable, i.e. are conjugate points, therefore OA = BI and OB = AI. For position A of the lens

$$u = OA = OI - AB - BI = d - l - u$$

$$\therefore \quad 2u = d - l \qquad \therefore \quad u = (d - l)/2$$

$$v = AI = OI + AB - AI = d + l - v$$

$$\therefore \quad 2v = d + l \qquad \therefore \quad v = (d + l)/2$$

Substituting for u and v in $1/v + 1/u = 1/f$ gives

$$\frac{1}{(d + l)/2} + \frac{1}{(d - l)/2} = \frac{1}{f}$$

Hence f follows. The method is suitable when u and v cannot be measured because the faces of the lens are inaccessible due, for example, to the lens being in a tube. *N.B.* A converging lens cannot form a real image on a screen if (i) the object is inside the principal focus or (ii) the distance between the object and the screen is less than $4f$ (this can be proved theoretically and experimentally). When the separation is $4f$, the object and image are then each distant $2f$ from the lens on opposite sides. In the 'two-position' method for f, the separation of the object and screen must exceed $4f$.

Magnification method. Using an object of known size (e.g. an illuminated transparent scale) direct measurement is made of the size of the image produced on a screen by the lens. The magnification m can thus be obtained directly.

Multiplying both sides of $1/v + 1/u = 1/f$ by v we get

$$1 + \frac{v}{u} = \frac{v}{f}$$

$$\therefore \quad 1 + m = \frac{v}{f}$$

$$\therefore \quad m = \frac{v}{f} - 1$$

192

A set of values of m and v are obtained and a graph of m against v plotted. It should be a straight line whose slope is $1/f$ and intercept on the v-axis (when $m = 0$) is f, Fig. 5.61.

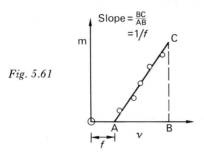

Fig. 5.61

(b) Diverging lens

Auxiliary converging lens method. A diverging lens normally forms a virtual image of a real object. Such an image cannot be located by a screen and is not easily found by a pin no parallax method. However with the help of a converging lens, a real image can be obtained.

In Fig. 5.62 the converging lens forms a real image at I' of an object O when the diverging lens is absent. This image is located and its position

Fig. 5.62

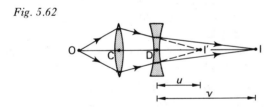

noted. The diverging lens is then placed between C and I' and the converging beam of light falling on it behaves as a virtual object at I'. The diverging lens forms a real image of I' at I, which is located. For the diverging lens $u = -I'D$ and $v = +ID$ whence f can be calculated.

Methods of measuring r for a lens surface

(*a*) *Converging lens* (Boys' method). The lens is floated on mercury and an object O moved until its image is also formed at O. Rays from O must be refracted at the top surface to be incident normally on the bottom surface and thence be reflected to retrace their own path to O, Fig. 5.63*a*.

If refraction was allowed to occur at *both* surfaces of the lens, the ray OAB, since it falls normally on the bottom surface, would pass straight through the lens along BD and form a *virtual* image of O at C, the centre of curvature of the bottom surface, Fig. 5.63*b*.

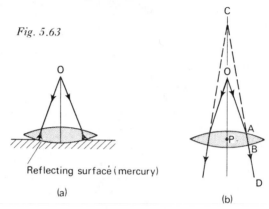

Fig. 5.63

Reflecting surface (mercury)

(a)

(b)

Knowing the focal length f of the lens and measuring u ($= +OP$), v ($= -CP$), which is the radius of curvature of the bottom surface, can be calculated from the simple *lens* equation $1/v + 1/u = 1/f$. The radius of the other surface is found in the same way by turning the lens over.

Notes. (*i*) Although use is made of *reflection* at the bottom surface of the lens, the calculation is based on what would occur if *refraction* through the lens took place.

(*ii*) The refractive index n of the lens material can be found if r_1, r_2 and f are known, using $1/f = (n - 1)(1/r_1 + 1/r_2)$.

(*b*) *Diverging lens.* The position is found in which an object coincides with its image formed by the weak *reflection* at the surface of the lens acting as a concave mirror. The object is then at the centre of curvature of the surface. The method is simply the 'self-conjugate point' method for a concave mirror (p. 162).

Defects in images

So far our discussion of the formation of images by spherical mirrors and lenses has been confined to paraxial rays; we have assumed that the mirror or lens had a small aperture and that object points were on or near the principal axis. In such cases it is more or less true to say that point images are formed of point objects. However, when rays are non-paraxial and objects are extended and mirrors and lens are of large aperture, the image can differ in shape, sharpness and colour from the object. Two of the most important image defects or *aberrations* will be considered.

OPTICAL PROPERTIES

(a) *Spherical aberration.* This arises with mirrors and lenses of large aperture and results in the image of an object point not being a point. The defect is due to the fact that the focal length of the mirror or lens for marginal rays is *less* than for paraxial rays—a property of a spherical surface.

Consider a point object at infinity (i.e. a long distance off) on the principal axis of a mirror or lens whose aperture is not small. The incident rays are parallel to the axis and are reflected or refracted so that the marginal rays farthest from the axis come to a focus at F_m whilst the paraxial rays give a point focus at F_p, Fig. 5.64a and b. All the reflected or refracted rays are tangents to a surface, called a *caustic surface*, which has an apex at F_p. (A caustic curve may be seen on the surface of a cup of tea in bright light, the inside of the cup acting as the mirror.) The nearest approach to a sharp image is the *circle of least confusion*, i.e. the smallest circle through which passes all the reflected or refracted rays.

In general, the image of any object point, on or off the axis, is a circular 'blur' and not a point. The distance $F_m F_p$ in Fig. 5.64 is the *longitudinal spherical aberration* of the mirror or lens for the particular object distance.

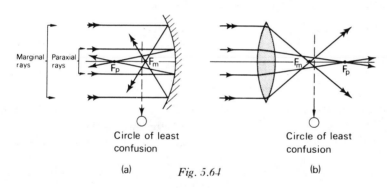

Marginal rays | Paraxial rays

Circle of least confusion

Circle of least confusion

(a)   *Fig. 5.64*   (b)

Whilst it is not possible to construct a mirror which always forms a point image of a point object on the axis, an ellipsoidal mirror achieves this for one definite point on the axis for both paraxial and marginal rays. In Fig. ⁻65a, ABC represents an ellipsoidal mirror with foci F_1 and F_2; *all* rays from F_1

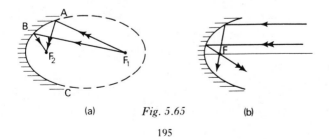

(a)   *Fig. 5.65*   (b)

195

are reflected through F_2. A parabola is an ellipse with one focus at infinity and so a paraboloidal mirror brings all rays from an object point *on* the axis at infinity to a point focus, thus accounting for its use as the objective in an astronomical telescope, Fig. 5.65b. (It should be noted, however, that it does not form a point image of a point object *off* the axis.) Searchlights and car headlamps have paraboloidal reflectors which produce a roughly parallel beam from a small light source at the focus. A perfectly parallel beam does not spread out as the distance from the reflector increases and its intensity does not therefore decrease on this account.

In a lens spherical aberration can be *minimized* if the angles of incidence at each refracting surface are kept small, thus, in effect, making all rays paraxial. This is achieved by sharing the deviation of the light as equally as possible between the surfaces. Fig. 5.66 shows parallel light falling on a

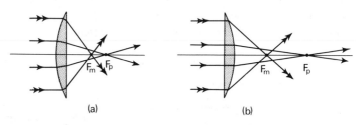

(a) (b)

Fig. 5.66

plano-convex lens; spherical aberration is smaller in (a) than in (b). Why? Why would it be better to have the convex side towards the object if the lens was used as a telescope objective but the other way round for a microscope objective?

Spherical aberration can also be reduced by placing a stop in front of the lens or mirror to cut off marginal rays but this has the disadvantage of making the image less bright.

(*b*) *Chromatic aberration.* This defect occurs only with lenses and causes the image of a white object to be blurred with coloured edges. A lens has a greater focal length for red light than for violet light, as shown in the exaggerated diagram of Fig. 5.67a. (This can be seen from $1/f = (n - 1)(1/r_1 + 1/r_2)$ bearing in mind that $n_{violet} > n_{red}$.) Thus a converging lens produces a series of coloured images of an extended white object, of slightly different sizes and at different distances from the lens, Fig. 5.67b. The eye, being most sensitive to yellow–green light, would focus the image of this colour on a screen but superimposed on it would be the other images, all out of focus.

Chromatic aberration can be eliminated for *two* colours (and reduced for all) by an *achromatic doublet*. This consists of a converging lens of crown glass

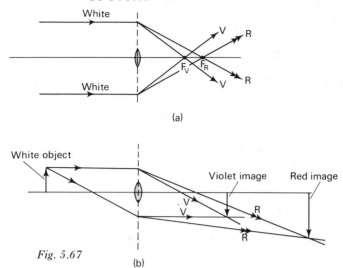

(a)

(b)

Fig. 5.67

combined with a diverging lens of flint glass. One surface of each lens has the same radius of curvature to allow them to be cemented together with Canada balsam and thereby reduce light loss by reflection, Fig. 5.68. The flint glass of the diverging lens produces the same dispersion as the crown glass of the converging lens but in the opposite direction and with less deviation of the light, so that overall the combination is converging. In Fig. 5.69, the dispersions (exaggerated) θ_1 and θ_2 are equal and opposite; for the deviations, $D_1 > D_2$.

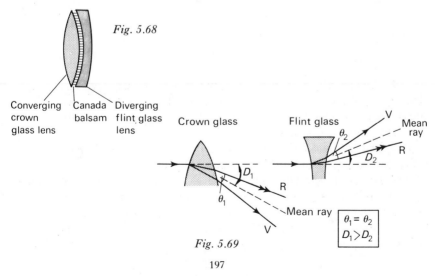

Fig. 5.68

Fig. 5.69

197

The eye and its defects

The construction of the human eye is shown in Fig. 5.70. The image on the retina is formed by successive refraction at the surfaces between the air, the cornea, the aqueous humour, the lens and the vitreous humour. The brain interprets the information transmitted to it as electrical impulses from the

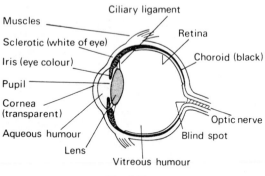

Fig. 5.70

retinal image and appreciates by experience that an inverted image means an upright object, Fig. 5.71. In good light the eye automatically focuses the image of an object on a very small region towards the centre of the retina called the fovea. The fovea permits the best observation of detail—to about two minutes of arc, i.e. to $\frac{1}{10}$ mm at about 20 cm. The periphery of the retina

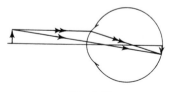

Fig. 5.71

can only detect much coarser detail but it is more sensitive to dim light. Objects at different distances are focused by the ciliary ligaments changing the shape of the lens—a process known as *accommodation*. It becomes more convex to view nearer objects.

The farthest point which can be seen distinctly by the unaided eye is called the *far point*—infinity for the normal eye; the nearest point that can be focused distinctly by the unaided eye is called the *near point*—25 cm for a normal adult eye but less for younger people. The distance of 25 cm is known as the *distance of most distinct vision*. The range of accommodation of the normal eye is thus from 25 cm to infinity and when relaxed it is focused on the latter point.

(*a*) *Short sight* (Myopia). The short-sighted person sees near objects clearly but his far point is closer than infinity. The image of a distant object is focused in front of the retina because the focal length of the eye is too short for the length of the eyeball, Fig. 5.72*a*. The defect is corrected by a *diverging* spectacle lens whose focal length *f* is such that it produces a virtual

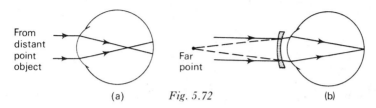

From
distant
point
object

Far
point

(a)   *Fig. 5.72*   (b)

image at the far point of the eye of an object at infinity, Fig. 5.72*b*. Thus if the far point is 200 cm, $v = -200$ cm, $u = \infty$ and from $1/f = 1/v + 1/u$ we get $1/f = -1/200 + 1/\infty$, therefore $f = -200$ cm.

(*b*) *Long sight* (Hypermetropia). The long-sighted person sees distant objects clearly but his near point is more than 25 cm from the eye. The image of a near object is focused behind the retina because the focal length of the eye is too long for the length of the eyeball, Fig. 5.73*a*. The defect is corrected by a *converging* spectacle lens of focal length *f* which gives a virtual image at the near point of the eye of an object at 25 cm, Fig. 5.73*b*. For example, if the near point is 50 cm, $u = +25$ cm, $v = -50$ cm and $1/f = -1/50 + 1/25 = +1/50$. Therefore $f = +50$ cm.

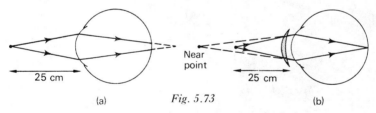

25 cm

Near
point

25 cm

(a)   *Fig. 5.73*   (b)

(*c*) *Presbyopia*. In this defect, which often develops with age, the eye loses its power of accommodation and two pairs of spectacles may be needed, one for distant objects and the other for reading. Sometimes 'bifocals' are used which have a diverging top part to correct for distant vision and a converging lower part for reading, Fig. 5.74.

Fig. 5.74

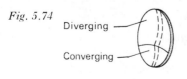

Diverging

Converging

Bifocal lens

(*d*) *Astigmatism*. If the curvature of the cornea varies in different directions, rays in different planes from an object are focused in different positions by the eye and the image is distorted. The defect is called astigmatism and anyone suffering from it will see one set of lines in Fig. 5.75 more sharply than the others. It may be possible to correct it with a non-spherical spectacle

Fig. 5.75

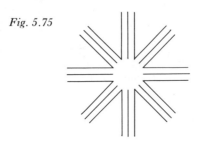

lens whose curvature increases the effect of that of the cornea in its direction of minimum curvature or decreases it in the maximum curvature direction.

(*e*) *Contact lenses*. These consist of tiny, unbreakable plastic lenses held to the cornea by the surface tension (p. 345) of eye fluid and in recent years they have increased in popularity. Fig. 5.76*a* shows one balanced on a finger tip. As well as being safer for sportsmen, they may help certain eye defects

Fig. 5.76*a*

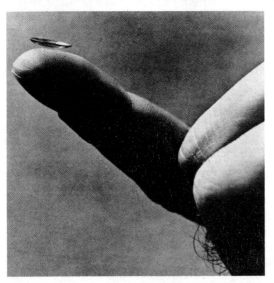

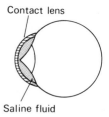

Contact lens

Saline fluid

Fig. 5.76*b*

which spectacles cannot. Thus if the cornea is conical-shaped, vision is very distorted but if a contact lens is fitted and the space between the lens and the cornea filled with a saline solution of the same refractive index as the cornea, normal vision results, Fig. 5.76*b*.

Magnifying power of optical instruments

Previously, when considering the magnification produced by mirrors and lenses, we used the idea of *linear magnification m* and showed that it was given by $m = v/u$. However if the image is formed at infinity, as it can be with some optical instruments, then m should be infinite! The difficulty is that we cannot get to the image to view it and in such cases m is therefore not a very helpful indication of the improvement produced by the instrument. A more satisfactory term is clearly necessary to measure this.

The *apparent* size of an object depends on the size of its image on the retina and, as Fig. 5.77 shows, this depends not so much on the actual size of the

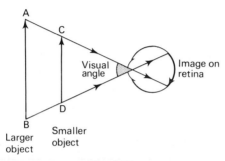

Fig. 5.77

object as on the angle it subtends at the eye, i.e. on the *visual angle*. Thus, AB is larger than CD but because it subtends the same visual angle as CD, it *appears* to be of equal size. The *angular magnification* or *magnifying power* M of an optical instrument is defined by the equation

$$M = \frac{\beta}{\alpha}$$

where β = angle subtended at the eye by the *image* formed when using the instrument, and α = angle subtended at the unaided eye (i.e. without the instrument) by the *object* at some 'stated distance'.

In the case of a telescope the 'stated distance' has got to be where the object (e.g. the moon) is; for a microscope it is usually taken to be the 'distance of most distinct vision' (i.e. 25 cm away) since it is at that distance the object is seen most distinctly by the normal, unaided eye.

The difference between the magnifying power M and the magnification m should be noted. M is the ratio of the *apparent* sizes of image and object and involves a comparison of visual angles; m is the ratio of the *actual* sizes of image and object. They do not necessarily have the same value but in some cases they do.

Magnifying glass

This is also called the simple microscope and consists of a converging lens forming a virtual, upright, magnified image of an object placed inside its principal focus, Fig. 5.78a. The image appears largest and clearest when it is at the near point.

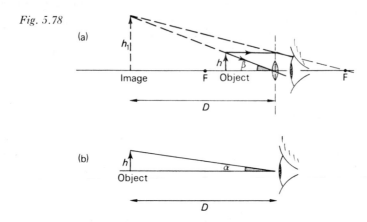

Fig. 5.78

(a)

(b)

Assuming rays are paraxial and that the eye is close to the lens, we can say $\beta = h_1/D$ where h_1 is the height of the image and D is the *magnitude* of the distance of most distinct vision (usually 25 cm). If the object is viewed at the near point by the unaided eye, Fig. 5.78b, we have $\alpha = h/D$ where h is the height of the object. Hence, the magnifying power M is given by

$$M = \frac{\beta}{\alpha} = \frac{h_1/D}{h/D} = \frac{h_1}{h}$$

In this case, $M = m$ where the linear magnification $m = v/u$; v and u being the image and object distances respectively.

$$\therefore \quad M = \frac{v}{u}$$

If $1/v + 1/u = 1/f$ is multiplied throughout by v, we get $v/u = v/f - 1$

$$\therefore \quad M = \frac{v}{f} - 1$$

It follows that a lens of short focal length has a large magnifying power. For example if $f = +5.0$ cm and $v = -D = -25$ cm (since image is virtual) then

$$M = \frac{v}{f} - 1 = \frac{-D}{f} - 1 = \frac{-25}{+5} - 1 = -6.0$$

The magnifying power is 6.0, i.e. $(D/f + 1)$; since M is a number the negative sign can be omitted.

You should draw a ray diagram for a magnifying glass forming a virtual image at *infinity* (where must the object be placed?) and use it to show that in this case $M = D/f$ numerically, i.e. M is one less than when the image is at the near point.

What is the effect, if any, on M if the eye is moved back from the lens when the image is at (i) the near point, (ii) infinity?

Compound microscope

The focal length of a lens can be decreased and its magnifying power thereby increased by making its surfaces more curved. However, serious distortion of the image results from excessive curvature and to obtain greater magnifying power a compound microscope is used consisting of two separated, converging lenses of short focal lengths.

The lens L_1 nearer to the object, called the 'objective', forms a real, magnified, inverted image I_1 of an object O placed just outside its principal focus F_o. I_1 is just inside the principal focus F_e of the second lens L_2, called the 'eyepiece', which acts as a magnifying glass and produces a magnified, virtual image I_2 of I_1. The microscope is said to be in 'normal adjustment' when I_2 is at the near point. Fig. 5.79 shows the usual constructional rays (p. 180), drawn to locate I_1 and I_2; note that the object is seen inverted.

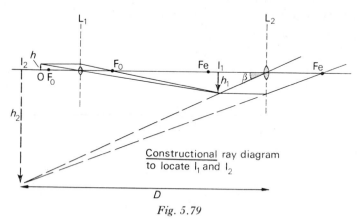

Constructional ray diagram
to locate I_1 and I_2

Fig. 5.79

(*a*) *Magnifying power.* We shall assume that (i) all rays are paraxial, (ii) the eye is close to the eyepiece and (iii) the microscope is in normal adjustment. $M = \beta/\alpha$ where, in this case,

OPTICAL PROPERTIES

β = angle subtended at the eye by I_2 *at the near point* = h_2/D (h_2 being the height of I_2 and D the distance of most distinct vision), and

α = angle subtended at the eye by O at the near point, without the microscope = h/D (h being the height of O).

Hence

$$M = \frac{\beta}{\alpha} = \frac{h_2/D}{h/D} = \frac{h_2}{h}$$

$$= \frac{h_2}{h_1} \times \frac{h_1}{h}$$

(where h_1 is the height of I_1). Now h_2/h_1 is the linear magnification m_e produced by the eyepiece and h_1/h is the linear magnification m_o due to the objective. Thus, $M = m_e \times m_o$, i.e. when the microscope is in normal adjustment with the final image at the near point, the magnifying power equals the linear magnification (as it does for a magnifying glass with the image at the near point). It follows that M will be large if f_o and f_e are small.

Many school-type microscopes have a low-power objective ($f_o \simeq 16$ mm) magnifying 10 times and a high-power objective ($f_o \simeq 4$ mm) magnifying 40 times. When used with a $\times 10$ eyepiece, the overall magnifying power is therefore 100 on low power and 400 on high power.

If prolonged observation is to be made it is more restful for the eye to view the final image I_2 at infinity instead of at the near point. The intermediate image I_1 must then be at the principal focus F_e of the eyepiece so that the emergent rays from the eyepiece are parallel. It can be shown that the magnifying power is then slightly less than for normal adjustment.

(b) *Resolving power*. This is the ability of an optical instrument to reveal detail, i.e. to form separate images of objects that are very close together. To increase the magnifying power unduly without also increasing the resolving power has been likened to stretching an elastic sheet on which a picture has been painted—the picture gets bigger but no more detail is seen.

It can be shown that the resolving power of a microscope is greater (*i*) the greater the angle θ subtended at the objective by a point in the object, Fig. 5.80, and (*ii*) the shorter the wavelength of the light used. Both these

Fig. 5.80

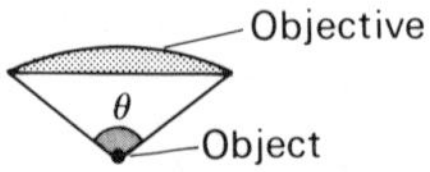

factors impose a definite limit on the resolving power and, having regard to this limit, the maximum useful magnifying power is about 600. In practice, in the interests of eye comfort, this value is often exceeded (over $\times 2000$ is attainable in the best instruments).

(c) *The eye ring.* The best position for an observer to place his eye when using a microscope is where it gathers most light from that passing through the objective—the image is then brightest and the field of view greatest.

In Fig. 5.81 the paths of two cones of rays are shown coming from the top and bottom of an object, filling the whole aperture of the objective and passing through the microscope. (Constructional rays, not shown here, are first

Fig. 5.81

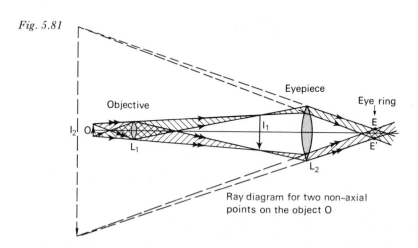

Ray diagram for two non–axial
points on the object O

drawn as in Fig. 5.79 to locate I_1 and I_2.) The only position at which the eye would receive both these cones (and also those falling on the objective from all other object points) is at EE′ where they cross. All light from the objective refracted by the eyepiece will pass through a small circle of diameter EE′ which must therefore be the image of the objective formed by the eyepiece. This image at EE′ is called the *eye ring* (or exit-pupil) and it is the best position for an observer's eye.

Ideally EE′ should equal the diameter of the average eye pupil and in a microscope a circular opening of this size is often fixed just beyond the eyepiece to indicate the eye ring position. If, for example, the objective is 15 cm from the eyepiece of focal length 1 cm then the distance v of the eye ring from the eyepiece is given by $1/v + 1/(+15) = 1/(+1)$, i.e. $v = 1.1$ cm.

(d) *A calculation.* You are strongly advised to work out numerical problems from first principles and not to quote formulae. It is also helpful to draw a diagram.

The objective and the eyepiece of a microscope may be treated as thin lenses with focal lengths of 2.0 cm and 5.0 cm respectively. If the distance between them is 15 cm and the final image is formed 25 cm from the eyepiece, calculate (i) the position of the object and (ii) the magnifying power of the microscope.

Let the positions of the object O, the first image I_1 and the final image I_2 be as in Fig. 5.82a. A ray from the top of O through the optical centre P_1 of the objective passes through the top of I_1 and a ray from the top of I_1

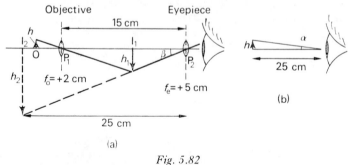

Fig. 5.82

through the optical centre P_2 of the eyepiece passes through the top of I_2 when produced backwards. Let h and h_1 be the heights of O and I_1 respectively.

(*i*) Consider the eyepiece. I_1 acts as the object and the final image I_2 is *virtual*. We have $v = -25$ cm and $f_e = +5.0$ cm

$$\therefore \quad \frac{1}{(-25)} + \frac{1}{u} = \frac{1}{(+5)}$$

$$\therefore \quad \frac{1}{u} = \frac{1}{5} + \frac{1}{25} = \frac{6}{25}$$

$$\therefore \quad u = I_1 P_2 = +4\tfrac{1}{6} \text{ cm}$$

Consider the objective. Image I_1 is *real* and at a distance $P_1 I_1$ from the objective where $P_1 I_1 = P_1 P_2 - I_1 P_2 = 15 - 4\tfrac{1}{6} = 10\tfrac{5}{6}$ cm. Hence $v = P_1 I_1 = +10\tfrac{5}{6} = +65/6$ cm. Also $f_o = +2.0$ cm

$$\therefore \quad \frac{1}{(+65/6)} + \frac{1}{u} = \frac{1}{(+2)}$$

$$\therefore \quad \frac{1}{u} = \frac{1}{2} - \frac{6}{65} = \frac{53}{130}$$

$$\therefore \quad u = OP_1 = +\frac{130}{53} \text{ cm}$$

$$= +2\tfrac{24}{53} \text{ cm} \quad (\simeq 2.5 \text{ cm})$$

The object O is about 2.5 cm from the objective.

(*ii*) Assuming the observer is close to the eyepiece the angle subtended at his eye is given by

$$\beta = \frac{h_1}{I_1P_2} = \frac{h_1}{(25/6)} = \frac{6h_1}{25}$$

The angle α subtended at the observer's eye when he views the object at his near point (assumed to be 25 cm away) without the microscope is given by $\alpha = h/25$, Fig. 5.82*b*. Therefore

$$\text{magnifying power } M = \frac{\beta}{\alpha}$$

$$= \frac{6h_1/25}{h/25} = \frac{6h_1}{h}$$

But

$$\frac{h_1}{h} = \frac{P_1I_1}{P_1O} = \frac{10\frac{5}{6}}{2\frac{24}{53}} = \frac{65/6}{130/53} = \frac{53}{12}$$

$$\therefore \quad M = 6 \times \frac{53}{12} = 27$$

Refracting astronomical telescope

A lens-type astronomical telescope consists of two converging lenses, one is an objective of long focal length and the other an eyepiece of short focal length. The objective L_1 forms a real, diminished, inverted image I_1 of a distant object at its principal focus F_o since the rays incident on L_1 from a *point* on such an object can be assumed parallel. The eyepiece L_2 acts as a magnifying glass and forms a magnified virtual image of I_1 and, when the telescope is in normal adjustment, this image is at infinity. I_1 must therefore be at the principal focus F_e of L_2, hence F_o and F_e coincide.

In Fig. 5.83 three *actual* rays are shown coming from the top of a distant

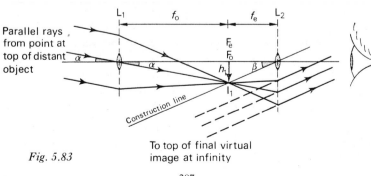

Fig. 5.83

Parallel rays from point at top of distant object

Construction line

To top of final virtual image at infinity

object and passing through the top of I_1 in the focal plane of L_1. They must emerge parallel from L_2 to appear to come from the top of the final image at infinity. They must also be parallel to the line joining the top of I_1 to the optical centre of L_2.

(a) *Magnifying power*. It will be assumed that (i) all rays are paraxial, (ii) the eye is close to the eyepiece and (iii) the telescope is in normal adjustment. Now $M = \beta/\alpha$ and in this case

$$\beta = \text{angle subtended at the eye by the}$$
final image at infinity
$$= \text{angle subtended at the eye by } I_1$$
$$= h_1/f_e$$

(h_1 being the height of I_1), and

$$\alpha = \text{angle subtended at the eye by the object}$$
without the telescope
$$= \text{angle subtended at the objective by the object}$$

(since the distance between L_1 and L_2 is very small compared with the distance of the object from L_1)

$$= h_1/f_o$$

Hence
$$M = \frac{\beta}{\alpha} = \frac{h_1/f_e}{h_1/f_o}$$

$$\therefore \quad M = \frac{f_o}{f_e}$$

Notes. (i) The above expression for M is true only for normal adjustment; the separation of the objective and eyepiece is then $f_o + f_e$.

(ii) A telescope is in normal adjustment when the final image is formed at infinity; a microscope is in normal adjustment with the final image at the near point.

For high magnifying power the objective should have a large focal length and the eyepiece a small one. The largest lens telescope in the world is at the Yerkes Observatory, U.S.A.; the objective has a focal length of about 20 metres and the most powerful eyepiece of about 6.5 mm. The maximum value of M is therefore $20 \times 10^3/6.5 \simeq 3000$.

If it is desired to form the final image at the near point, i.e. telescope not in normal adjustment, the eyepiece must be moved so that I_1 is closer to it than F_e. The magnifying power is then slightly greater than f_o/f_e.

(b) *Resolving power*. It can be shown that the ability of a telescope to reveal detail increases as the diameter of the objective increases. However, large lenses are not only difficult to make but they tend to sag under their own weight. The objective of the Yerkes telescope has a diameter of 1 metre which is about the maximum possible. There is no point in increasing the magnifying power of a telescope unduly if the resolving power cannot also be increased.

(c) *The eye ring.* As in the case of the microscope, the eye ring is in the best position for the eye and is the circular image of the objective formed by the eyepiece. All rays incident on the objective which leave the telescope pass through it. In Fig. 5.84a two cones of rays are shown coming from the top and bottom of a distant object and crossing at the eye ring EE′.

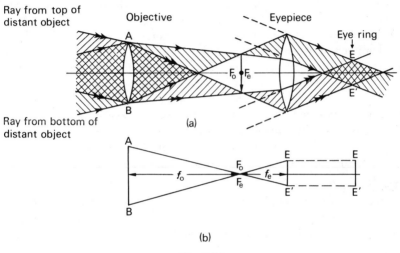

Fig. 5.84

If the telescope is in normal adjustment, the separation of the lenses is $f_o + f_e$ and from similar triangles in Fig. 5.84b

$$\frac{AB}{EE'} = \frac{f_o}{f_e}$$

But the magnifying power M for a telescope in normal adjustment is f_o/f_e, hence

$$M = \frac{\text{diameter of objective}}{\text{diameter of eye ring}}$$

This expression enables M to be found simply by illuminating the objective with a sheet of frosted glass and a lamp, locating the image of the objective formed by the eyepiece (i.e. the eye ring) on a screen and measuring the diameter.

(d) *Brightness of image.* A telescope increases the light-gathering power of the eye and in the case of a point object, such as a star, forms a brighter image. Thus, when the diameter of the objective is doubled, the telescope collects four times more light from a given star (why?) and since a point image is formed of a point object, whatever the magnifying power, the star appears brighter. Many more stars are therefore visible than would otherwise be seen and in fact the range of a telescope is proportional to the diameter of its objective.

The brightness of the background is not similarly increased because it acts as an extended object and, as we will now see, a telescope does not increase the brightness of such an object.

When the diameter of the eye ring equals the diameter of the pupil of the eye, almost all the light entering the telescope enters the eye. If M is the magnifying power of the telescope, the diameter of the objective is M times the diameter of the eye ring and the area of the objective is M^2 times greater. M^2 times more light enters the eye via the telescope than would enter it from the object directly. However the image has an area M^2 times that of the object since the telescope makes the object appear M times as high and M times as wide as it does to the unaided eye at the same distance. The brightness of the image cannot therefore exceed that of the object and is less, due to loss of light in the instrument. The contrast between a star and its background is thus increased by a telescope.

(e) *A calculation.*

An astronomical telescope has an objective of focal length 100 cm *and an eyepiece of focal length 5.0* cm. *Calculate the magnifying power when the final image of a distant object is formed (i) at infinity, (ii) 25.0* cm *from the eyepiece.*

(i) When the final image is at infinity, the telescope is in normal adjustment and the magnifying power is given by

$$M = \frac{f_o}{f_e} = \frac{100}{5.0} = 20$$

(ii) Let the position of I_1 be as shown in Fig. 5.85. A ray from the top of the distant object through the optical centre P_1 of the objective passes

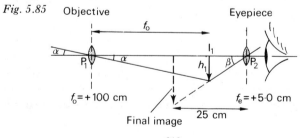

Fig. 5.85   Objective   Eyepiece

$f_o = +100$ cm $f_e = +5·0$ cm

Final image   25 cm

through the top of I_1. Also a ray from the top of I_1 through the optical centre P_2 of the eyepiece passes through the top of the final image when produced backwards.

For the eyepiece we can say $v = -25.0$ cm (final image *virtual*) and $f_e = +5.0$ cm.

$$\therefore \quad \frac{1}{(-25)} + \frac{1}{u} = \frac{1}{(+5)}$$

$$\therefore \quad \frac{1}{u} = \frac{1}{5} + \frac{1}{25} = \frac{6}{25}$$

$$\therefore \quad u = I_1P_2 = 4\tfrac{1}{6} \text{ cm}$$

If the eye is close to the eyepiece the angle β subtended at the eye is given by $\beta = h_1/I_1P_2 = h_1/(25/6) = 6h_1/25$. The angle α subtended at the unaided eye by the object = angle subtended at the objective by the object (see p. 208) $= h_1/f_o = h_1/100$

Hence
$$M = \frac{\beta}{\alpha} = \frac{6h_1/25}{h_1/100} = \frac{6 \times 100}{25}$$

$$= 24$$

Reflecting astronomical telescope

The largest modern astronomical telescopes use a concave mirror of long focal length as the objective instead of a converging lens but the principle is the same as the refracting telescope. One arrangement, called the Newtonian form after the inventor of the reflecting telescope, is shown in Fig. 5.86. Parallel rays from a distant point object on the axis are reflected first at

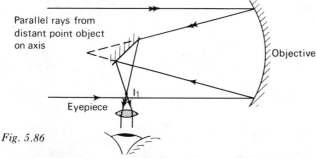

Parallel rays from
distant point object
on axis

Objective

Eyepiece

I_1

Fig. 5.86

the objective and then at a small plane mirror to form a real image I_1 which can be magnified by an eyepiece or photographed by having a film at I_1. The plane mirror, whose area is negligible compared with that of the concave mirror, deflects the light sideways without altering the effective focal length f_o of the objective. In normal adjustment the magnifying power is f_o/f_e where f_e is the focal length of the eyepiece.

The advantages of reflecting telescopes are

(*i*) no chromatic aberration since no refraction occurs at the objective,

(*ii*) no spherical aberration for a point object on the axis at infinity if a paraboloidal mirror is used (see p. 196),

(*iii*) a mirror can have a much larger diameter than a lens (since it can be supported at the back) thereby giving greater resolving power and a brighter image of a point object,

(*iv*) only one surface requires to be ground (compared with two for a lens), thus reducing costs.

The largest reflecting optical telescope in the world on Mount Palomar, California, Fig. 5.87, has a concave paraboloidal mirror of diameter 5 metres. It is made of low expansion glass which took six years to grind and the reflecting surface is coated with aluminium. Photographs of nebulae up to a distance of 10^{10} light-years away (1 light-year is the distance travelled by light in 1 year) can be taken. It is used in conjunction with spectrometers, cameras and other instruments in temperature-controlled, air-conditioned surroundings.

Fig. 5.87

For general astronomical work lens telescopes are more easily handled than large mirror telescopes; the latter are used only where high resolving power is required.

Other telescopes

(a) *Terrestrial telescope.* The final image in an astronomical telescope is inverted and whilst this is not a handicap for looking at a star, it is when viewing objects on earth.

A terrestrial telescope is a refracting astronomical telescope with an intermediate 'erecting' lens arranged as in Fig. 5.88 to be at a distance of $2f$

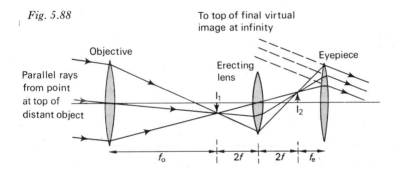

Fig. 5.88

(where f is the focal length of the erecting lens) from the inverted image I_1 formed by the objective. An erect image I_2 of the same size as I_1 is formed at $2f$ beyond the erecting lens and acts as an 'object' for the eyepiece in the usual way. A disadvantage of this arrangement is the increase in length of the telescope by $4f$.

(b) *Prism binoculars.* These consist of a pair of refracting astronomical telescopes with two totally reflecting prisms (angles 90°, 45° and 45°) between each objective and eyepiece as in Fig. 5.89. Prism A causes lateral

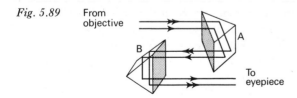

Fig. 5.89

inversion and prism B inverts vertically so that the final image is the same way round and the same way up as the object. Each prism reflects the light through 180° making the effective length of each telescope three times the distance between the objective and the eyepiece. Good magnifying power is thus obtained with compactness.

Prism binoculars marked '7 × 50' have a magnifying power of 7 and objectives of diameter 50 mm.

(c) . Galilean telescope. A final erect image is obtained in the Galilean telescope using only two lenses—a converging objective of large focal length f_o and a diverging eyepiece of small focal length f_e.

The image of a distant object would, in the absence of the eyepiece, be formed by the objective at I_1, where $P_1 I_1 = f_o$, Fig. 5.90. With the eyepiece

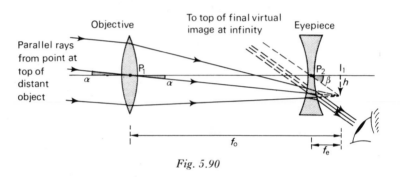

Fig. 5.90

in position at a distance f_e from I_1, the separation of the lenses is $f_o - f_e$ (numerically) and rays falling on the eyepiece emerge parallel, so that to the eye the top of the final image is *above* the axis of the telescope. An upright image at infinity is thus obtained. The converging light falling on the eyepiece behaves like a virtual object at I_1 and a virtual image of it is formed.

If the telescope is in normal adjustment, i.e. final image at infinity, the magnifying power is f_o/f_e, as for an astronomical telescope. In Fig. 5.90 the ray from the top of I_1 passing through the centre P_2 of the eyepiece goes to the top of the final image at infinity. It must therefore be parallel to the three parallel rays emerging from the eyepiece. The angle β subtended at the eye (close to the telescope) is thus given by $\beta = h/I_1 P_2 = h/f_e$ where h is the height of I_1. Also, the angle α subtended at the unaided eye by the object is very nearly equal to the angle subtended at the objective by the object, hence $\alpha = h/I_1 P_1 = h/f_o$. Thus, the magnifying power M is

$$M = \frac{\beta}{\alpha} = \frac{h/f_e}{h/f_o} = \frac{f_o}{f_e}$$

The Galilean telescope is shorter than the terrestrial telescope but the field of view is very limited because the eye ring is between the lenses (why?) and so inaccessible to the eye. Opera glasses consist of two telescopes of this type.

Spectrometer

The spectrometer is designed primarily to produce and make measurements on the spectra of light sources and is generally used with a diffraction grating but a prism can be employed. It also provides a very accurate method of measuring refractive index.

The instrument consists of (*i*) a fixed collimator with a movable slit of adjustable width (to produce a parallel beam of light from the source illuminating the slit), (*ii*) a turntable (having a circular scale) on which the grating or prism is placed and (*iii*) a telescope (with a vernier scale) rotatable about the same vertical axis as the turntable, Fig. 5.91. The converging lenses in the collimator and telescope are achromatic.

Four preliminary adjustments must first be made.

Fig. 5.91

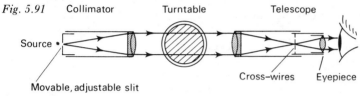

Collimator   Turntable   Telescope

Source

Movable, adjustable slit   Cross–wires   Eyepiece

(*a*) *Adjustments*

(*i*) *Eyepiece*. This is moved in the tube containing the cross-wires until the cross-wires are clearly seen. An image formed on the wires will then be distinct.

(*ii*) *Telescope*. A distant object (e.g. a vertical line of mortar between bricks in a building outside) is viewed through the telescope and the distance of the objective from the cross-wire eyepiece adjusted by a thumb-screw until there is no parallax between the image of the distant object and the cross-wires. Parallel rays entering the telescope are now brought to a focus at the cross-wires.

(*iii*) *Collimator*. The telescope is turned into line with the collimator and the slit, illuminated with sodium light, is moved in or out of the collimator tube until there is no parallax between the image of the slit and the cross-wires. The slit is then at the principal focus of the collimator lens which is producing a parallel beam.

(*iv*) *Levelling the table*. The method adopted depends on whether a grating or a prism is to be used; we will consider the latter at present. The prism is placed on the turntable with one face (AB in Fig. 5.92*a*) perpendicular to the lines on the table which join levelling screws S_1 and S_2. With the telescope at right angles to the collimator the table is rotated until a reflected image of the slit from AB enters the telescope. S_1 is then adjusted so that this image is in the centre of the field of view, as in Fig. 5.92*b*. Keeping the tele-

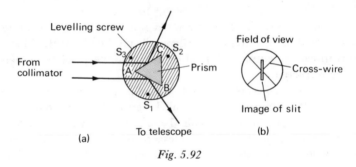

(a)

Levelling screw

From collimator

Prism

S_1

To telescope

Field of view

Cross-wire

Image of slit

(b)

Fig. 5.92

scope in the same position, the turntable is rotated to give a reflected image of the slit which is obtained in the telescope from face AC of the prism. S_3 *only* is adjusted to centralize the image of the slit in the field of view. The turntable is then level, i.e. the refracting edge of the prism is now parallel to the axis of rotation of the telescope.

(b) Measurement of refractive index

By finding the refracting angle A of a prism and its angle of minimum deviation D_{min}, the refractive index n of the material of the prism can be found for light of one colour (i.e. monochromatic light) from

$$n = \frac{\sin [(A + D_{min})/2]}{\sin (A/2)}$$ (p. 175)

To measure A, the slit is made as narrow as possible and the prism set on the turntable as in Fig. 5.93*a* so that the incident light is reflected from both faces AB and AC. The telescope is rotated in turn into the positions T_1 and

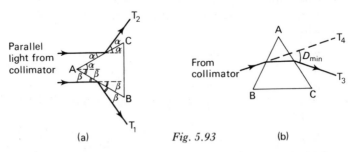

Parallel light from collimator

From collimator

(a)

Fig. 5.93

(b)

T_2 so that the images of the slit, *reflected* from AB and AC respectively, coincide with the intersection of the cross-wires. From the diagram we see that if three dotted lines are drawn parallel to the incident beam of light then $A = \alpha + \beta$ and the angle between T_1 and $T_2 = 2(\alpha + \beta) = 2A$. Hence A is half the angle read on the telescope scale between positions T_1 and T_2.

216

To measure D_{min} a monochromatic source, e.g. a sodium lamp or flame, must be used and the prism set on the turntable as in Fig. 5.93b. The minimum deviation position of the prism is found by rotating the table so that the telescope is as near the straight-through position as possible whilst still receiving a *refracted* image of the slit at the intersection of the cross-wires. The scale reading in this position T_3 is noted, the prism removed and the straight-through reading taken with the telescope and collimator in line, position T_4. D_{min} is the angle between T_3 and T_4. (Alternatively the minimum deviation position on each side can be found and the difference halved.)

Camera

A typical arrangement is shown in Fig. 5.94. The lens system has to have a field of view of about 50° (compared with 1° or so for an average microscope objective) and so the reduction of aberrations is a major consideration. Very

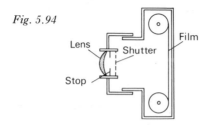

Fig. 5.94

large apertures give blurred images because of aberrations, so do very small apertures but due to the phenomenon called diffraction. The best images are therefore generally obtained with intermediate apertures. For some types of optical systems (e.g. eyes, cameras, enlargers) aberrations are more significant and the aperture has to be reduced to obtain clear images. For others (e.g. telescopes) diffraction is usually more significant and apertures have to be made as large as is practicable.

Cheap cameras use a meniscus lens, which is usually an achromatic doublet, and a stop to restrict the aperture. More expensive cameras have a lens system of several components designed to minimize the various aberrations. Focusing of objects at different distances is achieved by slightly altering the separation of the lens from the film.

In many cameras the amount of light passing through the lens can be altered by an aperture control or stop of variable width. This has a scale of f-numbers with all or some of the following settings—1.4, 2, 2.8, 4, 5.6, 8, 11, 16, 22, 32. These are such that reducing the f-number by one setting, say from 8 to 5.6, *doubles* the area of the aperture, i.e. the *smaller* the f-number the *larger* the aperture. An f-number of 4 means the diameter d of the aperture is $\frac{1}{4}$ the focal length f of the lens, i.e. $d = f/4$.

The aperture affects (*i*) the exposure time and (*ii*) the depth of field. Consider (*i*). Using the next lower *f*-number halves the exposure time needed to produce the same illumination on the film (since the area of the aperture has been doubled). The exposure required depends on the lighting conditions and must be brief if the object is moving. In better cameras exposure time can be varied.

The *depth of field* (often called the depth of focus) is the range of distances in which the camera can more or less focus objects simultaneously. A landscape photograph needs a large depth of field whilst in a family group it may be desirable to have the background out of focus. The depth of field is increased by reducing the lens aperture as can be seen from Fig. 5.95 in which

Fig. 5.95

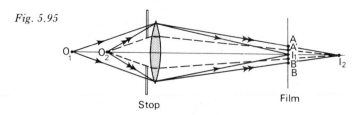

Stop Film

the images formed by a lens of point objects O_1 and O_2 are at I_1 and I_2 respectively. The diameter of the circular patch of light on a film in focus for I_1 is, for the out-of-focus I_2, AB if the whole aperture is used but only A′B′ if the lens is stopped down.

Projector

A projector is designed to throw on a screen a magnified image of a film or transparency. It consists of two main parts—an illumination system and a projection lens, Fig. 5.96.

The image on the screen is usually so highly magnified that very strong but uniform illumination of the film with white light is necessary if the image is also to be bright. This is achieved by directing light from a high power, intense, tungsten iodine filament lamp on to the film by means of a curved reflector and a condenser lens system arranged as shown. Since the screen

Fig. 5.96

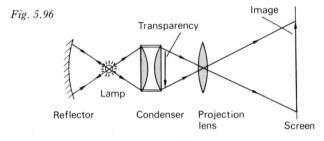

Image

Transparency

Lamp

Reflector Condenser Projection lens Screen

is generally a considerable distance away, the film, inverted, must be just outside the principal focus of the projection lens. Any chance of the image of the lamp appearing on the screen is also removed. To keep the projector as short as possible, the condenser has a short focal length f and the lamp is placed at a distance of $2f$ from it. What will be approximately (*i*) the separation of the condenser and the projection lens and (*ii*) the focal length of the projection lens if the film is close to the condenser?

Radio telescopes

Certain 'objects' in space, among them the sun, emit radio signals which can be picked up by huge aerials called radio telescopes. These do not give visual pictures as do optical telescopes but produce electrical signals which are often recorded graphically. There are various types of radio telescopes: the famous one at Jodrell Bank, Cheshire, Fig. 5.97, consists of a steerable metal reflector or 'dish', 75 metres in diameter; others have isolated aerials distributed over a large area.

The resolving power of any telescope depends on the diameter of the objective (lens, mirror or aerial) and the wavelength of the radiation from the object. The larger the diameter and the shorter the wavelength, the closer together can two distant points be and still be separated by the telescope. The Mount Palomar optical telescope has an objective (concave

Fig. 5.97

mirror) of diameter 5 metres but its resolving power is more than 1000 times that of the Jodrell Bank radio telescope with an aerial of diameter 75 metres. This is because the shortest wavelength of the radio signals from outer space that can penetrate the ionized layers in the upper atmosphere is about 1 cm, whereas the mean wavelength of light is 6×10^{-5} cm. (The longest wavelength that can pass through the earth's radio 'window' is about 30 metres.)

One of the strongest radio sources in our own star-system (the Galaxy) is the Crab nebula, a mass of luminous gas in the constellation of the Bull. This nebula is believed to be the remains of a star which underwent a tremendous explosion, becoming a *supernova* and shining so brightly that it was observed in 1054, according to Chinese records, in broad daylight for several months. The radio signals arise from the highly excited gas which is still expanding outwards from the explosion centre. Some other radio 'stars' may also be due to old supernovae but most lie outside our galaxy and cannot usually be identified with ordinary stars.

Whilst interstellar dust and gas stop light reaching us from distant stars, they do not block radio waves which can bring information about regions we cannot see. Very cold, rarefied hydrogen gas emits 21 cm-long waves and radio astronomers have established that invisible clouds of hydrogen in this state are very widely distributed in all space. By studying its distribution in our own galaxy, we now know that the latter is spiral-shaped and rotating like a huge Catherine-wheel.

Cosmology is concerned with how the universe began and there are two main theories. The 'evolutionary' (or 'big bang') theory suggests that millions of years ago all matter in the universe was concentrated into a very small volume of space, referred to as the 'primeval atom', which exploded, throwing out matter in all directions. From the debris the galaxies of stars gradually formed and the expanding universe is a result of this explosion. The 'steady-state' theory on the other hand proposes that new matter is continually being created out of nothing to fill up the empty space arising from the expansion of the universe. In this case the universe would always 'look' the same.

It is not easy to decide between the two theories but the problem may be resolved by radio astronomy. Some galaxies are known to be radio sources and the signals we receive give information about them, not as they are now, but as they were millions of years ago when the radio signals left them. If the evolutionary theory is correct we would expect the distant galaxies to be much closer together than those nearer to us; the more distant galaxies let us 'look back' through a longer time. If the steady-state theory applies there should be no difference in the average 'density' of galaxies near and far. Work done so far tends to support the evolutionary theory but the issue is still very open.

Two recent discoveries of radio astronomy are *quasars* and *pulsars*. Quasars (quasi-stellar) are very distant 'objects' that are small compared with ordinary galaxies but are very powerful radio sources. Pulsars also emit strong radio signals but in sharp regular pulses at rates varying from 30 pulses per second to 1 in 4 seconds. They are thought to be a very long way off, planet-sized and their source of energy is a mystery.

Electron microscopes

The electron microscope is analogous in principle to the optical microscope but its performance is far superior. The maximum magnifying power attainable with the best optical microscope is about 2000, for an electron microscope up to 100 000 is typical. The former can resolve detail about 10^{-6} m across, in the electron instrument it is very much smaller, about 10^{-9} m. Since atomic dimensions are of the order of 10^{-9} m (1 nm), this means that in some cases an electron microscope can reveal separate molecules.

The similarity between the paths of light in an optical microscope and the paths of electrons in an electron microscope can be seen from Fig. 5.98.

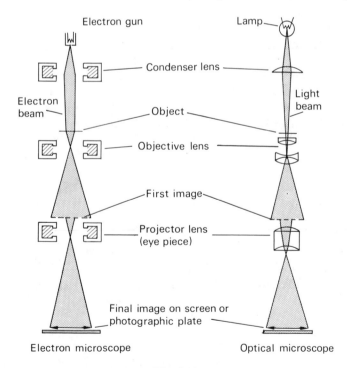

Electron gun   Lamp

Condenser lens

Electron beam

Object

Light beam

Objective lens

First image

Projector lens (eye piece)

Final image on screen or photographic plate

Electron microscope   Optical microscope

Fig. 5.98

221

In the latter, electrons are produced by an electron gun and the 'lenses' are electromagnets designed so that their fields focus the electron beam to give an image on a fluorescent screen or photographic plate. The focal lengths of the 'lenses' are variable and are determined by the current through the 'lens' coils.

When the object is struck by electrons, more penetrate in some parts than in others depending on the thickness and density of the part. The image is brightest where most electrons have been transmitted. The object must be very thin, otherwise too much electron scattering occurs and no image forms. Also, the whole arrangement is highly evacuated. Why? An air-lock device permits objects to be inserted and removed without loss of vacuum.

The high resolving power of an electron microscope arises from the fact that just as light can be considered to have both wave-like and particle-like properties, so *moving* electrons seem to have the characteristics of both particles and waves. (Electrons as well as light can give interference and diffraction effects.) The wavelength associated with a moving electron depends only on its speed which in turn depends on the p.d. accelerating it in the electron gun. It can be shown that for a p.d. of 100 kV, common in many electron microscopes, the wavelength is about 3.5×10^{-12} m. The resolving power of a microscope increases as the wavelength of the radiation falling on the object decreases (p. 204) and therefore, if we compare the above very small wavelength with that of light (say 6×10^{-5} m), we see why the electron microscope shows much greater detail.

Two recent developments in electron microscopy are the high voltage electron microscope and the scanning electron microscope. The first operates at 1 million volts and enables thicker specimens to be studied. In the second, the surface of a relatively large specimen is scanned by the electron beam; photographs taken with this instrument are shown on p. 5.

QUESTIONS

Spherical mirrors

1. If a concave mirror has a focal length of 10 cm, find the two positions where an object can be placed to give, in each case, an image twice the height of the object.

2. A convex mirror of radius of curvature 40 cm forms an image which is half the height of the object. Find the object and image positions.

3. A concave mirror of radius of curvature 25 cm faces a convex mirror of radius of curvature 20 cm and is 30 cm from it. If an object is placed midway between the mirrors find the nature and position of the image formed by reflection first at the concave mirror and then at the convex mirror.

4. Describe an experiment to determine the radius of curvature of a convex mirror by an optical method. Illustrate your answer with a ray diagram and explain how the result is derived from the observations.

A small convex mirror is placed 0.60 m from the pole and on the axis of a large concave mirror, radius of curvature 2.0 m. The position of the convex mirror is such that a real image of a distant object is formed in the plane of a hole drilled through the concave mirror at its pole. Calculate (a) the radius of curvature of the convex mirror, (b) the height of the real image if the distant object subtends an angle of 0.50° at the pole of the concave mirror. Draw a ray diagram to illustrate the action of the convex mirror in producing the image of a non-axial point of the object and suggest a practical application of this arrangement of mirrors. (J.M.B.)

Refraction at plane surfaces

5. A ray of light in air passes successively through parallel-sided layers of water and glass. If the angle of incidence in air is 60° and the refractive indices of water and glass are 4/3 and 3/2 respectively, calculate (a) the angle of refraction in the water, (b) the angle of incidence at the water–glass boundary and (c) the angle of refraction in the glass.

6. Find by calculation what happens to a ray of light which falls at an angle of (i) 30°, (ii) 60°, on a glass–water surface if the refractive index of the glass is 3/2 and of water 4/3.

7. The refractive indices of crown glass and of a certain liquid are 1.51 and 1.63 respectively. Determine the conditions under which total internal reflection can occur at a surface separating them.

8. (a) A ray of light is incident at 45° on one face of a 60° prism of refractive index 1.5. Calculate the total deviation of the ray.

(b) A ray of light just undergoes total internal reflection at the second face of a prism of refracting angle 60° and refractive index 1.5. What is its angle of incidence on the first face?

9. (a) A 60.0° prism is made of glass whose refractive index for a certain light is 1.65. At what angle of incidence will minimum deviation occur? Between what limits must the angle of incidence lie, if light is to pass through the prism by refraction at adjacent faces? (W. part qn.)

(b) A ray of monochromatic light is incident at an angle of 30.0° on a prism of which the refractive index for the given light is 1.52. What is the maximum refracting angle for the prism if the light is just to emerge from the opposite face? (A.E.B. part qn.)

10. If the refractive index of diamond for sodium light is 2.42, calculate the refracting angle of a diamond prism which will cause the greatest possible deviation of a beam of sodium light after two refractions (with no total internal reflection). Explain your reasoning. (C. part qn.)

Lenses: the eye

11. (a) The filament of a lamp is 80 cm from a screen and a converging lens forms an image of it on the screen, magnified three times. Find the distance of the lens from the filament and the focal length of the lens.

(b) An erect image 2.0 cm high is formed 12 cm from a lens, the object being 0.5 cm high. Find the focal length of the lens.

223

12. Explain what is meant by (a) a virtual image, (b) a virtual object, in geometrical optics. Illustrate your answer by describing the formation of (i) a virtual image of a real object by a thin converging lens, (ii) a real image of a virtual object by a thin diverging lens. In each instance draw a ray diagram showing the passage of two rays through the lens for a non-axial object point.   (J.M.B.)

13. A lens forms the image of a distant object on a screen 30 cm away. Where should a second lens, of focal length 30 cm, be placed so that the screen has to be moved 8.0 cm towards the first lens for the new image to be in focus?

14. The radii of curvature of the faces of a thin converging meniscus lens of glass of refractive index 3/2 are 15 cm and 30 cm. What is the focal length of the lens (a) in air, (b) when completely surrounded by water of refractive index 4/3?

15. Explain why a sign convention is adopted in geometrical optics. Describe a convention and explain its use in solving the following problem.

An equi-convex lens A is made of glass of refractive index 1.5 and has a power of 5.0 rad m^{-1}. It is combined in contact with a lens B to produce a combination whose power is 1.0 rad m^{-1}. The surfaces in contact fit exactly. The refractive index of the glass in lens B is 1.6. What are the radii of the four surfaces? Draw a diagram to illustrate your answer.   (W.)

16. Draw a diagram to explain what is meant by (a) the *principal axis*, (b) the *focal length* of a thin converging lens.

A small luminous object is placed on the axis of a thin plano-convex lens (made of glass of refractive index 1.6) on the side of the lens nearer to the plane face. When at a distance of 30 cm from the lens it coincides with the real inverted image formed by light which has undergone two refractions at the plane face and one reflection at the curved face. Find the position and nature of the image of this object formed by light transmitted directly by the lens.   (J.M.B.)

17. A person can focus objects between 60.0 cm and 500 cm from his eyes. What spectacles are needed to make his far point infinity? What is now his range of vision?

18. What spectacles are required by a person whose near and far points are 40.0 cm and 200 cm away respectively to bring his near point to a distance of 25.0 cm? Find his new range of vision.

Optical instruments

19. Explain the difference between the terms *magnifying power* and *magnification*, as used about optical systems. Illustrate this, by calculating both, in the case of an object placed 5.0 cm from a simple magnifying glass of focal length 6.0 cm, assuming that the minimum distance of distinct vision for the observer is 25 cm.   (S.)

20. (a) Explain the terms *magnifying power* and *resolving power* in connection with a microscope.

(b) A compound microscope is formed from two lenses of focal lengths 1.0 and 5.0 cm. A small object is placed 1.1 cm from the objective and the microscope adjusted so that the final image is formed 30 cm from the eyepiece. Calculate the angular magnification of the instrument. (Assume that the nearest distance of distinct vision is 25 cm.)

(A.E.B. part qn.)

21. Describe, with the help of diagrams, how (*a*) a single biconvex lens can be used as a magnifying glass, (*b*) two biconvex lenses can be arranged to form a microscope. State (*i*) one advantage, (*ii*) one disadvantage, of setting the microscope so that the final image is at infinity rather than at the near point of the eye.

A centimetre scale is set up 5.0 cm in front of a biconvex lens whose focal length is 4.0 cm. A second biconvex lens is placed behind the first, on the same axis, at such a distance that the final image formed by the system coincides with the scale itself and that 1.0 mm in the image covers 2.4 cm in the scale. Calculate the position and focal length of the second lens. (*O. and C.*)

22. State what is meant by *normal adjustment* in the case of an astronomical telescope.

Trace the paths of three rays from a distant non-axial point source through an astronomical telescope in normal adjustment.

Define the *magnifying power* of the instrument, and, by reference to your diagram, derive an expression for its magnitude.

A telescope consists of two thin converging lenses of focal lengths 100 cm and 10.0 cm respectively. It is used to view an object 2.00×10^3 cm from the objective. What is the separation of the lenses if the final image is 25.0 cm from the eye-lens? Determine the magnifying power for an observer whose eye is close to the eye-lens. (*J.M.B.*)

23. What should be the focal length of the objective of an astronomical telescope if the eyepiece is of focal length 5.0 cm and the lenses are to be fixed 85 cm apart in normal adjustment (final image at infinity)? What will then be the magnifying power obtained? (*No proofs required.*)

With such a telescope all the light received by the objective which also passes through the eyepiece eventually passes through a small circular region, called the exit pupil (eye ring), a short distance beyond the eyepiece. The exit pupil coincides in position and size with the image which would be formed by the eyepiece of the objective lens if the latter were a self-luminous object. Give arguments to justify these statements. How far behind the eyepiece will the exit pupil be in the case given above?

It is generally reckoned best to have an objective of such a size that the exit pupil can coincide in size and position with the pupil of the observer's eye. If his pupil (at night) has a diameter of 8.0 mm, what should be the diameter of the objective for this telescope?

What are the advantages of using an objective of as large a diameter as possible? What are the disadvantages of a large objective? (*S.*)

24. What is meant by the *f-number* of a camera lens? The stop of a camera lens is reduced from $f/8$ to $f/22$. State and explain in what ratio the illumination of the image on the film is changed. Explain also, with the aid of a diagram, why the 'depth of focus' is increased.

An image of a distant object is formed on a screen by an optical system consisting of a converging lens of focal length 15 cm placed co-axially 9.0 cm from a diverging lens of focal length 7.5 cm, the light being incident on the converging lens. Compare the size of this image with that produced of the same object by the converging lens alone. (*L.*)

Part 2 | MECHANICS

6 Statics and dynamics

Mechanics

Mechanics is concerned with the action of forces on a body. If the forces balance they are said to be in *equilibrium* and the branch of mechanics which deals with such cases is called *statics*—the subject reviewed in the first part of this chapter. In the second part of the chapter we will consider the effects of forces which are not in equilibrium—a study known as *dynamics*.

Many engineering and technological problems such as designing buildings, bridges (like the Forth road and rail bridges shown in Fig. 6.1), roads,

Fig. 6.1

reservoirs, jet engines and aircraft require the application of the principles of mechanics so that structures with the necessary strength are obtained using the minimum of material. Not only are these principles useful for dealing with the world of ordinary experience but, suitably supplemented, they enable us to deal with the physics of the atom on the one hand, and astronomy and space travel on the other.

Composition and resolution of forces

(a) *Scalar and vector quantities.* A scalar quantity has magnitude only and is completely described by a certain number of appropriate units. A vector quantity has both magnitude and direction; it can be represented by a straight line whose length represents the magnitude of the quantity on a particular scale and whose direction (shown by an arrow) indicates the direction of the quantity.

For example, if the points X and Y in Fig. 6.2a are 2 metres apart, the statement that XY = 2 m fully describes the *distance* between them;

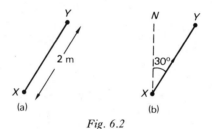

Fig. 6.2

distance is a scalar. However the *displacement* **XY** between the points is 2 m in a direction 30° east of north, Fig. 6.2b; displacement is a vector (like other vectors it is often printed in bold type). Other examples of scalars are mass, time, density, speed, energy. Force, velocity (displacement per unit time) and momentum are vectors. What kind of quantity is (i) temperature, (ii) acceleration?

(b) *Parallelogram of forces.* Scalars and vectors require different mathematical treatment. Thus scalars are added arithmetically but vectors are added geometrically by the *parallelogram law* which ensures their directions as well as their magnitudes are taken into account. The law will be illustrated by the addition of two displacements.

Suppose we walk from A to B and then from B to C as in Fig. 6.3a so that we suffer successive displacements **AB** and **BC**. The resultant displacement

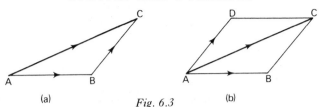

Fig. 6.3

is given by **AC** in magnitude and direction. The same resultant displacement (i.e. **AC**) would be obtained if we started from the same point A and drew AD equal to BC in magnitude and direction and then drew DC equal to AB, Fig. 6.3b. The sum of two vectors therefore equals the diagonal of the parallelogram of which the vectors are adjacent sides. How would you subtract two vectors?

The parallelogram law for the addition (composition) of forces is stated as follows.

If two forces acting at a point are represented in magnitude and direction by the sides of a parallelogram drawn from the point, their resultant is represented by the diagonal of the parallelogram drawn from the point.

(c) *Resolution of forces.* The reverse process to the addition of two vectors by the parallelogram law, is the splitting or *resolving* of one vector into two components. It is particularly useful in the case of forces when the components are taken at right angles to each other.

Suppose the force F is represented by OA in Fig. 6.4a and that we wish to find its components along OX and OY ($\angle XOY = 90°$). A perpendicular

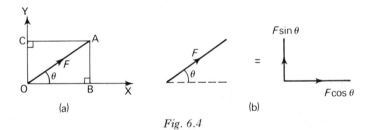

Fig. 6.4

AB is dropped from A on to OX and another AC from A on to OY, to give rectangle (parallelogram) OCAB. OB and OC are the required components or resolved parts. If $\angle AOB = \theta$ then

$$\cos \theta = OB/OA = OB/F$$
$$\therefore \quad OB = F \cos \theta$$
and
$$\sin \theta = AB/OA = OC/F$$
$$\therefore \quad OC = F \sin \theta$$

The two mutually perpendicular forces $F \cos \theta$ and $F \sin \theta$ are thus equivalent to F, Fig. 6.4b. The total effect of F along OX is represented by $F \cos \theta$. Also note that if $\theta = 0$, $F \cos \theta = F$ and $F \sin \theta = 0$, hence a force has no effect in a perpendicular direction. Resolving a force (or any vector) gives two quite independent forces and is a process we shall use frequently.

The component of a force of 10 N in a direction making an angle of 60° with it, is 10 cos 60 N (i.e. 5 N); in a direction perpendicular to this component the effective value of the force is 10 sin 60 N (i.e. $5\sqrt{3}$ N).

Moments and couples

A force applied to a hinged or pivoted body changes its rotation about the hinge or pivot. Experience shows that the turning effect or *moment* or *torque* of the force is greater the greater the magnitude of the force and the greater the distance of its point of application from the pivot.

The moment or torque of a force about a point is measured by the product of the force and the perpendicular distance from the line of action of the force to the point.

Thus in Fig. 6.5a if OAB is a trap-door hinged at O and acted on by forces P and Q as shown then

$$\text{moment of } P \text{ about } O = P \times OA$$

and, $$\text{moment of } Q \text{ about } O = Q \times OC$$

Fig. 6.5

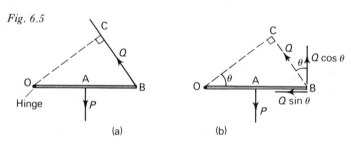

(a)   (b)

Note that the perpendicular distance must be taken. Alternatively, we can resolve Q into components $Q \cos \theta$ perpendicular to OB and $Q \sin \theta$ along OB, Fig. 6.5b. The moment of the latter about O is zero since its line of action passes through O. For the former we have

$$\text{moment of } Q \cos \theta \text{ about } O = Q \cos \theta \times OB$$
$$= Q \times OC \quad (\text{since } \cos \theta = OC/OB)$$

which is the same as before. Moments are measured in newton metres (N m) and are given a positive sign if they tend to produce clockwise rotation.

A *couple* consists of two equal and opposite parallel forces whose lines of action do not coincide; it always tends to change rotation. A couple is applied

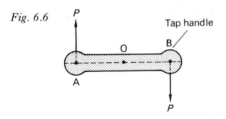

Fig. 6.6

Tap handle

to a water tap to open it. From Fig. 6.6 we can say that the moment or torque of the couple $P - P$ about O

$$= P \times \text{OA} + P \times \text{OB} \quad \text{(both are clockwise)}$$
$$= P \times \text{AB}$$

Hence,

   moment of couple = *one force* × *perpendicular distance between forces*

Equilibrium of coplanar forces

(*a*) *General conditions for equilibrium.* If a body is acted on by a number of coplanar forces (i.e. forces in the same plane) and is in equilibrium (i.e. there is rest or unaccelerated motion) then

(*i*) the components of the forces in both of any two directions (usually taken at right angles) must balance, and

(*ii*) the sum of the clockwise moments about any point equals the sum of the anticlockwise moments about the same point.

The first statement is a consequence of there being no translational motion in any direction and the second follows since there is no rotation of the body. In brief, if a body is in equilibrium the forces and the moments must *both* balance.

The following worked example (and also that on p. 237) shows how the conditions for equilibrium are used to solve problems.

(*b*) *Worked example. A sign of mass 5.0 kg is hung from the end B of a uniform bar AB of mass 2.0 kg. The bar is hinged to a wall at A and held horizontal by a wire joining B to a point C which is on the wall vertically above A. If angle ABC = 30°, find the force in the wire and that exerted by the hinge. (g = 10 m s^{-2}.)*

The weight of the sign will be 50 N and of the bar 20 N (since $W = mg$). The arrangement is shown in Fig. 6.7a. Let P be the force in the wire and suppose Q, the force exerted by the hinge, makes angle θ with the bar. The bar is uniform and so its weight acts vertically downwards at its centre G. Let the length of the bar be $2l$.

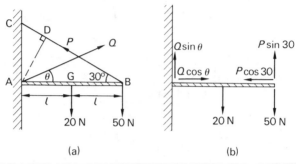

(a)   (b)

Fig. 6.7

(*i*) *There is no rotational acceleration,* therefore taking moments about A we have

clockwise moments = anticlockwise moment

i.e. $20 \times l + 50 \times 2l = P \times AD$ (AD is perpendicular to BC)

$$\therefore \quad 120l = P \times AB \sin 30 \quad \text{(sin 30} = AD/AB)$$

$$= P \times 2l \times 0.5$$

$$\therefore \quad P = 1.2 \times 10^2 \text{ N}$$

Note : by taking moments about A there is no need to consider Q since it passes through A and so has zero moment.

(*ii*) *There is no translational acceleration,* therefore the vertical components (and forces) must balance, likewise the horizontal components. Hence resolving Q and P into vertical and horizontal components (which now replace them, Fig. 6.7b) we have:

Vertically
$$Q \sin \theta + P \sin 30 = 20 + 50$$
$$\therefore \quad Q \sin \theta = 70 - 120(1/2)$$
$$Q \sin \theta = 10 \qquad \qquad (1)$$

Horizontally
$$Q \cos \theta = P \cos 30 = 120(\sqrt{3}/2)$$
$$\therefore \quad Q \cos \theta = 60\sqrt{3} \qquad \qquad (2)$$

Dividing (1) by (2)

$$\tan \theta = 10/(60\sqrt{3})$$

$$\therefore \quad \theta = 5.5°$$

Squaring (1) and (2) and adding

$$Q^2(\sin^2 \theta + \cos^2 \theta) = 100 + 10\,800$$

$$\therefore \quad Q^2 = 10\,900 \qquad (\sin^2 \theta + \cos^2 \theta = 1)$$

and

$$Q = 1.0(4) \times 10^2 \text{ N}$$

(c) *Structures.* Forces act at a joint in many structures and if these are in equilibrium then so too are the joints. The joint O in the bridge structure of Fig. 6.8 is in equilibrium under the action of forces P and Q exerted by

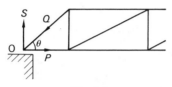

Fig. 6.8

the girders and the normal force S exerted by the bridge support at O. The components of the forces in two perpendicular directions at the joint must balance. Hence

$$S = Q \sin \theta \quad \text{and} \quad P = Q \cos \theta$$

If θ and S are known (the latter from the weight and loading of the bridge) then P and Q (which the bridge designer may wish to know) can be found. Other joints may be treated similarly (see question 4, p. 267).

Laws of friction

Frictional forces act along the surface between two bodies whenever one moves or tries to move over the other and in a direction so as to oppose relative motion of the surfaces. Sometimes it is desirable to reduce friction to a minimum but in other cases its presence is essential. For example, it is the frictional push of the ground on the soles of our shoes that enables us to walk. Otherwise our feet would slip backwards as they do when we try to walk on an icy road. The study of friction, wear and lubrication, now called *tribology*, is a matter of great importance to industry and is the subject of much research.

(a) *Coefficients of friction.* Friction between two solid surfaces can be studied using the apparatus of Fig. 6.9 in which the plank tends to move or does move, depending on the force applied to the crank, whilst the block remains at rest. The frictional force between the block and the plank is measured by the spring balance.

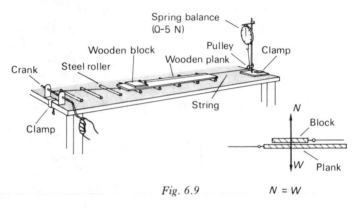

Fig. 6.9 $N = W$

As the crank is gently wound the spring balance reading increases and reaches a maximum value when the plank is about to move. This maximum force between the surfaces is called the *limiting* frictional force. When motion does start, the spring balance reading usually decreases slightly showing that the *sliding, kinetic* or *dynamic* frictional force (all terms are used) is rather smaller than the limiting value. The block can be set on edge to see if friction depends on the area of contact between the surfaces. The normal force N exerted by the plank on the block equals the weight W of the block. The effect on the frictional force of varying N can be found by placing weights on the block.

The results of such experiments are summarized in the following laws of friction, which hold approximately.

1. *The frictional force between two surfaces opposes their relative motion.*

2. *The frictional force does not depend on the area of contact of the surfaces if the normal reaction is constant.*

3. (a) *When the surfaces are at rest the limiting frictional force F is directly proportional to the normal force N, i.e. F $\propto$ N (or F/N = constant).*

(b) *When motion occurs the dynamic frictional force F' is directly proportional to the normal force N, i.e. F' $\propto$ N (or F'/N = constant) and is reasonably independent of the relative velocity of the surfaces.*

The coefficients of limiting and dynamic friction are denoted by μ and μ' respectively and are defined by the equations

$$\mu = F/N \quad \text{and} \quad \mu' = F'/N$$

For two given surfaces μ' is usually less than μ but they are often assumed equal. For wood on wood μ is about 0.2 to 0.5. In general a surface exerts a frictional force and the resultant force on a body on the surface has two components—a normal force N perpendicular to the surface and a frictional force F along the surface, Fig. 6.10. If the surface is smooth, as is sometimes

Fig. 6.10   Resultant force

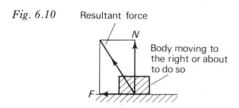

Body moving to
the right or about
to do so

assumed in mechanics calculations, $\mu = 0$ and so $F = 0$. Therefore a smooth surface only exerts a force at right angles to itself, i.e. a normal force N.

The coefficient of limiting friction, μ, can also be found by placing the block on the surface and tilting the latter to the angle θ at which the block is just about to slip, Fig. 6.11a. The three forces acting on the block are its weight mg, the normal force N of the surface and the limiting frictional force

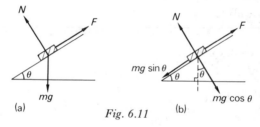

(a)

Fig. 6.11   (b)

$F(= \mu N)$. They are in equilibrium and if mg is resolved into components $mg \sin \theta$ along the surface and $mg \cos \theta$ perpendicular to the surface, Fig. 6.11b, then

$$F = \mu N = mg \sin \theta$$

and
$$N = mg \cos \theta$$

Dividing
$$\mu = \tan \theta$$

Hence μ can be found by measuring θ, called the *angle of friction*.

(b) *Worked example. A uniform ladder 4.0 m long, of mass 25 kg, rests with its upper end against a smooth vertical wall and with its lower end on rough ground. What must be the least coefficient of friction between the ground and the ladder for it to be inclined at 60° with the horizontal without slipping? (g =* 10 m s⁻².)

237

The weight (mg) of the ladder is 250 N and the forces acting on it are shown in Fig. 6.12. The wall is smooth and so the force S of the wall on the ladder is normal to the wall. Since the ladder is uniform its weight W

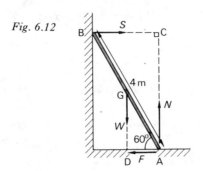

Fig. 6.12

can be taken to act at its mid-point G. If it is about to slip there will be a force exerted on it by the ground which can be resolved into a normal force N and a limiting frictional force $F = \mu N$, where μ is the required coefficient of friction.

The forces are in equilibrium.

Resolving vertically

$$N = W = 250 \text{ newtons}$$

Resolving horizontally

$$F = \mu N = S$$

Taking moments about A

$$S \times AC = W \times AD$$

$$S \times 4.0 \cos 30 = 250 \times 2.0 \sin 30 = 250$$

$$\therefore \quad S = 125/\sqrt{3} \text{ newtons}$$

Hence

$$\mu = \frac{S}{N} = \frac{125}{250\sqrt{3}}$$

$$= 0.29$$

Nature of friction

Close examination of the flattest and most highly polished surfaces shows that they have hollows and humps more than one hundred atoms high. When one solid is placed on another, contact therefore only occurs at a few places of small area, Fig. 6.13. From electrical resistance measurements of two

Fig. 6.13

metals in contact it is estimated that in the case of steel, the actual area that is touching may be no more than 1/10 000th of the apparent area.

The pressures at the points of contact are extremely high and cause the humps to flatten out (being plastically deformed) until the increased area of contact enables the upper solid to be supported. It is thought that at the points of contact small, cold welded 'joints' are formed by the strong adhesive forces between molecules which are very close together. These have to be broken before one surface can move over the other, thus accounting for law 1 (p. 236). Measurements show that changing the apparent area of contact of the bodies has little effect on the actual area for the same normal force, so explaining law 2. It is also found that the actual area is proportional to the normal force and since this theory suggests that the frictional force depends on the actual area we might expect the frictional force to be proportional to the normal force—as law 3 states.

Velocity and acceleration

(*a*) *Speed and velocity.* If a car travels from A to B along the route shown in Fig. 6.14*a*, its *average speed* is defined as *the actual distance travelled*, i.e. AXYZB, *divided by the time taken*. The speed at any instant is found by considering a very short time interval. Speed has magnitude only and is a scalar quantity.

Velocity is defined as *the distance travelled in a particular direction divided by the time taken*. The average velocity of the car in the direction AB is the distance between A and B, i.e. the length of the straight line AB, Fig. 6.14*b*, divided by the time actually taken for the journey from A to B. Thus AB is the displacement of the car, in this case it is not the actual path followed

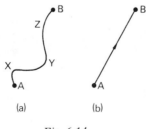

(a)   (b)

Fig. 6.14

from A to B, although of course it could be in other cases. Velocity can therefore also be defined as *the displacement in unit time*; it has both magnitude and direction and is a vector quantity.

The SI unit of velocity is 1 metre per second (1 m s^{-1}); 10 m s^{-1} = 36 km h^{-1}. In Fig. 6.14 if the route AXYZB is 40 kilometres and the car takes 1 hour for the journey, its average *speed* is 40 km h^{-1}. If B is 25 km north-east of A as the crow flies, its average *velocity* is only 25 km h^{-1} towards the north-east.

The velocity v of a body which undergoes a very small displacement δs in the very small time δt is given by the equation

$$v = \frac{\delta s}{\delta t}$$

Or more strictly, in calculus notation, the velocity v at an instant is defined by

$$v = \lim_{\delta t \to 0} \left(\frac{\delta s}{\delta t} \right) = \frac{ds}{dt}$$

Velocity is therefore the rate of change of displacement.

A body which covers equal distances in the same straight line in equal time intervals, no matter how short these are, is said to be moving with constant or *uniform velocity*. Only a body moving in a straight line can have uniform velocity. The direction of motion of a body travelling in a curved path is continually changing and so it cannot have uniform velocity even though its speed may be constant.

(*b*) *Acceleration.* A body is said to accelerate when its velocity changes. Thus if a very small velocity change δv occurs in a very small time interval δt, the acceleration a of the body is

$$a = \frac{\text{change in velocity}}{\text{time taken for change}} = \frac{\delta v}{\delta t}$$

More correctly, in calculus notation, the instantaneous acceleration a is defined by

$$a = \lim_{\delta t \to 0} \left(\frac{\delta v}{\delta t} \right) = \frac{dv}{dt}$$

In words, *acceleration is the rate of change of velocity*. For a car accelerating towards the north from 10 m s^{-1} to 20 m s^{-1} in 5.0 seconds we can say

$$\text{average acceleration} = \frac{(20 - 10) \text{ m s}^{-1}}{5.0 \text{ s}}$$

$$= 2.0 \text{ m s}^{-2} \text{ towards the north}$$

That is, on average, the velocity of the car increases by 2.0 m s^{-1} every second. Since 10 m s^{-1} = 36 km h^{-1} and 20 m s^{-1} = 72 km h^{-1}, we could also say the average acceleration is $(72 - 36)$ km h$^{-1}/5$ s = 7.2 km h^{-1} per second = 7.2 km h^{-1} s^{-1}.

Equations for uniform acceleration

The acceleration of a body is uniform if its velocity changes by equal amounts in equal times. We will now derive three useful equations for a body moving in a straight line with uniform acceleration.

Suppose the velocity of the body increases steadily from u to v in time t then the uniform acceleration a is given by

$$a = \frac{\text{change of velocity}}{\text{time taken}}$$

$$= \frac{v - u}{t}$$

$$\therefore \quad v = u + at \tag{1}$$

Since the velocity is increasing steadily, the average velocity is the mean of the initial and final velocities, i.e.

$$\text{average velocity} = \frac{u + v}{2}$$

If s is the displacement of the body in time t, then since average velocity = displacement/time = s/t, we can say

$$\frac{s}{t} = \frac{u + v}{2}$$

$$\therefore \quad s = \tfrac{1}{2}(u + v)t$$

But
$$v = u + at$$

$$\therefore \quad s = \tfrac{1}{2}(u + u + at)t$$

or
$$s = ut + \tfrac{1}{2}at^2 \tag{2}$$

If we eliminate t from (2) by substituting $t = (v - u)/a$ from (1), we get on simplifying

$$v^2 = u^2 + 2as \tag{3}$$

Knowing any three of u, v, a, s and t the others can be found.

Velocity–time graphs

Acceleration is rate of change of velocity (in calculus notation dv/dt) and equals at any instant the slope of the velocity–time graph. In Fig. 6.15, curve ① has zero slope and represents uniform velocity, curve ② is a straight line of constant slope and represents uniform acceleration, while curve ③ is for variable acceleration since its slope varies.

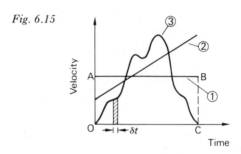

Fig. 6.15

The distance travelled by a body during any interval of time can also be found from a velocity–time graph, a fact which is especially useful in cases of non-uniform acceleration since the three equations of motion do not then apply. For the constant velocity case, curve ①, the distance travelled in time OC = velocity × time = OA × OC = area OABC. In curve ③, if we consider a small enough time interval δt, the velocity is almost constant and the distance travelled in δt will be the area of the very thin shaded strip. By dividing up the whole area under curve ③ into such strips it follows that the *total distance travelled in time OC equals the area between the velocity–time graph and the time-axis.*

Motion under gravity

(a) *Free fall.* Experiments show that at a particular place all bodies falling freely under gravity, in a vacuum or when air resistance is negligible, have the *same* constant acceleration irrespective of their masses. This acceleration towards the surface of the earth, known as the *acceleration due to gravity*, is denoted by g. Its magnitude varies slightly from place to place on the earth's surface and is approximately 9.8 m s^{-2}. The velocity of a freely falling body therefore increases by 9.8 m s^{-1} every second; in the equations of motion g replaces a.

A direct determination of g may be made using the apparatus of Fig. 6.16 in which the time for a steel ball-bearing to fall a known distance from rest is measured (to about 0.005 second) by an electric stop clock. When the two-way switch is changed to the 'down' position, the electromagnet releases the ball and simultaneously the clock starts. At the end of its fall the

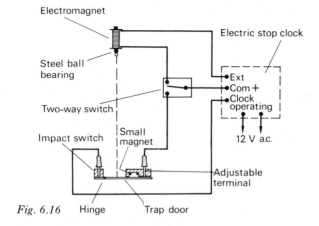

Fig. 6.16   Hinge   Trap door

ball opens the 'trap-door' on the impact switch and the clock stops. Air resistance is negligible for a dense object such as a ball-bearing. The result is found from $s = ut + \frac{1}{2}at^2$ since $u = 0$ and $a = g$.

The measurement of g using a simple pendulum is described on page 321.

(b) *Vertical projection.* The velocity of a body projected upwards from the ground decreases by 9.8 m s^{-1} (near enough to 10 m s^{-1}) every second —neglecting the effect of air resistance. Hence if a ball is thrown straight upward with an initial velocity of 30 m s^{-1} then in just over 3 seconds it will have zero velocity and be at its highest point.

(c) *Sign convention for displacement, velocity and acceleration.* The three equations for uniform acceleration can be used to solve problems on falling and rising bodies so long as a sign convention is adopted.

Vector quantities have magnitudes and directions. By their nature magnitudes are positive and direction is stated in terms of angles. However when we are dealing with one-dimensional effects, such as linear motion, the only way to indicate direction is by signs. If we take *downward as being positive,* a downward acceleration, velocity or displacement is positive. Hence $g = +10 \text{ m s}^{-2}$ and this is true whether the body is rising and slowing down or falling and speeding up.

Consider an example. Suppose a ball is thrown straight up with an initial velocity of 40 m s^{-1} and we wish to find its velocity and height after 2 seconds. We have $u = -40 \text{ m s}^{-1}$, $t = 2 \text{ s}$, $a = g = +10 \text{ m s}^{-2}$.

Since
$$v = u + at$$
$$v = -40 + 10 \times 2$$
$$= -20 \text{ m s}^{-1}$$

The ball has a velocity of 20 m s^{-1} upward. Also

$$s = ut + \tfrac{1}{2}at^2$$
$$= -40 \times 2 + \tfrac{1}{2} \times 10 \times 4$$
$$= -60 \text{ m}$$

It rises 60 m in 2 seconds.

Projectiles

In Fig. 6.17 a multiflash photograph is shown of the motion of two balls, one released from rest and the other projected simultaneously with a horizontal velocity. It is clear that the vertical motion of the projected ball (a constant acceleration $= g$) is unaffected by its horizontal motion (a constant velocity). The two motions are quite independent of each other.

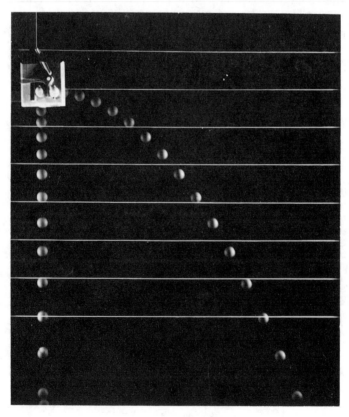

Fig. 6.17

STATICS AND DYNAMICS

Consider a body projected obliquely from O with velocity u at an angle θ to the horizontal, Fig. 6.18. Suppose we wish to know the height attained by the body and its horizontal range. If we resolve u into horizontal and vertical components $u \cos \theta$ and $u \sin \theta$ respectively, each component can be considered independently of the other.

Fig. 6.18

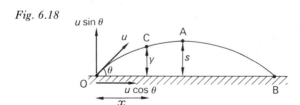

Vertical motion. Whilst rising, the body is subject to a constant acceleration $a = -g$. (Here it is convenient to take downward directed vectors as negative, which explains why g has a negative sign.) If s is the height attained then, since the initial velocity is $u \sin \theta$ and the final velocity zero, we have from the third equation of motion ($v^2 = u^2 + 2as$)

$$0 = u^2 \sin^2 \theta - 2gs$$

$$\therefore \quad s = \frac{u^2 \sin^2 \theta}{2g}$$

Also, if t is the time to reach the highest point A, it follows from the first equation of motion ($v = u + at$) that

$$0 = u \sin \theta - gt$$

$$\therefore \quad t = \frac{u \sin \theta}{g}$$

The time taken by the body to fall to the horizontal level of O is also t. Therefore

$$\text{time of flight} = 2t = 2\,\frac{u \sin \theta}{g}$$

Horizontal motion. Neglecting air resistance, the horizontal component $u \cos \theta$ remains constant during the flight since g has no effect in a horizontal direction. The horizontal distance travelled, OB,

$$= \text{horizontal velocity} \times \text{time of flight}$$

$$= \frac{u \cos \theta \times 2u \sin \theta}{g}$$

245

$$= \frac{2u^2 \sin \theta \cos \theta}{g}$$

$$= \frac{u^2 \sin 2\theta}{g} \qquad (\sin 2\theta = 2 \sin \theta \cos \theta)$$

For a given velocity of projection the range is a maximum when $\sin 2\theta = 1$, i.e. when $\theta = 45°$ and has the value u^2/g.

Trajectory. Let the body be at point C (co-ordinates x, y) at time t after projection from O. Therefore

$$x = tu \cos \theta$$

and

$$y = tu \sin \theta - \tfrac{1}{2}gt^2$$

Substituting for t in the second equation we get

$$y = \frac{x}{u \cos \theta} u \sin \theta - \frac{gx^2}{2u^2 \cos^2 \theta}$$

$$\therefore \quad y = x \tan \theta - \frac{gx^2}{2u^2 \cos^2 \theta}$$

This is of the form $y = ax + bx^2$ which is the equation of a parabola (a and b are constants for a given velocity and angle of projection). In practice air resistance causes slight deviation from a parabolic path.

Newton's laws of motion

Newton (1642–1727) studied and developed Galileo's (1564–1642) ideas about motion and subsequently stated the three laws which now bear his name. He established the subject of dynamics. His laws are a set of state-ments and definitions that we believe to be true because the results they predict are found to be in very exact agreement with experiment over a wide range of conditions. We do not regard them as absolutely true and more exact laws are required for certain extreme cases.

(a) *First law. If a body is at rest it remains at rest or if it is in motion it moves with uniform velocity* (i.e. constant speed in a straight line) *until it is acted on by a resultant force.*

The second part of the law appears to disagree with certain everyday experiences which suggest that a steady effort has to be exerted on a body, e.g. a bicycle, even to keep it moving with constant velocity (let alone to accelerate it), otherwise it comes to rest. The law on the other hand states that a moving body retains its motion naturally and if any change occurs (i.e. if it is accelerated) some outside agent—a force—must be responsible.

246

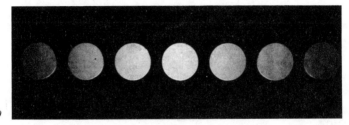

Fig. 6.19

It seems that the question to be asked about a moving body is not 'what keeps it moving' but 'what changes or stops its motion'. Frequently it is friction and if a body does move under near-frictionless conditions, its velocity is in fact almost uniform. This is shown in Fig. 6.19 which is a photograph of a puck, illuminated at equal time intervals by a flashing xenon lamp, moving on a cushion of carbon dioxide gas across a clean, level glass plate.

This law really defines a force as something which changes the state of rest or uniform motion of a body. Contact may be necessary, as when we push a body with our hands, or it may not be, as in the case of gravitational, electric and magnetic forces.

(*b*) *Mass.* The first law implies that matter has a built-in reluctance to change its state of rest or motion. This property, possessed by all bodies, is called *inertia*. Its effects are evident when a vehicle suddenly stops, causing the passengers to lurch forward (tending to keep moving), or starts, jerking the passengers backwards (since they tend to remain at rest).

The *mass* of a body is a measure of its inertia; a large mass requires a large force to produce a certain acceleration. The unit of mass is the *kilogram* (kg) and is the mass of a piece of platinum-iridium carefully preserved at Sèvres, near Paris. *In principle* the mass m of a body can be measured by comparing the accelerations a and a_0 produced by the same force in the body and the standard kilogram (m_0) respectively. The ratio of the two masses is then defined by

$$\frac{m}{m_0} = \frac{a_0}{a}$$

whence m can be calculated. *In practice* this is neither quick nor accurate and mass is most readily found using a beam balance to compare the weight of the body with that of a standard. It can be shown (p. 249) that the mass of a body is proportional to its weight and so a beam balance also compares masses.

The second law indicates how forces can be measured.

(*c*) *Second law. The rate of change of momentum of a body is proportional to the resultant force and occurs in the direction of the force.*

The momentum of a body of constant mass m moving with velocity u is, by definition, mu. That is

$$momentum = mass \times velocity$$

Suppose a force F acts on the body for time t and changes its velocity from u to v, then

$$change\ of\ momentum = mv - mu$$

$$\therefore \quad rate\ of\ change\ of\ momentum = \frac{m(v - u)}{t}$$

Hence, by the second law

$$F \propto \frac{m(v - u)}{t}$$

If a is the acceleration of the body then

$$a = \frac{v - u}{t}$$

$$\therefore \quad F \propto ma$$

or
$$F = kma$$

where k is a constant. Now, *one newton is defined as the force which gives a mass of 1 kilogram an acceleration of 1 metre per second per second.* Hence if $m = 1$ kg and $a = 1$ m s^{-2} then $F = 1$ N and substituting these values in $F = kma$ we obtain $k = 1$. Thus *with these units*

$$k = 1 \quad and \quad F = ma$$

This expression is one form of Newton's second law and it indicates that a force can be measured by finding the acceleration it produces in a known mass. It can be verified experimentally using, for example, a tickertape timer and trolleys on a runway. Two points should be noted when using $F = ma$ to solve numerical problems. First, F is the *resultant* (or unbalanced) force causing acceleration a in a certain direction and second, F must be in newtons, m in kilograms and a in metres per second squared.

(d) *Weight.* The weight W of a body is the force of gravity acting on it towards the centre of the earth. Weight is thus a *force*, not to be confused with mass which is independent of the presence or absence of the earth. If g is the acceleration of the body towards the centre of the earth then we can substitute F (force accelerating the body) $= W$ and $a = g$ in $F = ma$, hence

$$W = mg$$

Thus if $g = 9.8$ m s^{-2}, a body of mass 1 kg has a weight of 9.8 N (roughly 10 N). The mass m of a body is constant but its weight mg varies with position on the earth's surface since g varies from place to place. Weight can be measured by a calibrated spring balance.

Note. To be strictly accurate the value of mg recorded by a spring balance is not quite equal to W if W is the gravitational attraction on the body directed towards the *centre of the earth*. We will see later that due to the rotation of the earth the observed direction of g is not exactly towards the earth's centre and its observed value is slightly different from the true value in that direction (p. 297). Hence mg does not equal W in magnitude and direction but the differences are extremely small.

If two bodies of masses m_1 and m_2 have weights W_1 and W_2 at the same place then

$$W_1 = m_1 g \quad \text{and} \quad W_2 = m_2 g$$

$$\therefore \quad \frac{W_1}{W_2} = \frac{m_1}{m_2}$$

That is, the *weight of a body is proportional to its mass,* a fact we use when finding the mass of a body by comparing its weight with that of standard masses on a beam balance.

(*e*) *Third law. If body A exerts a force on body B, then body B exerts an equal but opposite force on body A.*

The law is stating that forces never occur singly but always in pairs as a result of the interaction between two bodies. For example, when you step forward from rest your foot pushes backwards on the earth and the earth exerts an equal and opposite force forward on you. Two bodies and two forces are involved. The comparatively small force you exert on the large mass of the earth produces no noticeable acceleration of the earth, but the equal force it exerts on your very much smaller mass causes you to accelerate. It is important to note that the equal and opposite forces *do not act on the same body*; if they did, there could never be any resultant forces and all acceleration would be impossible.

If you pull a string attached to a block with a force P to the right, Fig. 6.20, the string pulls you with an equal force P to the left. Generally we can

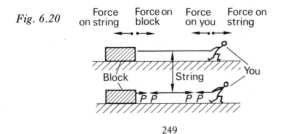

Fig. 6.20

assume the string transmits the force unchanged and so there is another pair of equal and opposite forces at the block. The string exerts a pull P to the right *on the block* and the block exerts an equal pull to the left *on the string*— one force acts on the block and the other on the string. The string is pulled outwards at both ends and is in a state of tension.

Two pairs of forces exist when a book lies at rest on a table; draw diagrams to show what they are.

$F = ma$ calculations

1. A Saturn V rocket develops an initial thrust of 3.3×10^7 N and has a lift-off mass of 2.8×10^6 kg. Find the initial acceleration of the rocket at lift-off. (Take $g = 10$ m s^{-2}.)

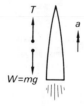

Fig. 6.21

Let T be the initial thrust on the rocket, m its mass and W its weight, Fig. 6.21. Then

$$W = mg = 2.8 \times 10^6 \text{ kg} \times 10 \text{ m s}^{-2}$$

$$= 2.8 \times 10^7 \text{ N}$$

The *resultant* upwards force *on the rocket* is $(T - W)$ and if a is the initial vertical acceleration, then from $F = ma$ we have

$$(T - W) = ma$$

$$\therefore \quad a = \frac{T - W}{m}$$

$$= \frac{3.3 \times 10^7 - 2.8 \times 10^7}{2.8 \times 10^6} \quad \frac{\text{N}}{\text{kg}}$$

$$= 1.8 \text{ m s}^{-2} \qquad (\text{N kg}^{-1})$$

Note. We can apply $F = ma$ here since the rocket is instantaneously at rest. In general this is not possible because the mass of the rocket changes.

2. Two blocks A and B are connected as in Fig. 6.22 on a horizontal friction-less floor and pulled to the right with an acceleration of 2.0 m s^{-2} by a force P. If $m_1 = 50$ kg and $m_2 = 10$ kg, what are the values of T and P?

Fig. 6.22

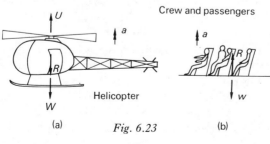

The forces acting *on the blocks* are shown. Apply $F = ma$ to each.

For B, $\qquad\qquad T = m_2a = 10 \times 2 = 20$ N

For A, $\qquad\quad P - T = m_1a = 50 \times 2 = 100$ N

$\qquad\qquad\qquad \therefore\ \ P = 120$ N

3. A helicopter of mass M and weight W rises with vertical acceleration, a, due to the upward thrust U generated by its rotor. The crew and passengers of total mass m and total weight w exert a combined force R on the floor of the helicopter. Write an equation for the motion of (a) the helicopter, (b) the crew and passengers.

The forces acting *on the helicopter* are the upwards force U due to the action of the rotor on the surrounding air, its weight W downwards due to the earth and the force R downwards exerted on the floor by the crew and passengers, Fig. 6.23a.

$\qquad \therefore\quad$ *Resultant* upwards force on helicopter $= U - W - R$

Hence, by the second law

$$U - W - R = Ma \qquad\qquad (1)$$

The forces acting *on the crew and passengers* are the upwards push of the floor of the helicopter (which by the third law must equal the downwards push R of the crew and passengers on the floor) and their weight w downwards, Fig. 6.23b.

$\qquad \therefore\quad$ *Resultant* upwards force on crew and passengers $= R - w$

Crew and passengers

Helicopter

(a) $\qquad$ *Fig. 6.23* $\qquad$ (b)

Hence, by second law

$$R - w = ma \qquad (2)$$

The required equations are (1) and (2).

Momentum

The *momentum* of a body was previously defined as the mass of the body multiplied by its velocity. If S I units are used, Newton's second law may be written

force = rate of change of momentum

In symbols

$$F = \frac{mv - mu}{t}$$

where F is the force acting on a body of mass m which increases its velocity from u to v in time t.

Hence

$$Ft = mv - mu$$

The quantity Ft is called the *impulse* of the force on the body. It is a vector and, like linear momenta, impulses in opposite directions must be given positive and negative signs. In words, the *impulse–momentum equation* is

impulse = change of momentum

The equation shows that impulse and momentum have the same units, i.e. N s or kg m s^{-1}.

These ideas are important in games. The good cricketer or tennis player 'follows through' with the bat or racquet when striking the ball. The force applied then acts for a longer time, the impulse is greater and so also is the change of momentum (and velocity) of the bat. On the other hand when a cricket ball is caught its momentum is reduced to zero. This is achieved by an impulse in the form of an opposing force acting for a certain time and whilst any number of combinations of force and time will give a particular impulse, the 'sting' can be removed from the catch by drawing back the hands as the ball is caught. A smaller force is thus applied for a longer time.

In collisions of this and other types, the force is not constant but builds up to a maximum value as the deformation of the colliding bodies increases. It does, however, have an average value.

Conservation of momentum

(a) *Principle.* Suppose a body A of mass m_1 and velocity u_1 collides with another body B of mass m_2 and velocity u_2 moving in the same direction, Fig. 6.24a. If A exerts a force F to the *right* on B for time t then by Newton's third law, B will exert an equal but opposite force F on A, also for time t (since the time of contact is the same for each) but to the *left*.

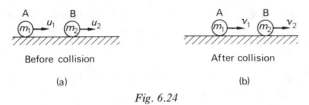

Before collision   After collision

(a)   (b)

Fig. 6.24

The bodies thus receive equal but opposite impulses Ft and so it follows from the impulse–momentum equation that the changes of momentum must be equal and opposite. The total momentum change of A and B is therefore zero, or in other words the *total momentum of* A *and* B *together remains* constant despite the collision. Thus if A has a reduced velocity v_1 after the collision and B has an increased velocity v_2, both in the same direction as before, Fig. 6.24b, then

$$m_1u_1 + m_2u_2 = m_1v_1 + m_2v_2$$

This important result, known as the *principle of conservation of momentum*, has been deduced from Newton's second and third laws and is a universal rule of the physical world which still holds even in certain extreme (relativistic) conditions where Newton's laws fail. It applies not only to collisions but to any interaction between two or more bodies. Thus in an explosion such as occurs when a gun is fired, the backward momentum component of the gun in a horizontal direction equals the component of the forward momentum of the shell and propellant gases so that the *total* momentum of the gun–shell system remains zero even though the momentum of each part changes.

No *external* agent must act on the interacting bodies otherwise momentum may be added to the system. Sometimes momentum does appear to be gained (or lost). For example a body falling towards the earth increases its downward momentum but the body is interacting with the earth (the external agent) which gains an equal amount of upward momentum from the attraction of the body on the earth. The complete system consists of the body *and* the earth and their total momentum remains constant. Similarly when a car comes to rest we believe that all the momentum it loses is transferred by the action of friction to the earth, although we cannot easily prove this.

The general statement of the principle is as follows.

When bodies in a system interact the total momentum remains constant provided no external force acts on the system.

(b) *Experimental test.* The principle can be investigated experimentally using a linear air track (which enables Perspex vehicles to move with negligible friction) and multiflash photography or electric stop clocks to measure velocities. Fig. 6.25a shows an air track (supplied with air by a domestic vacuum cleaner) and two vehicles with drinking straws attached so that a multiflash photograph can be taken using a xenon stroboscope which flashes at regular intervals.

In Fig. 6.25b a collision is shown between a vehicle of mass 'two' moving in from the left and one of mass 'three' from the right; the top markers give the velocities before the collision and the bottom ones after when the vehicles have 'rebounded' and are moving in opposite directions. (In Fig. 6.25a the hinged shutter is shown in the up position revealing to the camera the bottom half of one straw. At the exact instant of collision the shutter is rotated from one position to the other so that only one half of each straw is visible at any time. The velocities before and after the collision are thus obtained.)

Make measurements on Fig. 6.25b to see if momentum is conserved in this collision. In any attempt to verify the principle of conservation of momentum friction must be negligibly small. Why?

Fig. 6.25a

Fig. 6.25b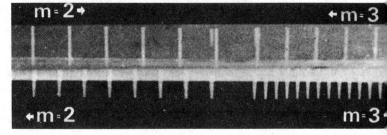

(c) *Speed of an air-rifle pellet.* An estimate is made by firing the pellet of mass m and speed v into 'Plasticine' on a model railway truck of total mass M on a friction-compensated runway, Fig. 6.26a, and measuring the speed V of the truck. Assuming conservation of momentum we have

momentum of pellet to the right before collision
= momentum of pellet and truck to the right after collision

$$\therefore \quad mv = (M + m)V$$

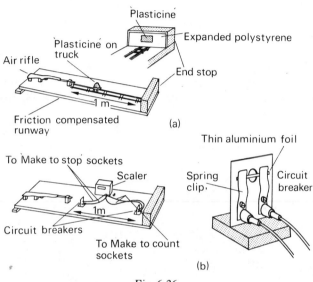

Fig. 6.26

V is found from the time taken by a card (say 10 cm long) attached to the truck to pass through a beam of light which switches on a millisecond scaler (or electric stop clock) when it interrupts the beam.

A check may be made on v and so also on the principle of conservation of momentum, by timing the pellet directly using the scaler and two aluminium foil 'circuit-breakers' 1 metre apart, Fig. 6.26b.

Rocket and jet propulsion

The principle of both is illustrated by the behaviour of an inflated balloon when released with its neck open. If the neck is closed there is a state of balance inside the balloon with equal pressure at all points, Fig. 6.27a.

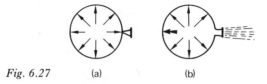

Fig. 6.27   (a)   (b)

When the neck is opened the pressure on the surface opposite the neck is now unbalanced and the balloon is forced to move in the opposite direction to that of the escaping air, Fig. 6.27*b*. According to the principle of conservation of momentum the air and the balloon have equal but opposite amounts of momentum, that is

$$m_{\text{air}} \times v_{\text{air}} = m_{\text{balloon}} \times v_{\text{balloon}}$$

In a rocket and a jet engine a stream of gas is produced at very high temperature and pressure and then escapes at high velocity through an exhaust nozzle. The thrust arises from the large increase in momentum of the exhaust gases. A rocket carries its own supplies of oxygen (liquid) and fuel (e.g. kerosene or liquid hydrogen), Fig. 6.28*a*. The mass of a rocket is not constant but decreases appreciably as it uses fuel (often at a rate of over 3000 kg s^{-1}). The acceleration consequently increases. A jet engine uses the surrounding air for its oxygen supply and so is unsuitable for space travel.

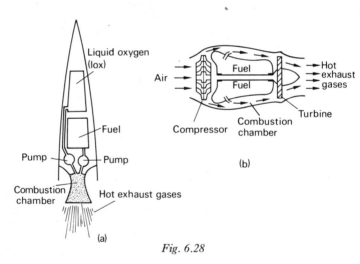

Fig. 6.28

Fig. 6.28*b* is a simplified drawing of one type of jet engine (gas turbine). The compressor draws in air at the front, compresses it, fuel (often paraffin) is injected and the mixture burns to produce hot exhaust gases which escape at high speed from the rear of the engine. These cause forward propulsion and drive the turbine which in turn rotates the compressor.

STATICS AND DYNAMICS

Momentum calculations

The concepts of impulse and momentum are useful when considering collisions and explosions, i.e. situations in which forces (called impulsive forces)
act for a short time.

*1. A jet of water emerges from a hose pipe of cross-section area 5.0×10^{-3}
m^2 with a velocity of 3.0 m s^{-1} and strikes a wall at right angles. Calculate the
force on the wall assuming the water is brought to rest and does not rebound.
(Density of water $= 1.0 \times 10^3$ kg m^{-3}.)*

If the water arrives with a velocity of 3.0 m s^{-1}, 3.0 m^3 hits every square
metre of the wall per second.

Hence volume of water striking wall per second

$$= 3.0 \text{ m s}^{-1} \times 5.0 \times 10^{-3} \text{ m}^2$$

$$= 1.5 \times 10^{-2} \text{ m}^3 \text{ s}^{-1}$$

Therefore mass of water striking wall per second

$$= 1.5 \times 10^{-2} \times 1.0 \times 10^3 \text{ kg s}^{-1}$$

$$= 15 \text{ kg s}^{-1}$$

Velocity change of water on striking wall
$$= 3.0 - 0 = 3.0 \text{ m s}^{-1}$$

Therefore momentum change per second of water on striking wall
$$= 15 \text{ kg s}^{-1} \times 3.0 \text{ m s}^{-1}$$

$$= 45 \text{ kg m s}^{-2}$$

But   force = momentum change per second

$$= 45 \text{ N}$$

(In practice the horizontal momentum of the water is seldom completely
destroyed and so the answer is only approximate.)

*2. A railway truck A of mass 2×10^4 kg travelling at 0.5 m s^{-1} collides
with another truck B of half its mass moving in the opposite direction with a
velocity of 0.4 m s^{-1}. If the trucks couple automatically on collision, find the
common velocity with which they move, Fig. 6.29.*

Fig. 6.29

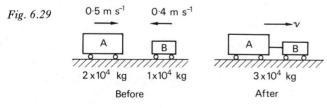

0·5 m s^{-1} 0·4 m s^{-1} v

A   B   A   B

2×10^4 kg 1×10^4 kg 3×10^4 kg

Before   After

Total momentum *to the right* of A and B before collision

$$= 2 \times 10^4 \times 0.5 - 1 \times 10^4 \times 0.4 \text{ kg m s}^{-1}$$

$$= 0.6 \times 10^4 \text{ kg m s}^{-1}$$

(If the momentum of A is taken as positive, that of B must be negative.)
Total momentum *to the right* of A and B after collision

$$= 3 \times 10^4 \times v \text{ kg m s}^{-1}$$

By the principle of conservation of momentum

$$3 \times 10^4 \times v = 0.6 \times 10^4$$

$$\therefore \quad v = 0.2 \text{ m s}^{-1}$$

3. A jet engine on a test bed takes in 20.0 kg of air per second at a velocity of 100 m s^{-1} *and burns 0.80 kg of fuel per second. After compression and heating the exhaust gases are ejected at 500* m s^{-1} *relative to the aircraft. Calculate the thrust of the engine.*

Velocity change of 20 kg of air $= (500 - 100) = 400 \text{ m s}^{-1}$

Therefore momentum change per second of 20 kg of air

$$= 20 \text{ kg s}^{-1} \times 400 \text{ m s}^{-1}$$

The initial velocity of the fuel is zero and so its velocity change is 500 m s^{-1}.

$\therefore$ Momentum change per second of 0.80 kg of fuel

$$= 0.80 \text{ kg s}^{-1} \times 500 \text{ m s}^{-1}$$

$\therefore$ Total momentum change per second of air and fuel

$$= (20 \times 400 + 0.80 \times 500) \text{ kg m s}^{-2}$$

$$= 8.4 \times 10^3 \text{ kg m s}^{-2}$$

But   force (thrust) = total change of momentum per second

$\therefore$ thrust of engine $= 8.40 \times 10^3 \text{ N}$ $(1 \text{ kg m s}^{-2} = 1 \text{ N})$

Note. If the engine is in an aircraft flying at 100 m s^{-1}, taking in air at this speed, the thrust would be about the same.

Work, energy and power

(*a*) *Work.* In science the term work has a definite meaning which differs from its everyday one. For example someone holding a heavy weight at rest may say and feel he is doing hard work but in fact none is being done on the weight in the scientific sense.

Work is done in science when a force moves its point of application along the direction of its line of action.

In the simple case of Fig. 6.30a, the constant force F and the displacement s are in the same direction and we define the work W done by the force on the body by

$$W = Fs$$

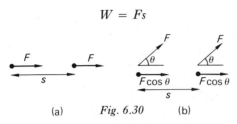

(a)   *Fig. 6.30*   (b)

If the force does not act in the direction in which motion occurs but at an angle θ to it as in Fig. 6.30b, then the work done is defined as the product of the component of the force in the direction of motion and the displacement in that direction. That is,

$$W = (F \cos \theta)s$$

When $\theta = 0$, $\cos \theta = 1$ and so $W = Fs$, in agreement with the first equation. When $\theta = 90°$, $\cos \theta = 0$ and F has no component in the direction of motion and so no work is done. Thus the work done by the force of gravity when a body is moved horizontally is zero.

If the force varies, the work done can be obtained from a force–displacement graph in which the component of the force in the direction of the displacement is plotted, Fig. 6.31. Suppose the force is F when the displacement is x, then the work done during a further, very small displacement

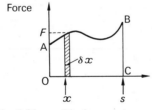

Fig. 6.31   Displacement

δx (which is so small that F can be considered constant during it) is $F\delta x$, i.e. the shaded area. By dividing up the whole area under the curve into narrow strips we see that the total work done during displacement s is represented by area OABC.

259

Work can be either positive or negative. It is positive if the force (or a component of it) acts in the same direction as the displacement (Figs. 6.30a and b), but negative if it is oppositely directed (θ is then $>90°$ and cos θ is negative). The work done by friction when it opposes one body sliding over another is negative.

The unit of work is *1 joule* and is *the work done by a force of 1 newton when its point of application moves through a distance of 1 metre in the direction of the force*. Thus

$$1 \text{ joule (J)} = 1 \text{ newton metre (N m)}$$

Work is a scalar although force and displacement are both vectors.

(*b*) *Energy.* When a body A does work by exerting a force on another body B, the body A is said to lose energy, equal in amount to the work it performs. Energy is therefore often defined as *that which enables a body to do work*; it is measured in joules, like work. When an interchange of energy occurs between two bodies we can look upon *the work done as measuring the quantity of energy transferred between them*. Thus if body A does 5 joules of work on body B then the energy transfer from A to B is 5 joules.

(*c*) *Power.* The power of a machine is *the rate at which it does work*, i.e. the rate at which it converts energy from one form to another. The unit of power is the *watt* (W) and equals a rate of working of 1 joule per second, i.e. $1 \text{ W} = 1 \text{ J s}^{-1}$.

The two basic reasons for bodies having mechanical energy will now be considered.

Kinetic and potential energy

(*a*) *Kinetic energy.* This is the energy a body has because of its motion. For example a moving hammer does work against the resistance of the wood into which a nail is being driven. An expression for kinetic energy can be obtained by calculating the amount of work the body will do while it is being brought to rest.

Consider a body of constant mass m moving with velocity u. Let a constant force F act on it and bring it to rest in a distance s, Fig. 6.32. Since the final velocity v is zero, from $v^2 = u^2 + 2as$ we have

$$0 = u^2 + 2as$$

$$\therefore \quad a = -\frac{u^2}{2s}$$

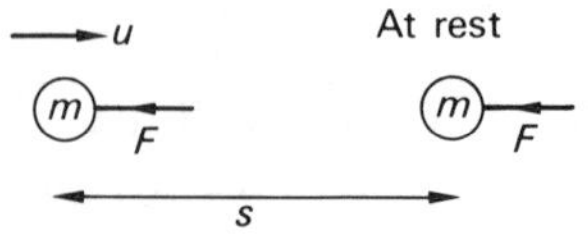

Fig. 6.32

The negative sign shows that the acceleration a is opposite in direction to u (as we would expect). The acceleration in the direction of F is thus $+u^2/2s$. The original kinetic energy of the body equals the work W it does against F, hence

$$\text{kinetic energy of body} = W = Fs$$

$$= mas \quad (\text{since } F = ma)$$

$$= ms\frac{u^2}{2s} \quad \left(\text{since } a = \frac{u^2}{2s}\right)$$

$$= \tfrac{1}{2}mu^2$$

Thence the kinetic energy of a body of mass m moving with speed u is $\tfrac{1}{2}mu^2$. Conversely if work is done on a body the gain of kinetic energy when its velocity increases from zero to u can be shown to be $\tfrac{1}{2}mu^2$.

In general if the velocity of a body of mass m increases from u to v when work is done on it by a force F acting over a distance s, then

$$Fs = \tfrac{1}{2}mv^2 - \tfrac{1}{2}mu^2$$

This is called the *work–energy equation* and may be stated

$$\frac{\text{work done by the}}{\text{forces acting on the body}} = \frac{\text{change in kinetic energy}}{\text{of the body}}$$

(b) *Potential energy*. This is the energy a system of bodies has because of the relative positions of its parts, i.e. due to its configuration. It arises when a body experiences a force in a field such as the earth's gravitational field. In that case the body occupies a position with respect to the earth and the potential energy is regarded as a joint property of the body–earth system and not of either body separately. The relative positions of the parts of the system, i.e. of the body and earth, determine its potential energy; the greater the separation the greater the potential energy.

Normally we are only concerned with differences of potential energy. In the gravitational case it is convenient to consider that the potential energy is zero when the body is at the surface of the earth. The potential energy when a body of mass m is at height h above ground level equals the work which must be done against the downward pull of gravity to raise the body to this height. A force, equal and opposite to mg, has to be exerted on the body over displacement h (assuming g is constant near the earth's surface). Therefore

$$\text{work done by external force against gravity} = \text{force} \times \text{displacement}$$

$$= mgh$$

$$\therefore \quad \text{potential energy} = mgh$$

On returning to ground level an amount of potential energy equal to *mgh* would be lost. A good example of this occurs when the water in a mountain reservoir falls to a lower level and does work by driving a power station turbine.

A stretched or compressed spring is also considered to have potential energy.

Conservation of energy

If a body of mass *m* is thrown vertically upwards with velocity *u* at A, it has to do work against the constant force of gravity, Fig. 6.33. When it has risen to B let its reduced velocity be *v*. By the definition of kinetic energy (k.e.)

    loss of k.e. between A and B = work done by body against *mg*

Fig. 6.33

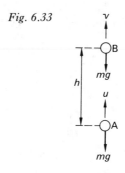

By the definition of potential energy (p.e.)

    gain of p.e. between A and B = work done by body against *mg*

$$\therefore \quad \text{loss of k.e.} = \text{gain of p.e.}$$

$$\therefore \quad \tfrac{1}{2}mu^2 - \tfrac{1}{2}mv^2 = mgh$$

This is called the *principle of conservation of mechanical energy* and may be stated as follows.

The total amount of mechanical energy (k.e. + p.e.) which the bodies in an isolated system possess is constant.

It applies only to frictionless motion, i.e. to *conservative systems*. Otherwise in the case of a rising body, work has to be done against friction as well as against gravity and the body gains less p.e. than when friction is absent. Furthermore, the gain of p.e. would depend on the path taken; it does not in a conservative system.

Work done against frictional forces is generally accompanied by a temperature rise. This suggests that we might include in our energy accountancy what we have called *internal energy*. This would then extend the energy conservation principle to non-conservative systems and we can then say, for example,

loss of k.e. = gain of p.e. + gain of internal energy

The mechanics of a body seen to be in motion has thus been related to a phenomenon which is apparently not mechanical and in which motion is not directly detected. (However, we *believe* that internal energy is random molecular kinetic and potential energy, see p. 115.) In a similar way, the idea of energy has been extended to other areas of physics and is now a unifying theme. In fact, physics is sometimes said to be the study of energy transformations, measured in terms of the work done by the forces created in the transformation.

The principle of conservation of mechanical energy is a special case of the more general *principle of conservation of energy*—one of the fundamental laws of science.

Energy may be transformed from one form to another, but it cannot be created or destroyed, i.e. the total energy of a system is constant.

Energy calculations

The work–energy equation $Fs = \frac{1}{2}mv^2 - \frac{1}{2}mu^2$ is useful for solving problems when the distance over which a force acts is known.

1. A car of mass 1.0×10^3 kg travelling at 72 km h^{-1} on a horizontal road is brought to rest in a distance of 40 m by the action of the brakes and frictional forces. Find (a) the average stopping force, (b) the time taken to stop the car.

A speed of 72 km h^{-1} = 72×10^3 m/3600 s = 20 m s^{-1}

(*a*) If the car has mass m and initial speed u, then

kinetic energy lost by car = $\frac{1}{2}mu^2$

If F is the average stopping force and s the distance over which it acts, then

work done by car against $F = Fs$

But
$$Fs = \tfrac{1}{2}mu^2$$

$$\therefore \quad F \times 40 \text{ m} = \tfrac{1}{2} \times (1.0 \times 10^3 \text{ kg}) \times (20 \text{ m s}^{-1})^2$$

$$\therefore \quad F = \frac{1.0 \times 10^3 \times 400}{2 \times 40} \quad \frac{\text{kg m}^2 \text{ s}^{-2}}{\text{m}}$$

$$= 5.0 \times 10^3 \text{ N}$$

263

(b) Assuming constant acceleration and substituting $v = 0, u = 20\,\text{m s}^{-1}$ and $s = 40\,\text{m}$ in $v^2 = u^2 + 2as$ we have

$$0 = 20^2 + 2a \times 40$$

$$\therefore \quad a = -5.0\,\text{m s}^{-2}$$

(the negative sign indicates the acceleration is in the opposite direction to the displacement).

Using $v = u + at$

$$0 = 20 - 5.0t$$

$$\therefore \quad t = 4.0\,\text{s}$$

2. *A bullet of mass 10 g travelling horizontally at a speed of 1.0×10^2 m s^{-1} embeds itself in a block of wood of mass 9.9×10^2 g suspended by strings so that it can swing freely. Find (a) the vertical height through which the block rises, (b) how much of the bullet's energy becomes internal energy. ($g = 10$ m s^{-2}.)*

(a) The bullet is brought to rest very quickly due to the resistance offered by the block and we shall assume that the block (with the bullet embedded) hardly moves until the bullet is at rest. Momentum is conserved in the collision and so

$$mu = (M + m)v$$

where m and M are the masses of the bullet and block respectively, u is the velocity of the bullet before impact and v is the velocity of the block + bullet as they move off.

$$\therefore \quad 10 \times 10^{-3}\,\text{kg} \times 1.0 \times 10^2\,\text{m s}^{-1} = (990 + 10) \times 10^{-3}\,\text{kg} \times v$$

$$\therefore \quad v = 1.0\,\text{m s}^{-1}$$

When the block has swung to its maximum height h, all its kinetic energy has become potential energy—if frictional forces are neglected. Conservation of energy therefore holds and we can say

$$\tfrac{1}{2}(M + m)v^2 = (M + m)gh$$

$$\therefore \quad h = \frac{v^2}{2g}$$

$$= \frac{(1\,\text{m s}^{-1})^2}{2 \times 10\,\text{m s}^{-2}} = \frac{1}{2 \times 10}\,\frac{\text{m}^2\,\text{s}^{-2}}{\text{m s}^{-2}}$$

$$\therefore \quad h = 5.0 \times 10^{-2}\,\text{m}$$

(b) Original kinetic energy of bullet = $\frac{1}{2}mu^2$

$$= \frac{1}{2} \times (10 \times 10^{-3} \text{ kg}) \times (1.0 \times 10^2 \text{ m s}^{-1})^2$$
$$= \frac{1}{2} \times 10 \times 10^{-3} \times 1.0 \times 10^4 \quad \text{kg m}^2 \text{ s}^{-2}$$
$$= 50 \text{ J}$$

Kinetic energy of block + bullet after impact

$$= \frac{1}{2}(M + m)v^2$$
$$= \frac{1}{2} \times (1000 \times 10^{-3} \text{ kg}) \times (1 \text{ m s}^{-1})^2$$
$$= 0.50 \text{ J}$$

Therefore

$$\text{internal energy produced} = \text{loss of kinetic energy}$$
$$= 50 - 0.50$$
$$= 49.5 \text{ J}$$

Elastic and inelastic collisions

Whilst momentum is always conserved in a collision, there is generally a change of some kinetic energy, usually to internal energy or to a very small extent to sound energy. Collisions in which a loss of kinetic energy occurs (see previous worked example) are said to be *inelastic*. In a *perfectly elastic* collision, kinetic energy is conserved.

The linear air track (or trolleys and ticker tape timers) can be used to investigate what happens to kinetic energy, as well as to momentum, in different collisions. Nearly perfect elastic collisions are obtained if a rubber band is fitted to the front of one vehicle and the other vehicle has a pointed end, Fig. 6.34a. By contrast, completely inelastic (i.e. no-bounce) collisions occur if a needle is fitted to one vehicle and 'Plasticine' inserted in a hole (in line with the needle) in the other, Fig. 6.34b. By making measurements on Fig. 6.25b (p. 254) find the total kinetic energy before and after the collision and say what kind of collision has occurred.

Fig. 6.34a

Fig. 6.34b

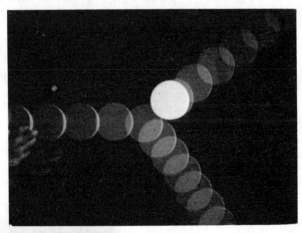

Fig. 6.35a

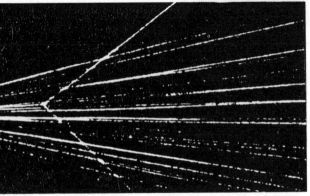

Fig. 6.35b

The head-on collisions of air–track vehicles are one dimensional. An oblique, two-dimensional collision between a moving, magnetic, 'dry-ice' puck and a stationary one of *equal mass* is shown in Fig. 6.35a. Make measurements to see if momentum is conserved (remember that momentum is a vector quantity). Compare the kinetic energy before and after the collision and comment on the result. What is the angle between the directions of motion of the pucks after the collision?

Collisions between atoms and other atomic particles were first studied in a cloud chamber; many such collisions are elastic. Fig. 6.35b shows a cloud chamber photograph of an alpha particle colliding with a helium nucleus. Compare this with Fig. 6.35a and assuming atomic particles and magnetic pucks behave similarly, comment on (*i*) the mass of an alpha particle compared with that of a helium nucleus, (*ii*) the type of collision which has occurred.

Electrons can have elastic or inelastic collisions with the atoms of a gas and the latter give information about the electronic structure of atoms.

266

STATICS AND DYNAMICS

QUESTIONS

Assume $g = 10$ m s^{-2}

Statics

1. Distinguish between scalar and vector quantities, giving two examples of each. How may the resultant of a number of vector quantities be obtained? Why are vector quantities frequently resolved into rectangular components?

It is possible for the product of two vector quantities to be a scalar. Give an example of two such quantities.

Show how, by considering the sail to be a flat plane and by resolving the thrust on the sail due to the wind into components, it is possible to explain how a yacht may make progress upwind. Why is it an advantage for a yacht to have a large and heavy keel? (*Hint :* see p. 376.)   (*A.E.B.*)

2. State the conditions of equilibrium of a body acted on by a system of coplanar forces.

An aerial attached to the top of a radio mast 20 m high exerts a horizontal force on it of 3.0×10^2 N. A stay-wire from the mid-point of the mast to the ground is inclined at $60°$ to the horizontal. Assuming the action of the ground on the mast can be regarded as a single force, find (*a*) the force exerted on the mast by the stay-wire, (*b*) the magnitude and direction of the action of the ground.

3. A uniform ladder 5.0 m long and having mass 40 kg rests with its upper end against a smooth vertical wall and with its lower end 3.0 m from the wall on rough ground. Find the magnitude and direction of the force exerted at the bottom of the ladder.

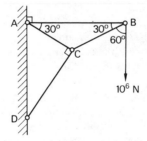

Fig. 6.36

4. Find the forces in the members of a pin-jointed structure shown in Fig. 6.36. State whether each force is tensile or compressive. Given that the maximum safe stress for the material used in each member is 8×10^7 N m^{-2}, for the member with the highest load calculate the minimum cross-sectional area.   (*J.M.B. Eng. Sc.*)

Dynamics

5. A dart player stands 3.00 m from the wall on which the board hangs and throws a dart which leaves his hand with a horizontal velocity at a point 1.80 m above the ground. The dart strikes the board at a point 1.50 m from the ground. Assuming air resistance to be negligible, calculate (*i*) the time of the flight of the dart, (*ii*) the initial speed of the dart and (*iii*) the speed of the dart when it hits the board. (*A.E.B. part qn.*)

6. A projectile is fired from ground level with a velocity of 500 m s^{-1} at 30° to the horizontal. Calculate its horizontal range, the greatest height it reaches and the time taken to rise to that height. (Neglect air resistance.)

7. A body slides, with constant velocity, down a plane inclined at 30° with the horizontal. Show in a diagram the forces acting on the body, and find the coefficient of kinetic friction between the body and the plane.

If the plane were now tilted so as to make an angle of 60° with the horizontal, with what acceleration would the body slide down the plane? What force, applied parallel to this plane, would be required to cause the body to move up the plane with a constant velocity? (W.)

8. An object of mass m rests on the floor of a lift which is ascending with acceleration a. Draw a diagram to show the external forces acting on the object, and write down its equation of motion. How do these forces arise? Show graphically how their magnitudes vary with the acceleration of the lift. What force constitutes the second member of the action–reaction pair in the case of each of these external forces?

9. Five identical cubes, each of mass m, lie in a straight line, with their adjacent faces in contact, on a horizontal surface, as shown in Fig. 6.37.

Fig. 6.37

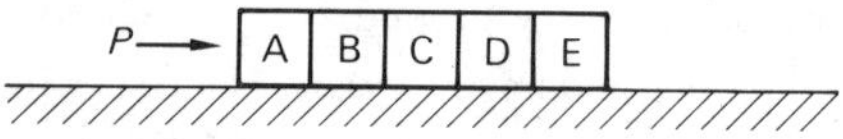

Suppose the surface is frictionless and that a constant force P is applied from left to right to the end face of A.

What is the acceleration of the system and what is the resultant force acting on each cube? What force does cube C exert on cube D?

If friction is present between the cubes and the surface, draw a graph to illustrate how the total frictional force varies as P increases uniformly from zero. (W.)

10. State Newton's second law of motion.

A stream of water travelling horizontally at 30 m s^{-1} is ejected from a hole of cross-sectional area 40 cm^2 and is directed against a vertical wall. Calculate the force exerted on the wall assuming that the water does not rebound.

What is the power of the pump needed to give the ejected water the necessary kinetic energy?

Density of water = 1.0 g cm^{-3} = 1.0 × 10^3 kg m^{-3} (J.M.B.)

11. Sketch a graph of the relationship between the kinetic energy E (plotted on the vertical axis) and the distance travelled x (plotted on the horizontal axis) for a body of mass m sliding from rest with negligible friction down a uniform slope which makes an angle of 30° with the horizontal. What is the gradient of the graph equal to? (S.)

12. Define linear momentum and state the principle of conservation of linear momentum. Explain briefly how you would attempt to verify this principle by experiment.

Sand is deposited at a uniform rate of 20 kilograms per second and with negligible kinetic energy on to an empty conveyor belt moving horizontally at a constant speed of 10 metres per minute. Find (a) the force required to maintain constant velocity, (b) the power required to maintain constant velocity and (c) the rate of change of kinetic energy of the moving sand. Why are the latter two quantities unequal? (O. and C.)

13. Write down an expression for the kinetic energy of a body.

(a) A car of mass 1.00×10^3 kg travelling at 20 m s^{-1} on a horizontal road is brought to rest by the action of its brakes in a distance of 25 m. Find the average retarding force.

(b) If the same car travels up an incline of 1 in 20 at a constant speed of 20 m s^{-1}, what power does the engine develop if the frictional resistance is 100 N?

14. State the principle of conservation of linear momentum and show how it follows from Newton's laws of motion.

A stationary radioactive nucleus of mass 210 units disintegrates into an alpha particle of mass 4 units and a residual nucleus of mass 206 units. If the kinetic energy of the alpha particle is E, calculate the kinetic energy of the residual nucleus.   (*J.M.B.*)

7 Circular motion and gravitation

Motion in a circle

In everyday life, in atomic physics and in astronomy and space travel there are many examples of bodies moving in paths which if not exactly circular are nearly so. In this chapter we will see how ideas developed for dealing with straight-line motion enable us to tackle circular motion.

A body which travels equal distances in equal times along a circular path has constant speed but *not* constant velocity. This is due to the way we have defined speed and velocity; speed is a scalar quantity, velocity is a vector quantity. Fig. 7.1 shows a ball attached to a string being whirled round in a

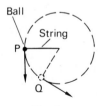

Fig. 7.1

horizontal circle. The velocity of the ball at P is directed along the tangent at P; when it reaches Q its velocity is directed along the tangent at Q. If the speed is constant the *magnitudes* of the velocities at P and Q are the same but their *directions* are different and so the velocity of the ball has changed. A change of velocity is an acceleration and a body moving uniformly in a circular path or arc is therefore accelerating.

In everyday language acceleration usually means going faster and faster, i.e. involves a change of speed. However, in physics it means a change of velocity and since the velocity changes not only when the speed changes, but also when the direction of motion changes, then, for example, a car rounding a bend (even at constant speed) is accelerating.

CIRCULAR MOTION AND GRAVITATION

Two useful expressions

We will use these from time to time when dealing with circular motion.

(a) *Angles in radians*: $s = r\theta$. Angles can be measured in radians as well as degrees. In Fig. 7.2 the angle θ, in radians, is defined by the equation

Fig. 7.2

$$\theta = \frac{s}{r}$$

If $s = r$ then $\theta = 1$ radian (rad). Therefore 1 radian is the angle subtended at the centre of a circle by an arc equal in length to the radius. When $s = 2\pi r$ (the circumference of a circle of radius r) then $\theta = 2\pi$ radians $= 360°$.

$$\therefore \quad 1 \text{ radian} = 360°/2\pi \simeq 57°$$

From the definition of a radian it follows that the length s of an arc which subtends an angle θ at the centre of a circle of radius r, is given by

$$s = r\theta$$

where θ is in radians.

(b) *Angular velocity*: $v = r\omega$. The speed of a body moving in a circle can be specified either by its speed along the tangent at any instant, i.e. by its linear speed, or, by its angular velocity. This is the angle swept out in unit time by the radius joining the body to the centre of the circle. It is measured in radians per second (rad s^{-1}).

We can derive an expression connecting angular velocity and linear speed. Consider a body moving uniformly from A to B in time t so that radius OA rotates through an angle θ, Fig. 7.2. The angular velocity ω of the body about O is

$$\omega = \frac{\theta}{t}$$

If arc AB has length s and if v is the constant speed of the body then

$$v = \frac{s}{t}$$

But from (a), $s = r\theta$ where r is the radius of the circle

$$\therefore \quad v = \frac{r\theta}{t}$$

But $\omega = \theta/t$

$$\therefore \quad v = r\omega$$

If $r = 3$ m and $\omega = 1$ revolution per second $= 2\pi$ rad s^{-1} then the linear speed $v = 6\pi$ m s^{-1}.

Deriving $a = v^2/r$

To obtain an expression for the acceleration of a small body (i.e. a particle) describing circular motion, consider such a body moving with *constant speed* v in a circle of radius r, Fig. 7.3a. If it travels from A to B in a short interval of time δt then, since distance = speed × time,

$$\text{arc AB} = v \, \delta t$$

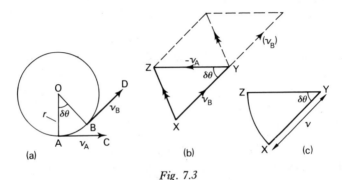

Fig. 7.3

Also, by the definition of an angle in radians

$$\text{arc AB} = r \, \delta\theta \qquad (\delta\theta = \angle\,\text{AOB})$$

$$\therefore \quad r \, \delta\theta = v \, \delta t$$

$$\therefore \quad \delta\theta = \frac{v \, \delta t}{r} \tag{1}$$

The vectors $\mathbf{v_A}$ and $\mathbf{v_B}$ drawn tangentially at A and B represent the velocities at these points. The *change* of velocity between A and B is obtained by subtracting $\mathbf{v_A}$ from $\mathbf{v_B}$. That is

$$\text{change of velocity} = \mathbf{v_B} - \mathbf{v_A}$$

But $$\mathbf{v_B} - \mathbf{v_A} = \mathbf{v_B} + (-\mathbf{v_A})$$

Hence, to subtract vector $\mathbf{v_A}$ from vector $\mathbf{v_B}$ we *add* vectors $\mathbf{v_B}$ and $(-\mathbf{v_A})$ by the parallelogram law.

In Fig. 7.3b, XY represents $\mathbf{v}_B$ in magnitude (v) and direction (BD); YZ represents ($-\mathbf{v}_A$) in magnitude (v) and direction (CA). The resultant, which gives the change of velocity, is then seen from the figure to be, in effect, vector XZ.

Since one vector ($-\mathbf{v}_A$) is perpendicular to OA and the other $\mathbf{v}_B$ is perpendicular to OB, $\angle XYZ = \angle AOB = \delta\theta$. If δt is very small, $\delta\theta$ will also be small and XZ in Fig. 7.3b will have almost same length as arc XZ in Fig. 7.3c which subtends angle $\delta\theta$ at the centre of a circle of radius v. Arc $XZ = v\,\delta\theta$ (from definition of radian) and so

$$XZ = v\,\delta\theta$$

But from (1)

$$\delta\theta = \frac{v\,\delta t}{r}$$

$$\therefore \quad XZ = \frac{v^2}{r}\,\delta t$$

The *magnitude* of the acceleration a between A and B is

$$a = \frac{\text{change of velocity}}{\text{time interval}} = \frac{XZ}{\delta t}$$

$$\therefore \quad a = \frac{v^2}{r}$$

If ω is the angular velocity of the body, $v = r\omega$ and we can also write

$$a = \omega^2 r$$

The *direction* of the acceleration is *towards the centre* O of the circle as can be seen if δt is made so small that A and B all but coincide; vector XZ is then perpendicular to $\mathbf{v}_A$ (or $\mathbf{v}_B$), i.e. along AO (or BO). We say the body has a *centripetal acceleration* (i.e. centre-seeking).

Does a body moving uniformly in a circle have *constant* acceleration? (Remember that acceleration is a vector.)

Centripetal force

Since a body moving in a circle (or a circular arc) is accelerating, it follows from Newton's first law of motion that there must be a force acting on it to cause the acceleration. This force, like the acceleration, will also be directed towards the centre and is called the *centripetal force*. It causes the body to deviate from the straight-line motion which it would naturally follow if the force were absent. The value F of the centripetal force is given by Newton's second law, that is

$$F = ma = \frac{mv^2}{r}$$

where m is the mass of the body and v is its speed in the circular path of radius r. If the angular velocity of the body is ω we can also say, since $v = r\omega$,

$$F = m\omega^2 r$$

When a ball attached to a string is swung round in a horizontal circle, the centripetal force which keeps it in a circular orbit arises from the tension in the string. We can think of the tension as tugging continually on the body and 'turning it in' so that it remains at a fixed distance from the centre. If the ball is swung round faster, a larger force is needed and if it is greater than the tension the string can bear, it breaks and the ball continues to travel along a tangent to the circle at the point of breaking, Fig. 7.4.

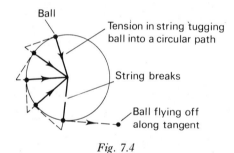

Ball

Tension in string tugging ball into a circular path

String breaks

Ball flying off along tangent

Fig. 7.4

Other examples of circular motion will be discussed presently but in all cases it is important to appreciate that the forces acting on the body must provide a resultant force of magnitude mv^2/r towards the centre. What is the nature of the centripetal force for (*a*) a car rounding a bend, (*b*) a space capsule circling the earth?

One arrangement for testing $F = mv^2/r$ experimentally is shown in Fig. 7.5. The turntable, driven by the electric motor, is *gradually* speeded up and the spring extends until the truck just reaches the stop at the end of the track. The speed v of the *truck* in orbit is found by measuring the time for one revolution of the turntable and then, with the turntable at rest, the radius r of the circle described by the truck (i.e. the distance from the centre of the turntable to the centre of the truck). Knowing the mass m of the truck, mv^2/r can be calculated.

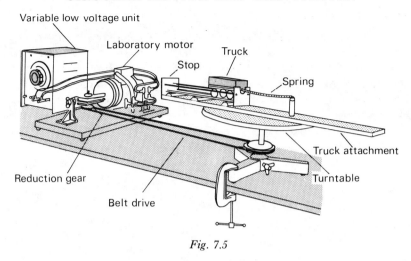

Fig. 7.5

The tension in the stretched spring is the centripetal force and this can be found by measuring with a spring balance, the tension required to extend the spring *by the same amount* as it is when the truck is at the end stop, Fig. 7.6. The value obtained should agree with the value of mv^2/r to within a few per cent.

The mass of the truck can be altered by loading it with lead plates and the experiment repeated for each mass.

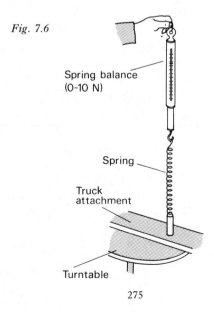

Fig. 7.6

Rounding a bend

If a car is travelling round a circular bend with uniform speed on a horizontal road, the resultant force acting on it must be directed to the centre of its circular path, i.e. it must be the centripetal force. This force arises from the interaction of the car with the air and the ground. The direction of the force exerted by the air on the car will be more or less opposite to the instantaneous direction of motion. The other and more important horizontal force is the frictional force exerted inwards by the ground on the tyres of the car, Fig. 7.7. The resultant of these two forces is the centripetal force.

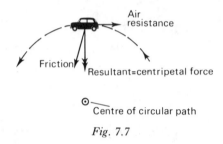

Fig. 7.7

The successful negotiation of a bend on a flat road therefore depends on the tyres and the road surface being in a condition that enables them to provide a sufficiently high frictional force—otherwise skidding occurs. Safe cornering that does not rely on friction is achieved by 'banking' the road.

The problem is to find the angle θ at which a bend should be banked so that the centripetal force acting on the car arises entirely from a component

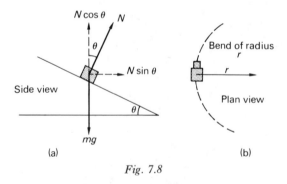

Fig. 7.8

of the normal force N of the road, Fig. 7.8a. Treating the car as a particle and resolving N vertically and horizontally we have, since $N \sin \theta$ *is* the centripetal force,

$$N \sin \theta = \frac{mv^2}{r}$$

where m and v are the mass and speed respectively of the car and r is the radius of the bend, Fig. 7.8b. Also, the car is assumed to remain in the same horizontal plane and so has no vertical acceleration, thus

$$N \cos \theta = mg$$

Hence, by division

$$\tan \theta = \frac{v^2}{gr}$$

The equation shows that for a given radius of bend, the angle of banking is only correct for one speed. In a race track, the banking becomes steeper towards the outside and the driver can select a position according to his speed, Fig. 7.9.

A bend in a railway track is also banked, in this case so that at a certain speed no lateral thrust has to be exerted by the outer rail on the flanges of the wheels of the train, otherwise the rails are strained. The horizontal component of the normal force of the rails on the train then provides the centripetal force.

Fig. 7.9

An aircraft in straight, level flight experiences a lifting force at right angles to the surface of its wings which balances its weight. To turn, the ailerons are operated so that the aircraft banks and the horizontal component of the lift supplies the necessary centripetal force, Fig. 7.10. The aircraft's weight

Fig. 7.10

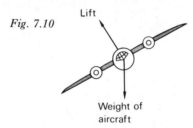

Lift

Weight of aircraft

is now opposed only by the vertical component of the lift and height will be lost unless the lift is increased by, for example, increasing the speed.

Other examples of circular motion

(*a*) *The rotor.* This device is sometimes present in amusement parks. It consists of an upright drum of diameter about 4 metres inside which people stand with their backs against the wall. The drum is spun at increasing speed about its central vertical axis and at a certain speed the floor is pulled downwards. The occupants do not fall but remain 'pinned' against the wall of the rotor.

The forces acting on a passenger of mass m are shown in Fig. 7.11. N is the normal force of the wall on the passenger and is the centripetal force

Fig. 7.11

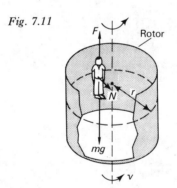

F
Rotor
N
r
mg
v

needed to keep him moving in a circle. Hence if r is the radius of the rotor and v the speed of the passenger then

$$N = \frac{mv^2}{r}$$

F is the frictional force acting upwards between the passenger and rotor wall and since there is no vertical motion of the passenger

$$F = mg$$

If μ is the coefficient of limiting friction between passenger and wall, we have $F = \mu N$

$$\therefore \quad \mu N = mg$$

$$\therefore \quad \mu = \frac{mg}{N} = \frac{mg}{mv^2/r}$$

$$\therefore \quad \mu = \frac{gr}{v^2}$$

This equation gives the minimum coefficient of friction required to prevent the passenger slipping; it does not depend on the passenger's weight. A typical value of μ between clothing and a rotor wall (of canvas) is about 0.40 and so if $r = 2$ m, v must be about 7 m s^{-1} (or more). What will be the angular velocity of the drum? How many revolutions will it make per minute?

(b) *Looping the loop.* A pilot who is not strapped into his aircraft can loop the loop without falling out at the top of the loop. A bucket of water can be swung round in a vertical circle without spilling. A ball-bearing can loop the loop on a length of curtain rail in a vertical plane. All three effects have similar explanations.

Consider the bucket of water when it is at the top of the loop, A in Fig. 7.12. If the weight mg of the water is *less than mv^2/r*, the normal force N of the bottom of the bucket on the water provides the rest of the force required to maintain the water in its circular path. However, if the bucket is swung more slowly then mg will be greater than mv^2/r and the 'unused' part of the weight causes the water to leave the bucket. What provides the centripetal force for the *water* when the bucket is at (*i*) B, (*ii*) C and (*iii*) D?

Fig. 7.12

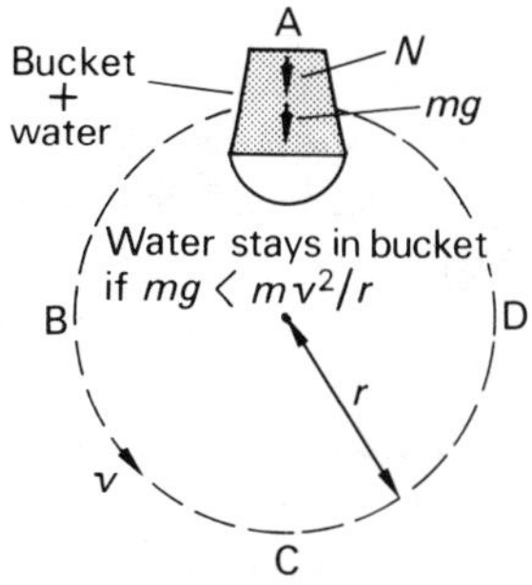

Bucket + water

A
N
mg

Water stays in bucket if $mg < mv^2/r$

B
D
r
v
C

(c) *Centrifuges.* These separate solids suspended in liquids or liquids of different densities. The mixture is in a tube, Fig. 7.13a, and when it is rotated at high speed in a horizontal circle the less dense matter moves towards the centre of rotation. On stopping the rotation, the tube returns to the vertical position with the less dense matter at the top. Cream is separated from milk in this way.

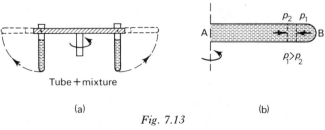

Tube + mixture

(a)   (b)

Fig. 7.13

The action uses the fact that if a horizontal tube of liquid is rotated, the force exerted by the closed end must be greater than when the tube was at rest so that it can provide the necessary centripetal force acting radially inwards. In Fig. 7.13b the liquid pressure at B is greater than at A and a pressure gradient exists along the tube. For any part of the liquid the force due to the pressure difference supplies exactly the centripetal force required. If this part of the liquid is replaced by matter of smaller density (and thus of smaller mass), the force is too large and the matter moves inwards.

During the launching and re-entry of space vehicles accelerations of about $8g$ occur and the resulting large forces which act on the surface of the astronaut's body cause blood to drain from some parts and congest others. If the brain is deprived, loss of vision and unconsciousness may follow. Tests with large man-carrying centrifuges in which passengers are subjected to high centripetal accelerations show that a person will tolerate $15g$ for a few minutes when his body is perpendicular to the direction of the acceleration but only $6g$ when in the direction of acceleration. What will be the best position for an astronaut to adopt at lift-off and re-entry?

Moment of inertia

In most of the cases of circular motion considered so far we have treated the body as a 'particle' so that all of it, in effect, revolves in a circle of the same radius. When this cannot be done we have to regard the rotating body as a system of connected 'particles' moving in circles of different radii. The way in which the mass of the body is distributed then affects its behaviour.

This may be shown by someone who is sitting on a freely rotating stool with a heavy weight in each hand. When he extends his arms, Fig. 7.14, the speed of rotation decreases but increases again when he brings them in. The

Heavy weight

Rotating stool

Fig. 7.14

angular velocity of the system clearly depends on how the mass is distributed about the axis of rotation. A concept is needed to express this property.

The mass of a body is a measure of its in-built opposition to any change of linear motion, i.e. mass measures inertia. The corresponding property for rotational motion is called the *moment of inertia*. The more difficult it is to change the angular velocity of a body rotating about a particular axis the greater is its moment of inertia about that axis. Experiment shows that a wheel with most of its mass in the rim is more difficult to start and stop than a uniform disc of equal mass rotating about the same axis; the former has a greater moment of inertia. Similarly the moment of inertia of the person on the rotating stool is greater when his arms are extended. It should be noted that moment of inertia is a property of a body rotating about a particular axis; if the axis changes so does the moment of inertia.

We now require a measure of moment of inertia which takes into account the distribution of mass about the axis of rotation and which plays a role in rotational motion, similar to that played by mass in linear, i.e. straight-line, motion.

Kinetic energy of a rotating body

Suppose the body of Fig. 7.15 is rotating about an axis through O with constant angular velocity ω. A particle A, of mass m_1, at a distance r_1 from O, describes its own circular path and if v_1 is its linear velocity along the tangent to the path at the instant shown, then $v_1 = r_1\omega$ and

Fig. 7.15

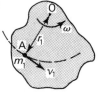

$$\text{kinetic energy of A} = \tfrac{1}{2}m_1v_1{}^2$$

$$= \tfrac{1}{2}m_1r_1{}^2\omega^2$$

281

The kinetic energy of the whole body is the sum of the kinetic energies of its component particles. If these have masses m_1, m_2, m_3, etc., and are at distances r_1, r_2, r_3, etc., from O then, since all particles have the same angular velocity ω (the body being rigid), we have

kinetic energy of whole body $= \frac{1}{2}m_1r_1{}^2\omega^2 + \frac{1}{2}m_2r_2{}^2\omega^2 + \frac{1}{2}m_3r_3{}^2\omega^2 + \cdots$

$$= \tfrac{1}{2}\omega^2(\textstyle\sum mr^2)$$

where $\sum mr^2$ represents the sum of the mr^2 values for all the particles of the body. The quantity $\sum mr^2$ depends on the mass and its distribution and is taken as a measure of the moment of inertia of the body about the axis in question. It is denoted by the symbol I and so

$$I = \sum mr^2$$

Therefore,

$$\text{kinetic energy of body} = \tfrac{1}{2}I\omega^2$$

Comparing this with the expression $\frac{1}{2}mv^2$ for linear kinetic energy we see that the mass m is replaced by the moment of inertia I and the velocity v is replaced by the angular velocity ω. The unit of I is kg m^2.

Values of I for regular bodies can be calculated (using calculus); that for a uniform rod of mass m and length l about an axis through its centre is $ml^2/12$. About an axis through its end it is $ml^2/3$.

It must be emphasized that rotational kinetic energy ($\frac{1}{2}I\omega^2$) is not a new type of energy but is simply the sum of the linear kinetic energies of all the particles of the body. It is a convenient way of stating the kinetic energy of a rotating rigid body.

The mass of a flywheel is concentrated in the rim, thereby giving it a large moment of inertia. When rotating, its kinetic energy is therefore large and explains why it is able to keep an engine (e.g. in a car) running at a fairly steady speed even though energy is supplied intermittently to it. Some toy cars have a small lead flywheel which is set into rapid rotation by a brief push across a solid surface. The kinetic energy of the flywheel will then keep the car in motion for some distance.

Work done by a couple

Rotation is changed by a couple, that is, by two equal and opposite parallel forces whose lines of action do not coincide. It is often necessary to find the work done by a couple so that the energy transfer occurring as a result of its action on a body is known.

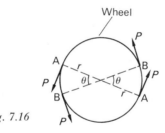

Fig. 7.16

Consider the wheel in Fig. 7.16 of radius r on which the two equal and opposite forces P act tangentially and rotation through angle θ (in radians) occurs.

$$\text{Work done by each force} = \text{force} \times \text{distance}$$
$$= P \times \text{arc AB} = P \times r\theta$$
$$\therefore \quad \text{total work done by couple} = Pr\theta + Pr\theta = 2Pr\theta$$

But, torque (or moment) of couple $= P \times 2r = 2Pr$ (p. 233). Therefore

$$\textit{work done by couple} = \textit{torque} \times \textit{angle of rotation}$$
$$= T\theta$$

For example, if $P = 2.0$ N, $r = 0.50$ m and the wheel makes 10 revolutions then $\theta = 10 \times 2\pi$ rad and $T = P \times 2r = 2.0$ N $\times 2 \times 0.50$ m $= 2$ N m. Hence, work done by couple $= T\theta = 2 \times 20\pi = 1.3 \times 10^2$ J.

In general, if a couple of torque T about a certain axis acts on a body of moment of inertia I through an angle θ about the same axis and its angular velocity increases from 0 to ω, then

$$\text{work done by couple} = \text{kinetic energy of rotation}$$
$$T\theta = \tfrac{1}{2}I\omega^2$$

Angular momentum

(*a*) *Definition*. In linear motion it is often useful to consider the (linear) momentum of a body. In rotational motion, *angular momentum* is important.

Consider a rigid body rotating about an axis O and having angular velocity ω at some instant, Fig. 7.17. Let A be a particle of this body, distant r_1 from

Fig. 7.17

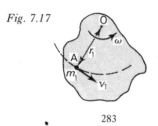

O and having linear velocity v_1 as shown, then the linear momentum of $A = m_1 v_1 = m_1 \omega r_1$ (since $v_1 = \omega r_1$).

The angular momentum of A about O is defined as the *moment of its momentum* about O. Hence

$$\text{angular momentum of A} = r_1 \times m_1 \omega r_1 = \omega m_1 r_1{}^2$$

$$\therefore \quad \text{total angular momentum of rigid body} = \sum \omega m r^2$$

$$= \omega \sum m r^2$$

$$= I \omega$$

where I is the moment of inertia of the body about O. Angular momentum is thus the analogue of linear momentum (mv), with I replacing m and ω replacing v.

(b) *Newton's second law.* A body rotates when it is acted on by a couple. The rotational form of Newton's second law of motion may be written (by analogy with $F = ma$),

$$T = I \alpha$$

where T is the torque or moment of the couple causing rotational acceleration α. In terms of momentum, the second law can be stated, for linear motion,

force = rate of change of linear momentum

i.e.
$$F = \frac{d(mv)}{dt}$$

and for rotational motion,

torque = rate of change of angular momentum

i.e.
$$T = \frac{d(I\omega)}{dt}$$

(c) *Conservation.* A similar argument to that used to deduce the principle of conservation of linear momentum from Newton's third law can be employed to derive the principle of conservation of angular momentum. It may be stated as follows.

The total angular momentum of a system remains constant provided no external torque acts on the system.

Ice skaters, ballet dancers, acrobats and divers use the principle. The diver in Fig. 7.18 leaves the high-diving board with outstretched arms and legs and some initial angular velocity about his centre of gravity. His angular

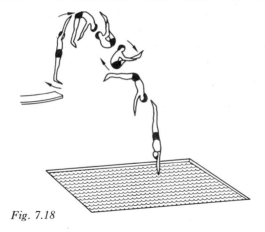

Fig. 7.18

momentum ($I\omega$) remains constant since no external torques act on him (gravity exerts no torque about his centre of gravity). To make a somersault he must increase his angular velocity. He does this by pulling in his legs and arms so that I decreases and ω therefore increases. By extending his arms and legs again, his angular velocity falls to its original value. Similarly a skater can whirl faster on ice by folding his arms.

The principle of conservation of angular momentum is useful for dealing with large rotating bodies such as the earth, as well as tiny, spinning particles such as electrons.

(*d*) *Worked example. A shaft rotating at 3.0 × 10³ revolutions per minute is transmitting a power of 10 kilowatts. Find the magnitude of the driving couple.*

Work done per second by driving couple = power transmitted by shaft.

Hence, since $1 \text{ W} = 1 \text{ J s}^{-1}$,

$$T\theta = 10 \times 10^3 \text{ J s}^{-1} \quad \text{(see p. 283)}$$

where T is the moment of the couple and θ is the angle through which the shaft rotates in 1 second. Now

$$3.0 \times 10^3 \text{ revs per minute} = 3.0 \times 10^3/60 \text{ revs per second}$$

$$= 50 \text{ revs per second}$$

$$\therefore \quad \theta = 50 \times 2\pi \text{ rad s}^{-1} \quad \text{(since } 2\pi \text{ rad} = 360° = 1 \text{ rev)}$$

$$\therefore \quad T = \frac{10 \times 10^3}{50 \times 2\pi} \text{ N m}$$

$$= 32 \text{ N m}$$

285

Kepler's laws

About 1542 the Polish monk Copernicus proposed that the earth, rather than being the centre of the universe as was generally thought, revolved round the sun, as did the other planets. This heliocentric (sun-centred) model was greatly developed by Kepler who, following on prolonged study of observations made by the Danish astronomer Tycho Brahé over a period of twenty years, arrived at a very complete description of planetary motion. He announced his first two laws in 1609 and the third in 1619.

1. *Each planet moves in an ellipse which has the sun at one focus.*
2. *The line joining the sun to the moving planet sweeps out equal areas in equal times.*
3. *The squares of the times of revolution of the planets (i.e. their periodic times T) about the sun are proportional to the cubes of their mean distances (r) from it (i.e. r^3/T^2 is a constant).*

In Fig. 7.19, if planet P takes the same time to travel from A to B as from C to D then the shaded areas are equal. Strictly speaking the distances in

Fig. 7.19

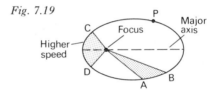

law 3 should be the semi-major axes of the ellipses but all the orbits are sufficiently circular for the mean radius to be taken. The third column of Table 7.1 shows the constancy of r^3/T^2.

Table 7.1

Planet	Mean radius of orbit r (metres)	Period of revolution T (seconds)	r^3/T^2
Mercury	5.79×10^{10}	7.60×10^6	3.36×10^{18}
Venus	1.08×10^{11}	1.94×10^7	3.35
Earth	1.49×10^{11}	3.16×10^7 (1 year)	3.31
Mars	2.28×10^{11}	5.94×10^7 (1.9 years)	3.36
Jupiter	7.78×10^{11}	3.74×10^8 (11.9 years)	3.36
Saturn	1.43×10^{12}	9.30×10^8 (29.5 years)	3.37
Uranus	2.87×10^{12}	2.66×10^9 (84.0 years)	3.34
Neptune	4.50×10^{12}	5.20×10^9 (165 years)	3.37
Pluto	5.90×10^{12}	7.82×10^9 (248 years)	3.36

Kepler's three laws enabled planetary positions, both past and future, to be determined accurately without the complex array of geometrical constructions used previously which were due to the Greeks. His work was also important because by stating his empirical laws (i.e. laws based on observation, not on theory) in mathematical terms he helped to establish the equation as a form of scientific shorthand.

Gravity and the moon

Kepler's laws summed up neatly *how* the planets of the solar system behaved without indicating *why* they did so. One of the problems was to find the centripetal force which kept a planet in its orbit round the sun, or the moon round the earth, in a way which agreed with Kepler's laws.

Newton reflected (perhaps in his garden when the apple fell) that the earth exerts an inward pull on nearby objects causing them to fall. He then speculated whether this same force of gravity might not extend out farther to pull on the moon and keep it in orbit. If it did, might not the sun also pull on the planets in the same way with the same kind of force? He decided to test the idea first on the moon's motion—as we will do now.

If r is the radius of the moon's orbit round the earth and T is the time it takes to complete one orbit, i.e. its period, Fig. 7.20, then using accepted values we have

Fig. 7.20

Earth

Moon

$$r = 3.84 \times 10^8 \text{ m}$$

$$T = 27.3 \text{ days}$$

$$= 27.3 \times 24 \times 3600 \text{ s}$$

(The time between full moons is 29.5 days but this is due to the earth also moving round the sun. The moon has therefore to travel a little farther to reach the same position relative to the sun. Judged against the background of the stars, the moon takes 27.3 days to make one complete orbit of the earth, which is its true period T.)

The speed v of the moon along its orbit (assumed circular) is

$$v = \frac{\text{circumference of orbit}}{\text{period}} = \frac{2\pi r}{T}$$

$$= \frac{2\pi \times 3.84 \times 10^8}{27.3 \times 24 \times 3600} \text{ m s}^{-1}$$

$$= 1.02 \times 10^3 \text{ m s}^{-1}$$

The moon's centripetal acceleration a will be

$$a = \frac{v^2}{r} = \frac{(1.02 \times 10^3 \text{ m s}^{-1})^2}{3.84 \times 10^8 \text{ m}}$$

$$= 2.72 \times 10^{-3} \text{ m s}^{-2}$$

The acceleration due to gravity at the earth's surface is 9.81 m s^{-2} and so if gravity is the centripetal force for the moon it must weaken between the earth and the moon. The simplest assumption would be that gravity halves when the distance doubles and at the moon it would be 1/60 of 9.81 m s^{-2} since the moon is 60 earth-radii from the centre of the earth and an object at the earth's surface is 1 earth-radius from the centre. But 9.81/60 = 1.64 × 10^{-1} m s^{-2}, which is still too large.

The next relation to try would be an inverse square law in which gravity is one-quarter when the distance doubles, one-ninth when it trebles and so on. At the moon it would be $1/60^2$ of 9.81 m s^{-2}, i.e. 9.81/3600 = 2.72 × 10^{-3} m s^{-2}—the value of the moon's centripetal acceleration.

Law of universal gravitation

Having successfully tested the idea of inverse square law gravity for the motion of the moon round the earth, Newton turned his attention to the solar system.

His proposal, first published in 1687 in his great work the *Principia* (Mathematical principles of natural knowledge), was that the centripetal force which keeps the planets in orbit round the sun is provided by the gravitational attraction of the sun for the planets. This, according to Newton, was the same kind of attraction as that of the earth for an apple. Gravity— the attraction of the earth for an object—was thus a particular case of gravitation. In fact, Newton asserted that every object in the universe attracted every other object with a gravitational force and that this force was responsible for the orbital motion of celestial (heavenly) bodies.

Newton's hypothesis, now established as a theory and known as the *law of universal gravitation,* may be stated quantitatively as follows.

Every particle of matter in the universe attracts every other particle with a force which is directly proportional to the product of their masses and inversely proportional to the square of their distances apart.

The gravitational attraction F between two particles of masses m_1 and m_2, distance r apart is thus given by

$$F \propto \frac{m_1 m_2}{r^2} \qquad \text{or} \qquad F = G \frac{m_1 m_2}{r^2}$$

where G is a constant, called the *universal gravitational constant*, and assumed to have the same value everywhere for all matter.

Newton believed the force was directly proportional to the mass of each particle because the force on a falling body is proportional to its mass ($F = ma = mg = m \times$ constant, therefore $F \propto m$), i.e. to the mass of the *attracted* body. Hence, from the third law of motion, he argued that since the falling body also attracts the earth with an equal and opposite force that is proportional to the mass of the earth, then the gravitational force between the bodies must also be proportional to the mass of the *attracting* body. The moon test justified the use of an inverse square law relation between force and distance.

The law applies to *particles* (i.e. bodies whose dimensions are very small compared with other distances involved), but Newton showed that the attraction exerted at an external point by a sphere of uniform density (or a sphere composed of uniform concentric shells) was the same as if its whole mass were concentrated at its centre. We tacitly assumed this for the earth in the previous section and will use it in future.

The gravitational force between two ordinary objects (say two 1 kg masses 1 metre apart) is extremely small and therefore difficult to detect. What does this indicate about the value of G in SI units? What will be the units of G in the SI system?

Testing gravitation

To test $F = Gm_1m_2/r^2$ for the sun and planets the numerical values of all quantities on both sides of the equation need to be known. Newton neither had reliable information about the masses of the sun and planets nor did he know the value of G and so he could not adopt this procedure. There are alternatives however.

(*a*) *Deriving Kepler's laws.* The behaviour of the solar system is summarized by Kepler's laws and any theory which predicts these would, for a start, be in agreement with the facts.

Suppose a planet of mass m moves with speed v in a circle of radius r round the sun of mass M, Fig. 7.21. Hence

Fig. 7.21

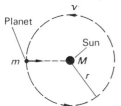

gravitational attraction
of sun for planet $= G\dfrac{Mm}{r^2}$

If this is the centripetal force keeping the planet in orbit then

$$G\frac{Mm}{r^2} = \frac{mv^2}{r}$$

$$\therefore \quad \frac{GM}{r} = v^2$$

If T is the time for the planet to make one orbit

$$v = \frac{2\pi r}{T}$$

$$\therefore \quad \frac{GM}{r} = \frac{4\pi^2 r^2}{T^2}$$

$$\therefore \quad GM = \frac{4\pi^2 r^3}{T^2}$$

Hence

$$\frac{r^3}{T^2} = \frac{GM}{4\pi^2}$$

Since GM is constant for any planet, r^3/T^2 is constant, which is Kepler's third law. We have considered a circular orbit but more advanced mathematics gives the same result for an elliptical one.

The first law can be derived by showing that if inverse square law gravitation holds, a planet moves in an orbit which is a conic section (i.e. a circle, ellipse, parabola or hyperbola) with the sun at one focus. Also, it may be shown that when a planet is acted on by *any* force, not just an inverse square law one, directed from the planet towards the sun, the radius covers equal areas in equal times—which is the second law.

(*b*) *Discovery of other planets.* Theories can never be proved correct, they are only disproved by making predictions which conflict with observations. A good theory should lead to new discoveries, as Newton's theory did.

The planets must exert gravitational pulls on one another, but except in the case of the larger planets like Jupiter and Saturn, the effect is only slight. The French scientist Laplace showed after Newton's time how to predict the effect of these disturbances (called perturbations) on Kepler's simple elliptical orbits.

The planet Uranus, discovered in 1781, showed small deviations from its expected orbit even after allowance had been made for the effects of known neighbouring planets. Two astronomers, Adams in England and Leverrier in France, working quite independently, predicted from the law of gravitation, the position, size and orbit of an unknown planet that could cause the observed perturbations. A search was made and the new planet located in 1846 in the predicted position by the Berlin Observatory. Thus Neptune was discovered.

In 1930, history was repeated when American astronomers discovered Pluto from perturbations of the orbit of Neptune.

Masses of the sun and planets

The theory of gravitation enables us to obtain information about the mass of any celestial body having a satellite. If the value of the gravitational constant G is known, the actual mass can be calculated. Otherwise only a comparison is possible. A determination of G was not made until after Newton's death.

The principle is simply to measure all the quantities in $F = Gm_1m_2/r^2$ except G which can then be calculated. The earliest determinations used a measured mountain as the 'attracting' mass and a pendulum as the 'attracted' one. The first laboratory experiment was performed by Cavendish in 1798. He measured the very small gravitational forces exerted on two small lead balls (m_1 and m_2) by two larger ones (M_1 and M_2) using a torsion balance, Fig. 7.22. In this, the force twists a calibrated wire. Modern measurements give the value

$$G = 6.7 \times 10^{-11} \text{ N m}^2 \text{ kg}^{-2}$$
$$(\text{or m}^3 \text{ s}^{-2} \text{ kg}^{-1})$$

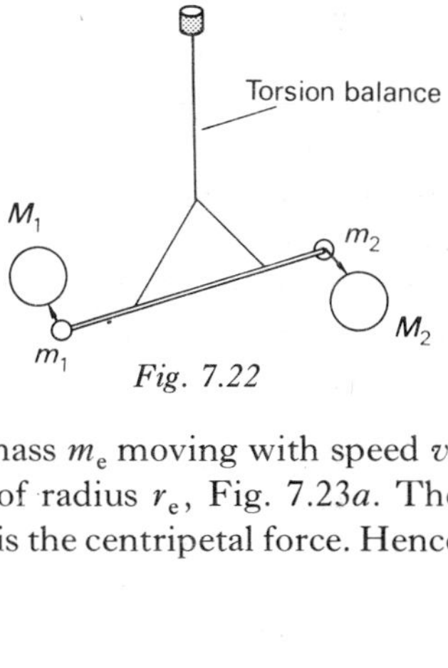

Torsion balance

M_1

m_2

M_2

m_1

Fig. 7.22

(a) *Mass of the sun.* Consider the earth of mass m_e moving with speed v_e round the sun of mass m_s in a circular orbit of radius r_e, Fig. 7.23a. The gravitational attraction of the sun for the earth is the centripetal force. Hence

$$G\frac{m_s m_e}{r_e^{2}} = \frac{m_e v_e^{2}}{r_e}$$

$$\therefore \quad m_s = \frac{v_e^{2} r_e}{G}$$

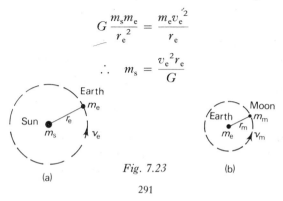

Earth

m_e

Sun

r_e

m_s

v_e

Moon

Earth

m_m

r_m

m_e

v_m

Fig. 7.23

(a)

(b)

291

If T_e is the time for the earth to make one orbit, then

$$v_e = \frac{2\pi r_e}{T_e} \quad \text{and} \quad m_s = \frac{4\pi^2}{G} \cdot \frac{r_e^3}{T_e^2}$$

Substituting for G, $r_e = 1.5 \times 10^{11}$ m and $T_e = 3.0 \times 10^7$ s (1 year), we get

$$m_s = 2.0 \times 10^{30} \text{ kg}$$

(b) *Mass of the earth.* Considering the moon of mass m_m moving with speed v_m round the earth of mass m_e in a circular orbit of radius r_m, Fig. 7.23b, we similarly obtain

$$G\frac{m_e m_m}{r_m^2} = \frac{m_m v_m^2}{r_m}$$

$$\therefore \quad m_e = \frac{v_m^2 r_m}{G}$$

Also if T_m is the period of the moon then $v_m = 2\pi r_m / T_m$ and so

$$m_e = \frac{4\pi^2}{G} \cdot \frac{r_m^3}{T_m^2}$$

Substituting for G, $r_m = 4.0 \times 10^8$ m and $T_m = 2.4 \times 10^6$ s (1·month) we find that

$$m_e = 6.0 \times 10^{24} \text{ kg}$$

The ratio of the mass of the sun to that of the earth is $2.0 \times 10^{30} : 6.0 \times 10^{24}$, i.e. 330 000:1. Table 7.2 gives the relative masses and densities of various bodies.

Table 7.2

	Mass (earth = 1)	Density (water = 1)
Sun	330 000	1.4
Moon	0.012	3.3
Mercury	0.056	6.1
Venus	0.82	5.1
Earth	1.0	5.5
Mars	0.11	4.1
Jupiter	320	1.4
Saturn	95	0.7
Uranus	15	1.6
Neptune	17	2.3
Pluto	0.8 (?)	?

Newton's work

(*a*) *Scientific explanation.* The charge is sometimes made that science does not get down to underlying causes and give the 'true' reasons. In many cases this is so. Newton's work raises the question of what is meant by scientific explanation.

Consider gravitation. Newton did not really explain why a body falls or why the planets move round the sun. He attributed these effects to something called 'gravitation' and this, like other basic scientific ideas, seems by its very nature to defy explanation in any simpler terms. It appears that we must accept it as a fundamental concept of science which is very useful because it enables us to regard apparently different phenomena—the falling of an apple and the motion of the planets—as having the same 'cause'.

A scientific explanation is very often an idea or concept that provides a connecting link between effects and so simplifies our knowledge. Explanations in terms of such concepts as energy, momentum, molecules, atoms, electrons, fall into this category. Concepts which do not cast their net wide are of little value in science.

(*b*) *Influence of Newton's work.* Starting from the laws of motion and gravitation Newton created a model of the universe which explained known facts, led to new discoveries and produced a unified body of knowledge. He united the physics of 'heaven and earth' by the same set of laws and so brought to a grand climax the work begun by Copernicus, Kepler and Galileo.

The success of Newtonian mechanics had a profound influence on both scientific and philosophical thought for 200 years. There arose a widespread belief that using scientific laws the future of the whole universe could be predicted if the positions, velocities and accelerations of all the particles in it were known at a certain time. This 'mechanistic' outlook regarded the universe as a giant piece of clockwork, wound up initially by the 'divine power' and now ticking over according to strict mathematical laws.

Today scientists are humbler and probability has replaced certainty. Also, whilst Newtonian mechanics is still perfectly satisfactory for the world of ordinary experience, it has been supplemented by two other theories. The *theory of relativity* has joined it for situations in which bodies are moving at very high speeds and *quantum mechanics* enables us to deal with the physics of the atom.

Earth's gravitational field

An action-at-a-distance effect in which one body A exerts a force on another body B not in contact with it, can be regarded as due to a 'field of force' in the region around A. The *field* may be considered to be the *interpretation* and the *force* on B the *observation*. (Body A will of course experience a force by being in the field due to B.)

We can think of the sun and all other celestial and terrestrial bodies as each having a gravitational field which exerts a force on any other body in the field. The *strength of a gravitational field is defined as the force acting on unit mass placed in the field*. Thus if a body of mass m experiences a force F when in the earth's field, the strength of the earth's field is F/m (in newtons per kilogram). Measurement shows that if $m = 1$ kg, then $F = 9.8$ N (at the earth's surface); the strength of the earth's field is therefore 9.8 N kg^{-1}. However if a mass m falls freely under gravity its acceleration g would be $F/m = 9.8$ m s^{-2} (since $F = ma = mg$).

We thus have two ways of looking at g. When considering bodies falling freely we can think of it as an acceleration (of 9.8 m s^{-2}), but when a body of known mass is *at rest* or is *unaccelerated* in the earth's field and we wish to know the gravitational force (in newtons) acting on it we regard g as the earth's gravitational field strength (of 9.8 N kg^{-1}).

At our level of study the *field* concept tends to be more useful when dealing with electric and magnetic effects whilst the *force* concept is generally employed for gravitational effects.

Acceleration due to gravity

(a) *Relation between g and G.* A body of mass m at a place on the earth's surface where the acceleration due to gravity is g, experiences a force $F = mg$ (i.e. its weight) due to its attraction by the earth, Fig. 7.24. Assuming the earth behaves as if its whole mass M were concentrated at its centre O, then, by the law of gravitation, we can also say that F is the gravitational pull of the earth on the body. Hence

$$F = G\frac{Mm}{r^2}$$

where r is the radius of the earth.

$$\therefore \quad mg = G\frac{Mm}{r^2}$$

$$\therefore \quad g = \frac{GM}{r^2}$$

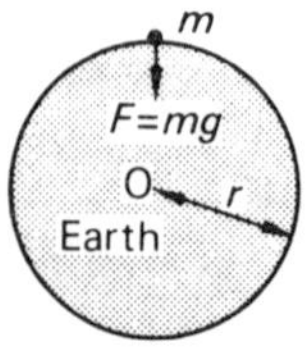

Fig. 7.24

It is worth noting that the mass m in $F = ma = mg$ is called the *inertial mass* of the body; it measures the opposition of the body to change of motion, i.e. its inertia. The mass of the same body when considering the law of gravitation is known as the *gravitational mass*. Experiments show that to a high degree of accuracy these two masses are equal for a given body and so we can, as we have done here, represent each by *m*.

(*b*) *Variation of g with height*. If g' is the acceleration due to gravity at a distance *a* from the centre of the earth where $a > r$, r being the earth's radius, then from (*a*),

$$g' = \frac{GM}{a^2} \quad \text{and} \quad g = \frac{GM}{r^2}$$

Dividing

$$\frac{g'}{g} = \frac{r^2}{a^2}$$

or

$$g' = \frac{r^2}{a^2} g$$

Above the earth's surface, the acceleration due to gravity g' thus varies inversely as the square of the distance *a* from the centre of the earth (since *r* and *g* are constant), i.e. it decreases with height as shown in Fig. 7.25.

Fig. 7.25   Using computed values of density

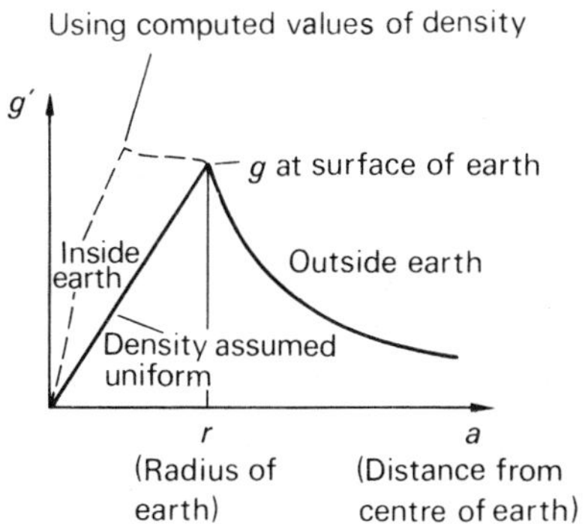

At height *h* above the surface, $a = r + h$

$$\therefore \quad g' = \frac{r^2}{(r + h)^2} g = \frac{1}{(1 + h/r)^2} g$$

$$= \left(1 + \frac{h}{r}\right)^{-2} g$$

If *h* is very small compared with *r* (6400 km) we can neglect powers of (*h/r*) higher than the first. Hence

$$g' = \left(1 - \frac{2h}{r}\right) g$$

(c) *Variation of g with depth.* At a point such as P below the surface of the earth, it can be shown that if the shaded spherical shell in Fig. 7.26 has uniform density, it produces no gravitational field inside itself. The gravi-

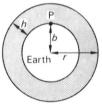

Fig. 7.26

tational acceleration g_1 at P is then due entirely to the sphere of radius b and if this is assumed to be of uniform density, then from (a)

$$g_1 = \frac{GM_1}{b^2} \quad \text{and} \quad g = \frac{GM}{r^2}$$

where M_1 is the mass of the sphere of radius b. The mass of a uniform sphere is proportional to its radius cubed, hence

$$\frac{M_1}{M} = \frac{b^3}{r^3}$$

But

$$\frac{g_1}{g} = \frac{M_1}{M} \cdot \frac{r^2}{b^2}$$

$$\therefore \quad \frac{g_1}{g} = \frac{b}{r} \quad \text{or} \quad g_1 = \frac{b}{r} g$$

Thus, assuming the earth has uniform density, the acceleration due to gravity g_1 is directly proportional to the distance b from the centre, i.e. it decreases linearly with depth, Fig. 7.25. At depth h below the earth's surface, $b = r - h$

$$\therefore \quad g_1 = \left(\frac{r - h}{r}\right) g = \left(1 - \frac{h}{r}\right) g$$

In fact, because the earth's density is not constant, g_1 actually *increases* for all depths now obtainable as shown by part of the dotted curve in Fig. 7.25.

(d) *Variation of g with latitude.* The observed variation of g over the earth's surface is largely due to (i) the equatorial radius of the earth exceeding its polar radius by about 21 km and thereby making g greater at the poles than at the equator where a body is farther from the centre of the earth, and (ii) the effect of the earth's rotation which we will now consider.

A body of mass m at any point of the earth's surface (except at the poles) must have a centripetal force acting on it. This force is supplied by part of the earth's gravitational attraction for it. On a stationary earth the gravitational pull of the earth on m would be mg where g is the acceleration due to gravity under such conditions. However, because of the earth's rotation, the observed gravitational pull is less than this and equals mg_0 where g_0 is the *observed* acceleration due to gravity. Hence

$$\text{centripetal force on body} = mg - mg_0$$

At the equator, the body is moving in a circle of radius r where r is the earth's radius and it has the same angular velocity ω as the earth. The centripetal force is then $m\omega^2 r$ and so

$$mg - mg_0 = m\omega^2 r$$

$$\therefore \quad g - g_0 = \omega^2 r$$

Substituting $r = 6.4 \times 10^6$ m and $\omega = 1$ revolution in 24 hours $= 2\pi/(24 \times 3600)$ rad s^{-1}, we get $g - g_0 = 3.4 \times 10^{-2}$ m s^{-2}. Assuming the earth is perfectly spherical this is also the difference between the polar and equatorial values of the acceleration due to gravity. (At the poles $\omega = 0$ and so $g = g_0$.) The observed difference is 5.2×10^{-2} m s^{-2}, of which 1.8×10^{-2} m s^{-2} arises from the non-sphericity of the earth.

At latitude θ on an assumed spherical earth, the body describes a circle of radius $r \cos \theta$, Fig. 7.27a. The magnitude of the centripetal force required is thus $m\omega^2 r \cos \theta$ and is smaller than at the equator since ω has the same value. However its direction is along PQ whereas mg acts along PO towards

Fig. 7.27

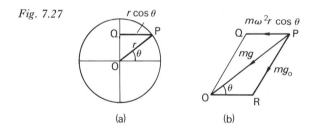

(a) (b)

the centre of the earth. The observed gravitational pull mg_0 is therefore less than mg by a force $m\omega^2 r \cos \theta$ along PQ and will be in a different direction from mg. The value and direction of mg_0 must be such that when it is compounded by the parallelogram law with $m\omega^2 r \cos \theta$ along PQ, it gives mg along PO, Fig. 7.27b. The direction of g_0 as shown by a falling body or a plumb line is not exactly towards the centre of the earth except at the poles and the equator.

CIRCULAR MOTION AND GRAVITATION

Artificial satellites

(a) *Satellite orbits*. The centripetal force which keeps an artificial satellite in orbit round the earth is the gravitational attraction of the earth for it. For a satellite of mass m travelling with speed v in a circular orbit of radius R (measured from the centre of the earth), we have

$$\frac{mv^2}{R} = \frac{GMm}{R^2}$$

where M is the mass of the earth.

$$\therefore \quad v^2 = \frac{GM}{R}$$

But
$$g = \frac{GM}{r^2} \qquad \text{(see p. 294)}$$

where r is the radius of the earth and g is the acceleration due to gravity at the earth's surface.

$$\therefore \quad v^2 = \frac{gr^2}{R}$$

If the satellite is close to the earth, say at a height of 100–200 km, then $R \simeq r$ and

$$v^2 = gr$$

Substituting $r = 6.4 \times 10^6$ m (6400 km) and $g = 9.8$ m s^{-2}

$$v = \sqrt{gr} = \sqrt{9.8 \text{ m s}^{-2} \times 6.4 \times 10^6 \text{ m}}$$

$$= 7.9 \times 10^3 \text{ m s}^{-1}$$

$$\therefore \quad v \simeq 8 \text{ km s}^{-1}$$

The time for the satellite to make one complete orbit of the earth, i.e. its period T, is

$$T = \frac{\text{circumference of earth}}{\text{speed}} = \frac{2\pi r}{v}$$

$$= \frac{2\pi \times 6.4 \times 10^6 \text{ m}}{7.9 \times 10^3 \text{ m s}^{-1}}$$

$$\therefore \quad T \simeq 5000 \text{ s} \simeq 83 \text{ minutes}$$

We can regard a satellite in orbit as being continually pulled in by gravity from a straight-line tangent path to a circular path, Fig. 7.28. It 'falls' again

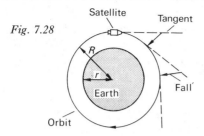

Fig. 7.28

and again from the tangents instead of continuing along them; its horizontal speed is such that it 'falls' by the correct distance to keep it in a circle. Although the satellite has an acceleration towards the centre of the earth, i.e. in a vertical direction, it has no vertical velocity because it 'falls' at the same rate as the earth's surface falls away underneath it. With respect to the earth's surface its velocity in a vertical direction is zero since the distance between the satellite and the earth's surface remains constant. In practice it is very difficult to achieve an exactly circular orbit.

(*b*) *Launching a satellite.* To be placed in orbit a satellite must be raised to the desired height and given the correct speed and direction by the launching rocket. A typical launching sequence using a two-stage rocket might be as follows.

At lift-off, the rocket, with a manned or unmanned space capsule on top, is held down by clamps on the launching pad for a few seconds until the exhaust gases have built up an upward thrust which exceeds the rocket's weight. The clamps are then removed by remote control and the rocket accelerates upwards. Fig. 7.29 shows the lift-off of a Saturn V rocket and a manned Apollo space capsule on their way to earth orbit before a trip to land astronauts on the moon. To penetrate the dense lower part of the atmosphere by the shortest possible route, the rocket rises vertically initially and after this is gradually tilted by the guidance system. The first-stage rocket, which may burn for about 2 minutes producing a speed of 3 km s^{-1} or so, lifts the vehicle to a height of around 60 km, then separates and falls back to earth, landing many kilometres from the launching site.

The vehicle now coasts in free flight (unpowered) to its orbital height, say 160 km, where it is momentarily moving horizontally (i.e. parallel to earth's surface immediately below). The second-stage rocket then fires and increases the speed to that required for a circular orbit at this height (about 8 km s^{-1}). By firing small rockets, the capsule is separated from the second stage which follows behind, also in orbit.

The equation $v^2 = gr^2/R$ (p. 298) for circular orbits shows that each orbit requires a certain speed and the greater the orbit radius R the smaller the speed v.

Fig. 7.29

Fig. 7.30

Some notable space flights are given in Table 7.3. Synchronous satellites have a period of 24 hours, exactly the same as that of the earth, and so remain in the same position above the earth, apparently stationary. By acting as relay stations, they make continuous, world-wide communications (e.g. of television programmes) possible. Fig. 7.30 shows an American earth-orbiting space station (Skylab), from which manned vehicles will be used for journeys to and from the earth.

Table 7.3

Name	Launch date	Descent date or lifetime	Period (min)	Height (km)	Notes
Sputnik 1	4 Oct 1957	4 Jan 1968 (?)	96.2	215–939	Mass 83.6 kg First artificial earth satellite
Explorer 1	31 Jan 1958	11 years	114.8	356–2548	Mass 14.0 kg Discovered inner Van Allen belt
Vostok 1	12 Apr 1961	12 Apr 1961	89.3	169–315	First manned space-flight. One orbit by Yuri Gagarin
Mercury 6	20 Feb 1962	20 Feb 1962	88.6	159–265	First U.S. manned spaceflight. Three orbits by John Glenn
Early Bird	6 Apr 1965	10^6 years	1437	35 000–36 000	First commercial synchronous communications satellite: 'stationary' between Africa and South America
Pageos A	24 June 1966	50 years	181.4	4207–4271	Inflated sphere 30 metres in diameter. Visible to the naked eye at times given in national daily newspapers
Apollo 11	16 July 1969	24 July 1969			First men to land on the moon and return to earth with 'moon samples'

Weightlessness

An astronaut orbiting the earth in a space vehicle with its rocket motors off is said to be 'weightless'. If weight means the pull of the earth on a body, then the statement, although commonly used, is misleading. A body is not truly weightless unless it is outside the earth's (or any other) gravitational field, i.e. at a place where $g = 0$. In fact it is gravity which keeps an astronaut and his vehicle in orbit. To appreciate what 'experiencing the sensation of weightlessness' means we will consider similar situations on earth.

We are made aware of our weight because the ground (or whatever supports us) exerts an *upward* push on us as a result of the *downward* push our feet exert on the ground. It is thus upward push which makes us 'feel' the force of gravity. When a lift suddenly starts upwards the push of the floor on our feet increases and we feel heavier. On the other hand if the support is reduced we seem to be lighter. In fact *we judge our weight from the upward push exerted on us by the floor*. If our feet are completely unsupported we experience weightlessness. Passengers in a lift which has a continuous downward acceleration equal to g would get no support from the floor since they, too, would be falling with the same acceleration as the lift. There is no upward push on them and so no sensation of weight is felt. The condition is experienced briefly when we jump off a wall or dive into a swimming pool.

An astronaut in an orbiting space vehicle is not unlike a passenger in a freely falling lift. The astronaut is moving with *constant speed* along the orbit, but since he is travelling in a circle he has a centripetal acceleration— of the same value as that of his space vehicle and equal to g at that height. The walls of the vehicle exert no force on him, he is unsupported, the physiological sensation of weight disappears and he floats about 'weightless'. Similarly any object released in the vehicle does not 'fall'; anything not in use must be firmly fixed and liquids will not pour. Summing up, to be strictly correct we should not use the term 'weightless' unless by weight we mean the force exerted on (or by) a body on its support and generally we do not.

It is important to appreciate that although 'weightless' a body still has mass and it would be just as difficult to push it in space as on earth. An astronaut floating in his vehicle could still be injured by hitting a hard but weightless object. Fig. 7.31 is an illustration from Jules Verne's book *From the earth to the moon*, published in 1865; it shows spaceship passengers in a truly weightless condition. They were not in orbit round the earth or moon, where *exactly* were they?

Speed of escape

The faster a ball is thrown upwards the higher does it rise before it is stopped and pulled back by gravity. To escape from the earth into outer space we will show shortly that an object must have a speed of just over 11 km s^{-1} —called the *escape speed*.

The multi-stage rockets in use at present burn their fuel in a comparatively short time to obtain the best performance. They behave rather like objects thrown upwards, i.e. like projectiles. For a journey to, say, the moon, they therefore have to attain the escape speed. In many cases this is done by

Fig. 7.31

first putting the final vehicle into a 'parking' orbit round the earth with a speed of 8 km s^{-1} and then firing the final rocket again to reach escape speed in the appropriate direction.

The attainment of the escape speed is not a necessary condition however. The essential thing is that a certain amount of *energy* is required to escape from the earth and if rockets were available which could develop large power over a long time, escape would still be possible without ever achieving escape speed. In fact if we had a long enough ladder and the necessary time and energy we could walk to the moon!

The escape speed is obtained from the fact that the potential energy gained by the body equals its loss of kinetic energy, if air resistance is neglected. The work done measures the energy change. Let m be the mass of the escaping body and M the mass of the earth. The force F exerted on the object by the earth when it is distance x from the centre of the earth is

$$F = G\frac{Mm}{x^2}$$

Therefore work done δW by gravity when the body moves a further short distance δx upwards is

$$\delta W = -F\,\delta x = -G\frac{Mm}{x^2}\,\delta x$$

(the negative sign shows the force acts in the opposite direction to the displacement, see p. 260). Therefore

total work done while body escapes $= \displaystyle\int_r^\infty -G\frac{Mm}{x^2}\,dx$ (r = radius of earth)

$$= -GMm\left[-\frac{1}{x}\right]_r^\infty = GMm\left[\frac{1}{x}\right]_r^\infty$$

$$= -\frac{GMm}{r}$$

If the body leaves the earth with speed v and just escapes from its gravitational field

$$\tfrac{1}{2}mv^2 = \frac{GMm}{r}$$

$$\therefore\ v = \sqrt{\frac{2GM}{r}}$$

But $\qquad\qquad g = \dfrac{GM}{r^2}$ (p. 294)

$$\therefore\ v = \sqrt{2gr}$$

Substituting $r = 6.4 \times 10^6$ m and $g = 9.8$ m s^{-2}, we get $v \simeq 11$ km s^{-1}.

Possible paths for a body projected at different speeds from the earth are shown in Fig. 7.32.

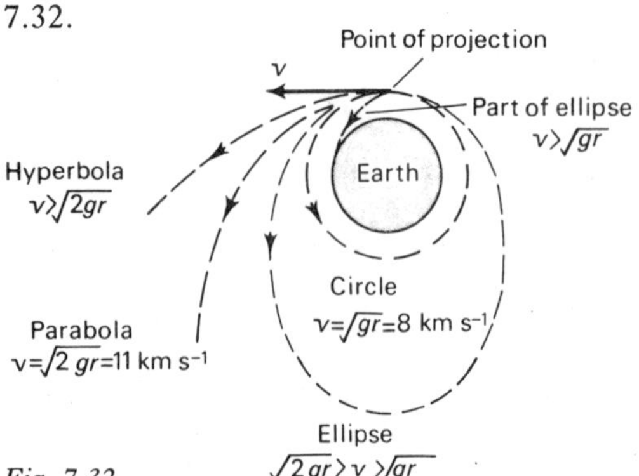

Fig. 7.32

304

Air molecules at s.t.p. have an average speed of about 0.5 km s^{-1} which, being much less than the escape speed, ensures that the earth's gravitational field is able to maintain an atmosphere of air round the earth. The average speed of hydrogen molecules at s.t.p. is more than three times that of air molecules and explains their rarity in the earth's atmosphere. The moon has no atmosphere. Can you suggest a possible reason?

QUESTIONS

Assume $g = 10$ m s^{-2} unless stated otherwise.

Circular motion

1. A particle moves in a semicircular path AB of radius 5.0 m with constant speed 11 m s^{-1}, Fig. 7.33. Calculate (a) the time taken to travel from A to B (take $\pi = 22/7$), (b) the average velocity, (c) the average acceleration.

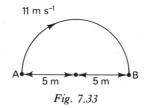

11 m s^{-1}

A 5 m   5 m B

Fig. 7.33

2. The turntable of a record player makes 45 revolutions per minute. Calculate (a) its angular velocity in rad s^{-1}, (b) the linear velocity of a point 0.12 m from the centre.

3. What is meant by a *centripetal force*? Why does such a force do no work in a circular orbit?

   (a) An object of mass 0.50 kg on the end of a string is whirled round in a horizontal circle of radius 2.0 m with a constant speed of 10 m s^{-1}. Find its angular velocity and the tension in the string.

   (b) If the same object is now whirled in a vertical circle of the same radius with the same speed, what are the maximum and minimum tensions in the string?

4. Explain exactly how the centripetal acceleration is caused in the following cases:

   (a) a train on a circular track;

   (b) a conical pendulum.

   A small bob of mass 0.1 kg is suspended by an inextensible string of length 0.5 m and is caused to rotate in a horizontal circle of radius 0.4 m which has its centre vertically below the point of suspension. Show on a sketch the two forces acting on the bob as seen by an outside observer. Find (i) the resultant of these forces, (ii) the period of rotation of the bob.　　　　　　　　　　　　　　　　　　　　　　　　(C. *part qn.*)

5. Explain why a particle moving with uniform speed in a circle must be acted upon by a centripetal force. Derive from first principles an expression for the magnitude of this force.

A pilot of mass 84.0 kg loops the loop (i.e. executes a vertical circle) in his aircraft at a steady speed of 300 km hr^{-1}. Account for the centripetal force acting on him at (a) the highest, (b) the lowest points of the circle and calculate its value. Also calculate the magnitude of the force with which the pilot is pressed into the seat at the highest and lowest points of the loop if the radius of the circle is 0.580 km. (J.M.B.)

Moment of inertia

6. A bicycle wheel has a diameter of 0.50 m, a mass of 0.80 kg and a moment of inertia about its axle of 4.0×10^{-2} kg m^2.

Assuming $\pi = 22/7$ find the values of the following quantities when the wheel rolls, at 7 rotations per second without slipping, over a horizontal surface:

(a) the angular velocity in radians per second,

(b) the linear velocity of the centre of gravity,

(c) the instantaneous linear velocity of the topmost point on the wheel,

(d) the total kinetic energy of the wheel. (*Hint*: the wheel has both rotational and translational kinetic energy.) (J.M.B.)

7. Explain, in non-mathematical terms, the physical significance of *moment of inertia*. Illustrate your answer by reference to two examples in which moments of inertia are involved.

A flywheel of moment of inertia 6.0×10^{-3} kg m^2 is rotating with an angular velocity of 20 rad s^{-1}. Calculate the steady couple required to bring it to rest in 10 revolutions. (*L. part qn.*)

8. Derive an expression for the kinetic energy of a rotating rigid body.

An electric motor supplies a power of 5×10^2 W to drive an unloaded flywheel of moment of inertia 2 kg m^2 at a steady speed of 6×10^2 revolutions per minute. How long will it be before the flywheel comes to rest after the power is switched off assuming the frictional couple remains constant?

9. (a) Define *angular momentum*. State the principle of conservation of angular momentum.

A figure skater is spinning about a vertical axis with his arms extended vertically upwards. Will he spin faster or slower when he allows his arms to fall until they are horizontal? Has his kinetic energy been increased or decreased? How do you account for the change?

(b) A horizontal disc rotating freely about a vertical axis makes 90 revolutions per minute. A small piece of putty of mass 2.0×10^{-2} kg falls vertically on to the disc and sticks to it at a distance of 5.0×10^{-2} m from the axis. If the number of revolutions per minute is thereby reduced to 80, calculate the moment of inertia of the disc.

Gravitation

10. The mass of the earth is 5.98×10^{24} kg and the gravitational constant is 6.67×10^{-11} m^3 kg^{-1} s^{-2}. Assuming the earth is a uniform sphere of radius 6.37×10^6 m, find the gravitational force on a mass of 1.00 kg at the earth's surface.

CIRCULAR MOTION AND GRAVITATION

11. Define moment of inertia and angular momentum.

A small planet, mass m, moves in an elliptical orbit round a large sun, mass M, which is at the focus F_1 of the ellipse, Fig. 7.34.

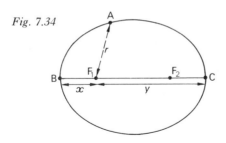

Fig. 7.34

Write down an expression for the force acting on the planet when it is at a position A, at a distance r from the sun. Indicate the direction of this force on a copy of Fig. 7.34, and also mark on it the directions of the planet's velocity and acceleration when at A. What is the magnitude of this acceleration?

According to Kepler's second law, the planet is moving faster at B than at C. Account for this with reference to the principle of conservation of energy.

What is the moment of inertia of the planet, when at B, about F_1? The velocities at B and C are v_B and v_C. Use the principle of conservation of angular momentum to deduce the ratio v_B/v_C. (*S.*)

12. It is proposed to place a communications satellite in a circular orbit round the equator at a height of 3.59×10^7 m above the earth's surface. Find the period of revolution of the satellite in hours and comment on the result. (Use the values given in *Question 10* for the radius and mass of the earth and the constant of gravitation.)

13. State Newton's Law of Gravitation. If the acceleration due to gravity, g_m, at the Moon's surface is 1.70 m s^{-2} and its radius is 1.74×10^6 m, calculate the mass of the Moon.

To what height would a signal rocket rise on the Moon, if an identical one fired on Earth could reach 200 m? (Ignore atmospheric resistance.) Explain your reasoning.

Explain, using algebraic symbols and stating which quantity each represents, how you could calculate the distance D of the Moon from the Earth (Mass M_e) if the Moon takes t seconds to move once round the Earth.

What is meant by 'weightlessness', experienced by an astronaut orbiting the Earth, and how is it caused? Explain also whether he would have the same experience when falling freely back to Earth in his capsule just prior to re-entry in the Earth's atmosphere. (Gravitational constant $= 6.67 \times 10^{-11}$ m^3 kg^{-1} s^{-2}; acceleration due to gravity $= 9.81$ m s^{-2}.) (*S.*)

14. Explaining each step in your calculation and pointing out the assumptions you make, use the information below to estimate the mean distance of the moon from the earth.

Period of rotation of the moon around the earth $= 27.3$ days
Radius of earth $= 6.37 \times 10^3$ km
Acceleration due to gravity at earth's surface, $g = 9.81$ m s^{-2} (*J.M.B. part qn.*)

15. The graph (Fig. 7.35) shows how the force of attraction on a 1 kg mass towards the earth varies with its distance from the centre of the earth. It shows that the force at a distance equal to the radius R of the earth is 10 newtons.

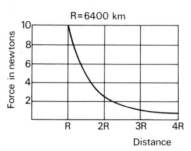

Fig. 7.35

(*a*) Calculate the force on the mass at distances of (*i*) $9R$ (*ii*) $10R$ from the earth's centre.

(*b*) Shade in on (a copy of) the graph an area which gives the energy change in moving the mass from $3R$ to $2R$.

(*c*) Make a rough estimate of the increase in kinetic energy when the mass falls from $10R$ to $9R$. (*O. and C. Nuffield*)

8 Mechanical oscillations

Introduction

In previous chapters linear and circular motion were considered. Another common type of motion is the to-and-fro repeating movement called a *vibration* or *oscillation*.

Examples of oscillatory motion are provided by a swinging pendulum, the balance wheel of a watch, a mass on the end of a vibrating spring, the strings and air columns of musical instruments when producing a note.

Fig. 8.1a

Fig. 8.1b

Sound waves are transmitted by the oscillation of the particles of the medium in which the sound is travelling. We also believe that the atoms in a solid vibrate about fixed positions in their lattice.

Engineers as well as scientists need to know about vibrations. They can occur in turbines, aircraft, cars, tall buildings and chimneys and were responsible for the collapse of the Tacoma Narrows suspension bridge in America in 1940 when a moderate gale set the bridge oscillating until the main span broke up, Fig. 8.1 *a* and *b*. In metal structures they can cause fatigue failure (p. 43). ·

In a mechanical oscillation there is a continual interchange of potential and kinetic energy due to the system having (*i*) *elasticity* (or springiness) which allows it to store p.e. and (*ii*) *mass* (or inertia) which enables it to have k.e. Thus, when a body on the lower end of a spiral spring, Fig. 8.2, is pulled

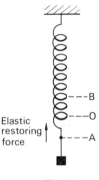

Elastic restoring force

- - - B
- - - O
- - - A

Fig. 8.2

down and released, the elastic restoring force pulls the body up and it accelerates towards its equilibrium position O with increasing velocity. The accelerating force decreases as the body approaches O (since the spring is stretched less) and so the *rate* of change of velocity (i.e. the acceleration) decreases.

At O the restoring force is zero but because the body has inertia it overshoots the equilibrium position and continues to move upwards. The spring is now compressed and the elastic restoring force acts again but downwards towards O this time. The body therefore slows down and at an increasing rate due to the restoring force increasing at greater distances from O. The body eventually comes to rest above O and repeats its motion in the opposite direction, p.e. stored as elastic energy of the spring being continually changed to k.e. of the moving body and vice versa. The motion would continue indefinitely if no energy loss occurred, but energy is lost. Why?

The time for a complete oscillation from A to B and back to A, or from O to A to O to B and back to O again is the *period T* of the motion. The *frequency f* is the number of complete oscillations per unit time and a little thought (perhaps using numbers) will indicate that

$$f = \frac{1}{T}$$

An oscillation (or cycle) per second is a *hertz*. The maximum displacement OA or OB is called the *amplitude* of the oscillation.

Some other simple oscillatory systems are shown in Fig. 8.3. It is worth

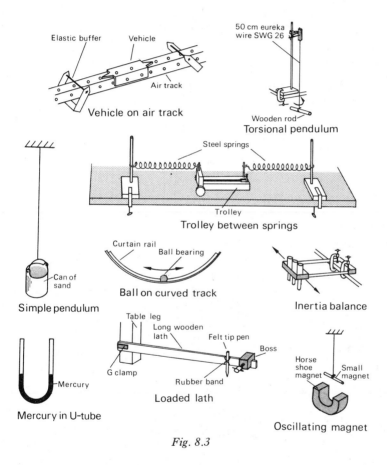

Fig. 8.3

trying to discover experimentally (*i*) which have a constant period (compared with the oscillating balance wheel of a watch), (*ii*) what factors determine the period (or frequency) of the oscillation and (*iii*) whether 'time-traces' of their motions can be obtained and what they look like.

Simple harmonic motion

In Fig. 8.4, N is a body oscillating in a straight line about O, between A and B; N could be a mass hanging from a spiral spring. Previously, in linear motion we considered accelerations that were constant in magnitude and

Fig. 8.4

direction and in circular motion the accelerations (centripetal) were constant in magnitude if not in direction. In oscillatory motion, the accelerations, like the displacements and velocities, change periodically in both magnitude and direction.

Consider first displacements and velocities. When N is below O, the displacement (measured from O) is downwards; the velocity is directed downwards when N is moving away from O but upwards when it moves towards O and is zero at A and B. When N is above O, the displacement is upwards and the velocity upwards or downwards according to whether N is moving away from or towards O.

The variation of acceleration can be seen by considering a body oscillating on a spiral spring. The magnitude of the elastic restoring force increases with displacement but always acts towards the equilibrium position (i.e. O); the resulting acceleration must therefore behave likewise, increasing with displacement but being directed to O whatever the displacement. Thus if N is below O, the displacement is downwards and the acceleration upwards, but if the displacement is upwards the acceleration is downwards. If we adopt the sign convention that quantities acting downwards are positive and those acting upwards are negative then acceleration and displacement always have opposite signs in an oscillation. Fig. 8.5 summarizes these facts and should be studied carefully.

The simplest relationship between the magnitudes of the acceleration and the displacement would be one in which the acceleration a of the body is directly proportional to its displacement x. Such an oscillation is said to be a *simple harmonic motion* (s.h.m.) and is defined as follows.

If the acceleration of a body is directly proportional to its distance from a fixed point and is always directed towards that point, the motion is simple harmonic.

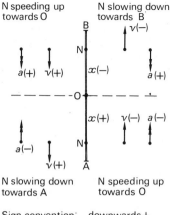

N speeding up
towards O

N slowing down
towards B

Fig. 8.5

Sign convention: downwards +
upwards −

The equation relating acceleration and displacement can be written

$$a \propto -x$$

or

$$a = -\text{constant}.x$$

The negative sign indicates that although the acceleration is larger at larger displacements it is always in the opposite direction to the displacement, i.e. towards O. What kind of motion would be represented by the above equation if a positive sign replaced the negative sign? (It would not be an oscillation.)

In practice many mechanical oscillations are nearly simple harmonic, especially at small amplitudes, or are combinations of such oscillations. In fact any system which obeys Hooke's law will exhibit this type of motion when vibrating. The equation for s.h.m. turns up in many problems in sound, optics, electrical circuits and even in atomic physics. In calculus notation it is written

$$\frac{d^2x}{dt^2} = -\text{constant}.x$$

where $a = dv/dt = d^2x/dt^2$. Using calculus this second order differential equation can be solved to give expressions for displacement and velocity. However, in the next section we shall use a simple geometrical method which links circular motion and simple harmonic motion.

Equations of s.h.m.

Suppose a point P moves round a circle of radius r and centre O with uniform angular velocity ω, its speed v round the circumference will be constant and equal to ωr, Fig. 8.6a. As P revolves, N, the foot of the perpendicular from

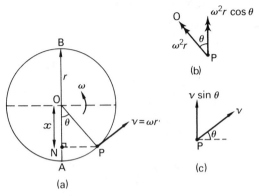

Fig. 8.6

P on to the diameter AOB, moves from A to O to B and returns through O to A as P completes each revolution. Let P and N be in the positions shown at time t after leaving A, with radius OP making angle θ with OA and distance ON being x. We will now show that N describes s.h.m. about O.

(a) *Acceleration.* The motion of N is due to that of P, therefore the acceleration of N is the *component* of the acceleration of P parallel to AB. The acceleration of P is $\omega^2 r$ (or v^2/r) along PO and so the component of this parallel to AB is $\omega^2 r \cos \theta$, Fig. 8.6b. Hence the acceleration a of N is

$$a = -\omega^2 r \cos \theta$$

The negative sign, as explained before, indicates mathematically that a is always directed towards O. Now $x = r \cos \theta$

$$\therefore \quad a = -\omega^2 x$$

Since ω^2 is a positive constant, this equation states that the acceleration of N towards O is directly proportional to its distance from O. N thus describes s.h.m. about O as P moves round the circle—called the *auxiliary circle*—with constant speed.

The table below gives values of a for different values of x and we see that a is zero at O and a maximum at the limits A and B of the oscillation where the direction of motion changes.

x	0	$+r$	$-r$
a	0	$-\omega^2 r$	$+\omega^2 r$

Using the arrangement of Fig. 8.7 the shadow of a ball moving steadily in a circle can be viewed on a screen. The shadow moves with s.h.m. and represents the *projection* of the ball on the screen.

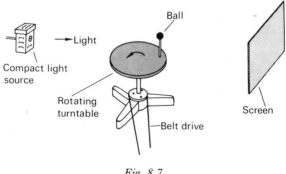

Fig. 8.7

(b) *Period.* The period T of N is the time for N to make one complete oscillation from A to B and back again. In the same time P will travel once round the auxiliary circle and therefore

$$T = \frac{\text{circumference of auxiliary circle}}{\text{speed of P}}$$

$$= \frac{2\pi r}{v} = \frac{2\pi}{\omega} \qquad (\text{since } v = \omega r)$$

For a particular s.h.m. ω is constant and so T is constant and independent of the amplitude r of the oscillation. If the amplitude increases, the body travels faster and so T remains unchanged. A motion which has a constant period whatever the amplitude is said to be *isochronous* and this property is an important characteristic of s.h.m.

(c) *Velocity.* The velocity of N is the *component* of P's velocity parallel to AB which

$$= -v \sin \theta \qquad (\text{see Fig. 8.6}c)$$

$$= -\omega r \sin \theta \qquad (\text{since } v = \omega r)$$

Since $\sin \theta$ is positive when $0° < \theta < 180°$, i.e. N moving upwards, and negative when $180° < \theta < 360°$, i.e. N moving downwards, the negative sign ensures that the velocity is negative when acting upwards and positive when acting downwards (see Fig. 8.5). The variation of the velocity of N with time t (assuming P, and so N, start from A at zero time)

$$= -\omega r \sin \omega t \qquad (\text{since } \theta = \omega t)$$

315

The variation of the velocity of N with displacement x

$$= -\omega r \sin \theta$$

$$= \pm \omega r \sqrt{1 - \cos^2 \theta} \quad \text{(since } \sin^2 \theta + \cos^2 \theta = 1\text{)}$$

$$= \pm \omega r \sqrt{1 - (x/r)^2}$$

$$= \pm \omega \sqrt{r^2 - x^2}$$

Hence the velocity of N is

$$\pm \omega r \text{ (a maximum) when } x = O$$

$$\text{zero when } x = \pm r$$

(d) *Displacement.* This is given by

$$x = r \cos \theta$$

$$= r \cos \omega t$$

The graph of the variation of the displacement of N with time (i.e. its 'time-trace') is shown in Fig. 8.8a and like those for velocity and acceleration in Figs. 8.8b and c it is sinusoidal. Note that when the velocity is zero the acceleration is a maximum and vice versa. We say there is a *phase difference* of a quarter of a period (i.e. $T/4$) between the velocity and the acceleration. What is the phase difference between the displacement and the acceleration?

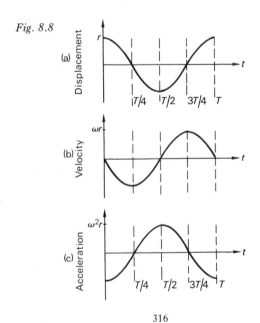

Fig. 8.8

316

(e) *Summary*

$$\text{acceleration} = -\omega^2 x = -\omega^2 r \cos \omega t$$

$$\text{velocity} = \pm \omega \sqrt{r^2 - x^2} = -\omega r \sin \omega t$$

$$\text{displacement} = r \cos \omega t$$

$$\text{period} = 2\pi/\omega$$

These equations are true for *any* s.h.m.

Expression for ω

Consider the equation $a = -\omega^2 x$. We can write (ignoring signs)

$$\omega^2 = \frac{a}{x} = \frac{ma}{mx} = \frac{ma/x}{m}$$

where m is the mass of the system. The force causing the acceleration a at displacement x is ma, therefore ma/x is the force per unit displacement. Hence

$$\omega = \sqrt{\frac{force\ per\ unit\ displacement}{mass\ of\ oscillating\ system}}$$

The period T of the s.h.m. is given by

$$T = \frac{2\pi}{\omega}$$

$$= 2\pi \sqrt{\frac{mass\ of\ oscillating\ system}{force\ per\ unit\ displacement}}$$

This expression shows that T increases if (*i*) the *mass* of the oscillating system *increases* and (*ii*) the *force per unit displacement decreases*, i.e. if the elasticity factor decreases.

An oscillation is simple harmonic if its equation of motion can be written in the form

$$a = -(\text{positive constant}).x$$

For convenience the 'positive constant' is usually represented by ω^2 since $T = 2\pi/\omega$. Hence ω is the square root of the 'positive constant' in the acceleration–displacement equation.

MECHANICAL OSCILLATIONS

Mass on a spring

(a) *Period of oscillations*. The extension of a spiral spring which obeys Hooke's law is directly proportional to the extending tension. A mass m attached to the end of a spring exerts a downward tension mg on it and if it stretches it by an amount l as in Fig. 8.9a, then if k is the tension required to

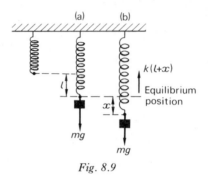

Fig. 8.9

produce unit extension (called the *spring constant* and measured in N m^{-1}) the stretching tension is also kl and so

$$mg = kl$$

Suppose the mass is now pulled down a further distance x below its equilibrium position, the stretching tension acting downwards is $k(l + x)$ which is also the tension in the spring acting upwards, Fig. 8.9b. Hence resultant restoring force *upwards* on mass

$$= k(l + x) - mg$$
$$= kl + kx - kl \qquad \text{(since } mg = kl\text{)}$$
$$= kx$$

When the mass is released it oscillates up and down. If it has an acceleration a at extension x then by Newton's second law

$$-kx = ma$$

The negative sign indicates that at the instant shown a is upwards (negative on our sign convention) while the displacement x is downwards (i.e. positive).

$$\therefore \quad a = -\frac{k}{m}x = -\omega^2 x$$

where $\omega^2 = k/m = $ a positive constant since k and m are fixed. The motion is therefore simple harmonic about the equilibrium position so long as Hooke's law is obeyed. The period T is given by

$$T = \frac{2\pi}{\omega} = 2\pi\sqrt{\frac{m}{k}}$$

If follows that $T^2 = 4\pi^2 m/k$. If the mass m is varied and the corresponding periods T found, a graph of T^2 against m is a straight line but it does not pass through the origin as we might expect from the above equation. This is due to the mass of the spring itself being neglected in the above derivation. Its effective mass and a value of g can be found experimentally.

(b) *Measurement of g and effective mass of spring.* Let m_s be the effective mass of the spring then

$$T = 2\pi\sqrt{\frac{m + m_s}{k}}$$

But $$mg = kl$$

Substituting for m in the first equation and squaring, we get

$$T^2 = \frac{4\pi^2}{k}\left(\frac{kl}{g} + m_s\right)$$

$$\therefore \quad l = \frac{g}{4\pi^2} \cdot T^2 - \frac{g\,m_s}{k}$$

By measuring (i) the static extension l and (ii) the corresponding period T, using several different masses in turn, a graph of l against T^2 can be drawn. It is a straight line of slope $g/4\pi^2$ and intercept gm_s/k on the l axis, Fig. 8.10.

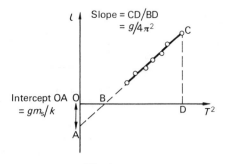

Fig. 8.10

Thus g and m_s can be found. Theory suggests that the effective mass of a spring is about one-third of its actual mass.

Simple pendulum

(a) *Period of oscillations.* The simple pendulum consists of a small bob (in theory a 'particle') of mass m suspended by a light inextensible thread of length l from a fixed point B, Fig. 8.11. If the bob is drawn aside slightly

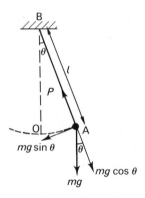

Fig. 8.11

and released, it oscillates to-and-fro in a vertical plane along the arc of a circle. We shall show that it describes s.h.m. about its equilibrium position O.

Suppose at some instant the bob is at A where arc OA $= x$ and $\angle$ OBA $= \theta$. The forces on the bob are P and the weight mg of the bob acting vertically downwards. Resolving mg radially and tangentially at A we see that the tangential component $mg \sin \theta$ is the unbalanced restoring force acting towards O. (The component $mg \cos \theta$ balances P.) If a is the acceleration of the bob along the arc at A due to $mg \sin \theta$ then the equation of motion of the bob is

$$-mg \sin \theta = ma$$

The negative sign indicates that the force is towards O while the displacement x is measured along the arc from O in the opposite direction.

When θ is small, sin $\theta = \theta$ in radians (e.g. if $\theta = 5°$, sin $\theta = 0.0872$ and $\theta = 0.0873$ rad) and $x = l\theta$ (see p. 271). Hence

$$-mg\,\theta = -mg\frac{x}{l} = ma$$

$$\therefore \quad a = -\frac{g}{l}x = -\omega^2 x \quad \text{(where } \omega^2 = g/l)$$

The motion of the bob is thus simple harmonic *if the oscillations are of small amplitude,* i.e. θ does not exceed $10°$. The period T is given by

$$T = \frac{2\pi}{\omega} = \frac{2\pi}{\sqrt{g/l}}$$

$$= 2\pi\sqrt{\frac{l}{g}}$$

T is therefore independent of the amplitude of the oscillations and at a given place on the earth's surface where g is constant, it depends only on the length l of the pendulum.

A multiflash photograph of a single swing of a simple pendulum is shown in Fig. 8.12.

Fig. 8.12

(b) *Measurement of g.* A fairly accurate determination of g can be made by measuring T for different values of l and plotting a graph of l against T^2. A straight line AB is then drawn so that the points are evenly distributed about it, Fig. 8.13. It should pass through the origin and its slope BC/CA gives an average value of l/T^2 from which g can be calculated since

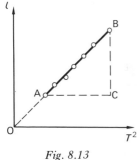

Fig. 8.13

$$T = 2\pi\sqrt{\frac{l}{g}}$$

$$\therefore \quad T^2 = 4\pi^2\frac{l}{g}$$

$$\therefore \quad g = 4\pi^2\frac{l}{T^2} = 4\pi^2\frac{BC}{CA}$$

The experiment requires (*i*) 100 oscillations to be timed, (*ii*) an angle of swing less than $10°$, (*iii*) the length l to be measured to the centre of the bob, (*iv*) the oscillations to be counted as the bob passes the equilibrium position O. Why?

MECHANICAL OSCILLATIONS

S.H.M. calculations

1. A particle moving with s.h.m. has velocities of 4 cm s^{-1} and 3 cm s^{-1} at distances of 3 cm and 4 cm respectively from its equilibrium position. Find (a) the amplitude of the oscillation, (b) the period, (c) the velocity of the particle as it passes through the equilibrium position.

Fig. 8.14

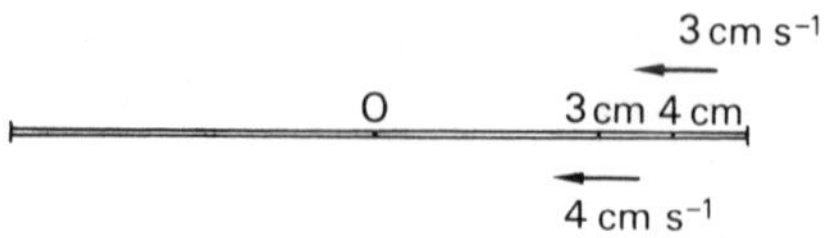

(a) Using the previous notation and taking the case shown in Fig. 8.14, the equation for the velocity is

$$\text{velocity} = -\omega\sqrt{r^2 - x^2} \qquad (\text{p. 316})$$

if we take velocities and displacements to the left as being negative and those to the right positive.

When $x = +3$ cm, velocity $= -4$ cm s^{-1}; therefore

$$-4 = -\omega\sqrt{r^2 - 9}$$

When $x = +4$ cm, velocity $= -3$ cm s^{-1}; therefore

$$-3 = -\omega\sqrt{r^2 - 16}$$

Squaring and dividing these equations we get

$$\frac{16}{9} = \frac{r^2 - 9}{r^2 - 16}$$

Hence
$$r = \pm 5 \text{ cm}$$

(b) Substituting for r in one of the velocity equations we find

$$\omega = 1 \text{ s}^{-1}$$

$$\therefore \ T = \frac{2\pi}{\omega} = 2\pi \text{ s}$$

(c) At the equilibrium position $x = 0$

$$\therefore \quad \text{velocity} = \pm\omega\sqrt{r^2 - x^2}$$

$$= \pm\omega r$$

$$= \pm 5 \text{ cm s}^{-1}$$

2. A light spiral spring is loaded with a mass of 50 g and it extends by 10 cm. Calculate the period of small vertical oscillations. (g = 10 m s⁻²)

The period T of the oscillations is given by

$$T = 2\pi \sqrt{\frac{m}{k}}$$

where
$$m = 50 \times 10^{-3} \text{ kg}$$

and
$$k = \text{force per unit displacement}$$

$$= \frac{50 \times 10^{-3} \times 10 \text{ N}}{10 \times 10^{-2} \text{ m}} = 5.0 \text{ N m}^{-1}$$

$$\therefore \quad T = 2\pi \sqrt{\frac{50 \times 10^{-3}}{5}} = 2\pi \sqrt{10^{-2}} \text{ s}$$

$$= 2\pi \times 10^{-1} \text{ s}$$

$$= 0.63 \text{ s}$$

3. A simple pendulum has a period of 2.0 s and an amplitude of swing 5.0 cm. Calculate the maximum magnitudes of (a) the velocity of the bob, (b) the acceleration of the bob.

(*a*) $T = 2\pi/\omega$, therefore $\omega = 2\pi/T = 2\pi/2 = \pi$ s⁻¹. Velocity is a maximum at the equilibrium position where $x = 0$ and

$$= \pm\omega\sqrt{r^2 - x^2}$$

$$= \pm\pi\sqrt{25} \qquad \text{(since } r = \pm 5 \text{ cm)}$$

$$= \pm 5\pi \text{ cm s}^{-1}$$

$$= \pm 16 \text{ cm s}^{-1}$$

(*b*) Acceleration is a maximum at the limits of the swing where $x = r = \pm 5.0$ cm and

$$= -\omega^2 r$$

$$= -\pi^2 \times 5 \text{ cm s}^{-2}$$

$$= -50 \text{ cm s}^{-2}$$

Energy of s.h.m.

In an oscillation there is a constant interchange of energy between the kinetic and potential forms and if the system does no work against resistive forces (i.e. is undamped) its total energy is constant, as we shall now show.

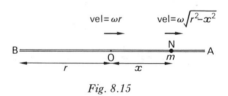

Fig. 8.15

(a) *Kinetic energy.* The velocity of a particle N of mass m at a distance x from its centre of oscillation O, Fig. 8.15,

$$= +\omega\sqrt{r^2 - x^2}$$

∴ k.e. at displacement $x = \frac{1}{2}m\omega^2(r^2 - x^2)$

(b) *Potential energy.* As N moves out from O towards A (or B) work is done against the force (e.g. the tension in a stretched spring) trying to restore it to O. Thus N loses k.e. and gains p.e. When $x = 0$, the restoring force is zero; at displacement x, the force is $m\omega^2 x$ (since the acceleration has magnitude $\omega^2 x$). Hence

average force on N while moving to displacement $x = \frac{1}{2}m\omega^2 x$

∴ work done = average force × displacement in direction of force

$$= \frac{1}{2}m\omega^2 x \times x = \frac{1}{2}m\omega^2 x^2$$

∴ p.e. at displacement $x = \frac{1}{2}m\omega^2 x^2$

(c) *Total energy.* At displacement x we have

total energy = k.e. + p.e.

$$= \frac{1}{2}m\omega^2(r^2 - x^2) + \frac{1}{2}m\omega^2 x^2$$

$$= \frac{1}{2}m\omega^2 r^2$$

This is constant, does not depend on x and is directly proportional to the product of (*i*) the mass, (*ii*) the square of the frequency and (*iii*) the square of the amplitude. Fig. 8.16 shows the variation of k.e., p.e. and total energy with displacement.

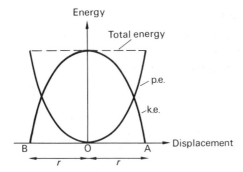

Fig. 8.16

In a simple pendulum all the energy is kinetic as the bob passes through the centre of oscillation and at the top of the swing it is all potential.

Damped oscillations

The amplitude of the oscillations of, for example, a pendulum gradually decreases to zero due to the resistive force that arises from the air. The motion is therefore not a perfect s.h.m. and is said to be *damped* by air resistance; its energy becomes internal energy of the surrounding air.

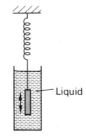

Fig. 8.17

The behaviour of a mechanical system depends on the extent of the damping. The damping of the mass on the spring in Fig. 8.17 is greater than when it is in air. Undamped oscillations are said to be *free*, Fig. 8.18a. If a system is slightly damped oscillations of decreasing amplitude occur, Fig. 8.18b. When heavily damped no oscillations occur and the system re-

Fig. 8.18

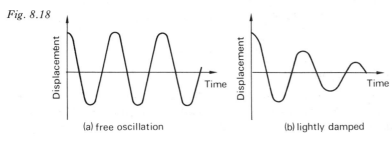

(a) free oscillation   (b) lightly damped

325

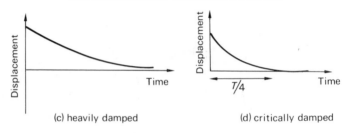

Fig. 8.18 cont.

turns very slowly to its equilibrium position, Fig. 8.18c. When the time taken for the displacement to become zero is a minimum, the system is said to be *critically damped*, Fig. 8.18d.

The motion of many devices is critically damped on purpose. Thus the shock absorbers on a car critically damp the suspension of the vehicle and so resist the setting up of vibrations which could make control difficult or cause damage. In the shock absorber of Fig. 8.19 the motion of the suspension up or down is opposed by viscous forces when the liquid passes through

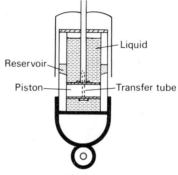

Fig. 8.19

the transfer tube from one side of the piston to the other. The damping of a car can be tested by applying your weight to the suspension momentarily; the car should rapidly return to its original position without vibrating.

Instruments such as balances and electrical meters are critically damped (i.e. dead-beat) so that the pointer moves quickly to the correct position without oscillating. The damping is often produced by electromagnetic forces.

Forced oscillation and resonance

(*a*) *Barton's pendulums.* The assembly consists of a number of paper cone pendulums (made by folding paper circles of about 5 cm diameter) of lengths varying from $\frac{1}{4}$ m to $\frac{3}{4}$ m, each loaded with a plastic curtain ring. All

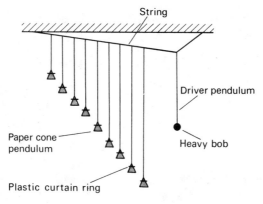

Fig. 8.20

are suspended from the same string as a 'driver' pendulum which has a heavy bob and a length of $\frac{1}{2}$ m, Fig. 8.20.

The driver pendulum is pulled well aside and released so that it oscillates in a plane perpendicular to that of the diagram. The motion settles down after a short time and all the pendulums oscillate with very nearly the *same* frequency as the driver but with *different* amplitudes. This is a case of *forced oscillation*.

The pendulum whose length equals that of the driver has the greatest amplitude; its natural frequency of oscillation is the same as the frequency of the driving pendulum. This is an example of *resonance* and the driving oscillator then transfers its energy most easily to the other system, i.e. the paper cone pendulum of the same length.

The amplitudes of oscillations also depend on the extent to which the system is damped. Thus removing the rings from the paper cone pendulums reduces their mass and so increases the damping. All amplitudes are then found to be reduced and that of the resonant frequency is less pronounced. These results are summarized by the graphs in Fig. 8.21 which indicate that

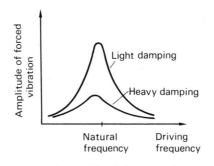

Fig. 8.21

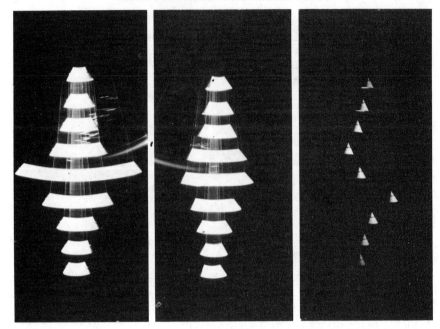

Fig. 8.22a   Fig. 8.22b   Fig. 8.22c

the sharpest resonance is given by a lightly damped system. Fig. 8.22a and b are time-exposure photographs taken with the camera looking along the line of swinging pendulums towards and at the same level as the bob of the driver. Is (a) more or less damped than (b)?

Careful observation shows that the resonant pendulum is always a quarter of an oscillation behind the driver pendulum, i.e. there is a phase difference of a quarter of a period. The shorter pendulums are nearly in phase with the driver, while those that are longer than the driver are almost half a period behind it. This is evident from the instantaneous photograph of Fig. 8.22c, taken when the driver is at maximum displacement to the left.

(b) *Hacksaw blade oscillator.* The arrangement, shown in Fig. 8.23, provides another way of finding out what happens when one oscillator (a loaded hacksaw blade) is driven by another (a heavy pendulum), as often occurs in practice. The positions of the mass on the blade and the pendulum bob can both be adjusted to alter the natural frequencies. By using different rubber bands the degree of coupling may be varied, as can the damping, by turning the postcard. The motion of the driver is maintained by gentle, timely taps just below its support.

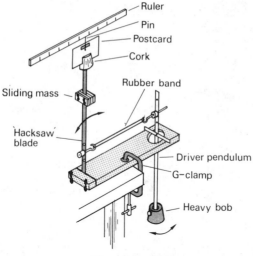

Fig. 8.23

There is scope for investigating (*i*) the transient oscillations that occur as the motion starts and before the onset of steady conditions, (*ii*) resonance, (*iii*) phase relationships, (*iv*) damping and (*v*) coupling.

(*c*) *Examples of resonance.* These are common throughout science and are generally useful. Thus in the production of musical sounds from air columns in wind instruments resonance occurs, in many cases, between the vibrations of air columns and of small vibrating reeds. Electrical resonance occurs when a radio circuit is tuned by making its natural frequency for electrical oscillations equal to that of the incoming radio signal.

Information about the strength of chemical bonds between ions can be obtained by a resonance effect. If we regard electromagnetic radiation (e.g. light, infrared, etc.) as a kind of oscillating electrical disturbance which, when incident on a crystal, subjects the ions to an oscillating electrical force, then, with radiation of the correct frequency, the ions could be set into oscillation by resonance, Fig. 8.24. Energy would be absorbed from the radiation and the absorbed frequency could be found using a suitable spectrometer. With sodium chloride absorption of infrared radiation occurs.

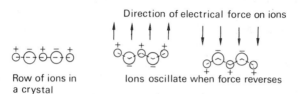

Fig. 8.24

Fig. 8.25

Resonance in mechanical systems is a source of trouble to engineers. The Tacoma Narrows Bridge disaster (p. 310) was really caused by the wind producing an oscillating resultant force in resonance with a natural frequency of the bridge. An oscillation of large amplitude built up and destroyed the structure. Fig. 8.25 shows the tail of a wind tunnel model of an aircraft being tested for resonance, i.e. shaken at different frequencies. It is important that the natural frequencies of vibration of an aircraft do not equal any that may be produced by the forces experienced in flight. Otherwise resonance might occur and undue stress result.

S.H.M.—a mathematical model

Real oscillators such as a motor cycle on its suspension, a tall chimney swaying in the wind, atoms (or ions) vibrating in a crystal, only approximate to the ideal type of motion we call s.h.m.

S.H.M. describes in mathematical terms a kind of motion which is not fully realized in practice. It is a mathematical model and is useful because it represents well enough many real oscillations. This is due to its simplicity; complications such as damping, variable mass and variable stiffness (elastic modulus) are omitted and the only conditions imposed on the system are

MECHANICAL OSCILLATIONS

that the restoring force should be directed towards the centre of the motion and be proportional to the displacement.

A more complex model might, for example, take damping into account and as a result be a better description of a particular oscillator, but it would probably not be so widely applicable. On the other hand if a model is too simple it may be of little use for dealing with real systems. A model must have just the correct degree of complexity. The mathematical s.h.m. model has this and so is useful in practice.

Film loops (8 mm)

1. *Tacoma Narrows Bridge collapse*—Ealing Scientific Ltd, Bushey Mill Lane, Watford, Herts.
2. *Wind-induced oscillations*—Penguin Education, Harmondsworth, Middlesex.

QUESTIONS

1. Write a short account of *simple harmonic motion* explaining the terms *amplitude, time period* and *frequency*.

A particle of mass m moves such that its displacement from the equilibrium position is given by $y = a \sin \omega t$ where a and ω are constants. Derive an expression for the kinetic energy of the particle at a time t and show that its value is a maximum as the particle passes through the equilibrium position.

A steel strip clamped at one end vibrates with a frequency of 30 Hz, and an amplitude of 4.0 mm at the free end. Find (a) the velocity of the free end as it passes through the equilibrium position and (b) the acceleration at the maximum displacement.

(*A.E.B.*)

2. A simple pendulum of length 80 cm is oscillating with an amplitude of 4.0 cm. Calculate the velocity of the bob as it passes through the mid-point of its oscillation. Explain why the tension in the string as it passes through this point is different from that in the string when the pendulum hangs vertically at rest and state which tension has the greater value. (*L. part qn.*)

3. Define *simple harmonic motion*.

Describe an experiment to measure the acceleration due to gravity at the earth's surface, using a simple pendulum. Derive the equation used.

A pendulum of length 130 cm has a periodic time of T_1. The bob is now pulled aside and made to move as a conical pendulum in a horizontal circle of radius 50.0 cm. The period of rotation is T_2. Find the ratio $T_1 : T_2$. (*L.*)

331

MECHANICAL OSCILLATIONS

4. Define simple harmonic motion and explain the meaning of the terms amplitude, period, and frequency.

A body of mass 0.10 kg hangs from a long spiral spring. When pulled down 10 cm below its equilibrium position A and released, it vibrates with simple harmonic motion with a period of 2.0 s.

(a) What is its velocity as it passes through A?

(b) What is its acceleration when it is 5.0 cm above A?

(c) When it is moving upwards, how long a time is taken for it to move from a point 5.0 cm below A to a point 5.0 cm above?

(d) What are the maximum and minimum values of its kinetic energy, and at what points of the motion do they occur?

(e) What is the value of the total energy of the system and does it vary with time?

(W.)

5. Describe an experiment to demonstrate the effects of damping on the oscillatory motion of a vibrating system undergoing (a) free and (b) forced harmonic oscillations. Draw labelled diagrams to illustrate the results you would expect to obtain.

What is the physical origin of the damping mechanisms in the case of (a) the oscillations of a simple pendulum in air, (b) the vibrations of a bell sounding in air and (c) oscillatory currents in an electrical circuit?

A spring is supported at its upper end. When a mass of 1.0 kg is hung on the lower end the new equilibrium position is 5.0 cm lower. The mass is then raised 5.0 cm to its original position and released. Discuss as fully as you can the subsequent motion of the system.

(O. and C.)

6. A mass of 2.0 kg is hung from the lower end of a spiral spring and extends it by 0.40 m. When the mass is displaced a further short distance x and released, it oscillates with acceleration a towards the rest position. If $a = -kx$ and if the tension in the spring is always directly proportional to its extension, what is the value of the constant k? (Earth's gravitational field strength is 9.8 N kg^{-1}.)

7. Define *simple harmonic motion* and state where the magnitude of the acceleration is (a) greatest, (b) least.

Some sand is sprinkled on a horizontal membrane which can be made to vibrate vertically with simple harmonic motion. When the amplitude is 0.10 cm the sand just fails to make continuous contact with the membrane. Explain why this phenomenon occurs and calculate the frequency of vibration. ($g = 10$ m s^{-2}) (J.M.B.)

8. If a mass hanging on a vertical spring which obeys Hooke's Law is given a small vertical displacement it will oscillate vertically, above and below its equilibrium position. Why is the motion of the mass simple harmonic motion?

Explain why the oscillations of a simple pendulum, consisting of a mass m hanging from a thread of length l, are almost perfectly simple harmonic, and why the deviations from perfect simple harmonic motion become greater if the amplitude of the swing is increased.

A metre ruler is clamped to the top of a table so that most of its length overhangs and is free to vibrate with vertical simple harmonic motion. Calculate the maximum velocity of the tip of the ruler if the amplitude of the vibration is 5.0 cm and the frequency is 4.0 Hz. What is the maximum possible amplitude if a small object placed on the ruler at the vibrating end is to maintain contact with it throughout the vibration?

If this amplitude were very slightly exceeded, at what stage in the vibration would contact between the object and the ruler be broken? Explain clearly why it would happen at this point. ($g = 10 \text{ m s}^{-2}$) (*S.*)

9. By reference to a particular system explain what is meant by (*a*) forced vibrations, (*b*) resonance. What are the effects of damping? (*J.M.B.*)

10. Because of the effect on the comfort of the ride and the noise experienced by drivers, a manufacturer wishes to investigate the various vibrations (of the body, wheels, springs, door panels and so on) which can arise in motor cars. The manufacturer has two cars, one of which is reported to be much more uncomfortable and noisy than the other.

Draft an outline plan of a programme of tests to:

(*a*) identify and compare vibrations in the cars,

(*b*) investigate the reasons for discomfort and noise,

(*c*) provide guidance for designers of new models.

You may assume that any test equipment needed can be made available but you have to say for what the equipment is needed. Your plan should include an explanation of why the various tests are proposed.

You are asked to plan the tests and *not* to predict their results nor to explain how the car design might be improved. (*O. and C. Nuffield*)

9 Fluids at rest

Introduction

Liquids and gases can flow and are called *fluids*. Before considering their behaviour at rest, certain basic terms will be defined, by way of revision.

The *density* ρ of a sample of a substance of mass m and volume V is defined by the equation

$$\rho = \frac{m}{V}$$

In words, density is the *mass per unit volume*. The density of water (at 4 °C) is 1.00 g cm^{-3} or, in SI units, 1.00×10^3 kg m^{-3}; the density of mercury (at room temperature) is about 13.6 g cm^{-3} or 13.6×10^3 kg m^{-3}.

The term *relative density* is sometimes used and is given by

$$relative\ density = \frac{density\ of\ material}{density\ of\ water}$$

It is a ratio and has no unit. The relative density of mercury is thus 13.6.

If a force acts on a surface (like the weight of a brick on the ground) it is often more useful to consider the *pressure* exerted rather than the force. The pressure p caused by a force F acting normally on a surface of area A is defined by

$$p = \frac{F}{A}$$

Pressure is therefore *force per unit area*; the SI unit is the *pascal* (Pa) which equals a pressure of 1 newton per square metre.

Pressure in a liquid

In designing a dam like that in Fig. 9.1 the engineer has to know, among other things, about the size and point of action of the resultant force exerted on the dam by the water behind it. This involves making calculations based on the expression for liquid pressure.

Fig. 9.1

(a) *Expression for liquid pressure.* The pressure exerted by a liquid is experienced by any surface in contact with it. It increases with depth because the liquid has weight and we define the *pressure at a point in a liquid as the force per unit area on a very small area round the point.* Thus if the force is δF and the small area δA then the pressure p at the point is given by

$$p = \frac{\text{force}}{\text{area}} = \frac{\delta F}{\delta A}$$

More exactly, in calculus terms, the defining equation is

$$p = \underset{\delta A \to 0}{\text{limit}} \frac{\delta F}{\delta A}$$

An expression for the pressure p at a depth h in a liquid of density ρ can be found by considering an extremely small horizontal area δA, Fig. 9.2. The force δF acting vertically downwards on δA equals the weight of the liquid column of height h and uniform cross-section area δA above it. We can say

Fig. 9.2

volume of liquid column $= h\, \delta A$

mass of liquid column $\quad = h\, \delta A\, \rho$

weight of liquid column $= h\, \delta A\, \rho\, g$

335

where g is the acceleration due to gravity (or the strength of the earth's gravitational field). Hence

$$\delta F = h \, \delta A \, \rho \, g$$

and

$$p = \frac{\delta F}{\delta A} = \frac{h \, \delta A \, \rho \, g}{\delta A}$$

$$\therefore \quad p = h\rho g$$

Thus the pressure at a point in a liquid depends only on the depth and the density of the liquid. If h is in m, ρ in kg m^{-3} and g in m s^{-2} (or N kg^{-1}) then p is in Pa.

Notes. (*i*) In the above derivation, δA was considered to be horizontal but it can be shown that the same result is obtained for any other orientation of δA, i.e. *the pressure at a point in a liquid acts equally in all directions*—as experiment confirms.

(*ii*) The force exerted on a surface in contact with a liquid at rest is *perpendicular* to the surface at all points. Otherwise, the equal and opposite force exerted by the containing surface on the liquid would have a component parallel to the surface which would cause the liquid to flow.

(*b*) *Transmission of pressure.* A liquid can transmit any external pressure applied to it to all its parts. Use is made of this property in the hydraulic press to produce a large force from a small one. In its simplest form it consists of a narrow cylinder connected to a wide cylinder, both containing liquid (usually oil) and fitted with pistons A and B as in Fig. 9.3a. If, for example,

Fig. 9.3a

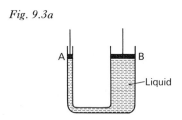

A has a cross-sectional area of 1×10^{-4} m^2 (i.e. 1 cm^2) and a downwards force of 1 N is applied to it, a pressure of 1×10^{-4} Pa (i.e. 1 N cm^{-2}) in excess of atmospheric pressure is transmitted through the liquid. If B has a cross-sectional area of 100×10^{-4} m^2 (i.e. 100 cm^2), it experiences an upwards force of 100 N. Piston B acts against a fixed plate and in a large press like that in Fig. 9.3b, it is used to forge motor car parts from steel.

The same principle operates in hydraulic jacks for lifting cars in a garage and also in the hydraulic braking system of a car. In the latter, the force applied to the brake pedal causes a piston to produce an increase of pressure

Fig. 9.3b

Fig. 9.3c

in an oil-filled cylinder and this is transmitted through oil-filled pipes to four other pistons which apply the brake-shoes or discs to the car wheels. This results in the same pressure being applied to all wheels and minimizes the risk of the car pulling to one side or skidding.

(c) *High pressure water jet cutting*. This is a new cutting technique which is entirely dust-free. It can be used with a wide range of materials such as slate, stone, brake-lining material, Formica, rubber, foams and is especially advantageous for materials like asbestos, the dust from which can cause respiratory disease. In the equipment shown, Fig. 9.3c, the cutting jet is guarded by a Perspex cover, the jet pressure is about 3.5×10^8 Pa (3500 times normal atmospheric pressure), the water flow is 3.0 litres per minute and the width of cut can be varied from about 0.3 mm to 0.9 mm.

Liquid columns

(a) U-*tube manometer*. An open U-tube containing a suitable liquid can be used to measure pressures, for example, the pressure of the gas supply, and is called a U-tube manometer. It uses the fact that the pressure in a column of liquid is directly proportional to the height of the column. In the U-tube manometer of Fig. 9.4a the pressure p to be measured acts on the surface of the liquid at A and balances the pressure of the liquid column BC of height h, *plus* atmospheric pressure P, acting on B. Hence

$$p = P + h\rho g$$

where ρ is the density of the liquid in the manometer. For small pressures, water or a light oil is used, for medium pressures mercury is suitable. The amount by which p exceeds atmospheric pressure, i.e. $(p - P)$, equals the pressure due to the column of liquid BC, i.e. $h\rho g$. Consequently it is often convenient to state a pressure as a number of mm of water or mercury rather than in Pa.

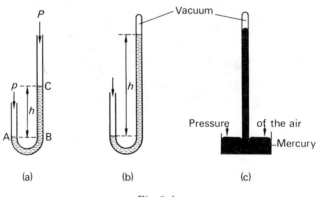

Fig. 9.4

When the absolute pressure is required and not the excess over atmospheric, the limb of the tube open to the atmosphere is replaced by a closed, evacuated one, Fig. 9.4b. The height h of the liquid column then gives the absolute pressure directly. The principle is used in the measurement of atmospheric pressure by a mercury barometer, the short limb being replaced by a reservoir of mercury, Fig. 9.4c. The column of mercury is supported by the pressure of the air on the surface of the mercury in the reservoir and any change in this causes the length of the column to vary.

(b) *Balancing liquid columns.* If a U-tube contains two immiscible liquids of different densities the surfaces of the liquids are not level. In Fig. 9.5 the column of water AB is balanced by the column of paraffin CD. Hence

$$\text{pressure at B} = \text{pressure at D}$$

$$\therefore \quad h_1 \rho_1 g = h_2 \rho_2 g$$

where h_1, ρ_1 and h_2, ρ_2 are the heights and densities of the water and paraffin respectively. It follows that

$$\frac{\rho_2}{\rho_1} = \frac{h_1}{h_2}$$

Measurement of h_1 and h_2 thus gives a simple way of finding the relative density of paraffin. The limbs of the U-tube need not have the same diameter (if not too small). Why not?

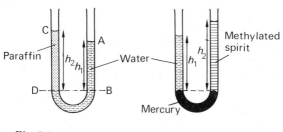

Fig. 9.5   Fig. 9.6

Miscible liquids can be separated by mercury as in Fig. 9.6. In this case the heights of the columns are adjusted by adding liquid until the mercury surfaces are exactly level. Measurements of h_1 and h_2 are then made from this level.

Archimedes' principle

When a body is immersed in a liquid it is buoyed up and appears to lose weight. The upward force is called the *upthrust* of the liquid on the body and is due to the pressure exerted by the liquid on the lower surface of the body being greater than that on the top surface since pressure increases with depth. The law summarizing such effects was discovered over 2000 years ago by Archimedes. It also applies to bodies in gases and is stated as follows.

When a body is completely or partly immersed in a fluid it experiences an upthrust, or apparent loss in weight, which is equal to the weight of fluid displaced.

A numerical case is illustrated in Fig. 9.7*a*; more briefly we can say: *upthrust = weight of fluid displaced.*

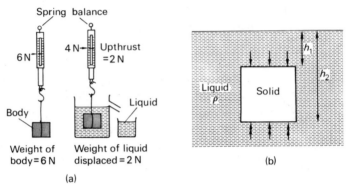

Fig. 9.7

The principle can be verified experimentally, or deduced theoretically by considering the pressures exerted by a liquid on the top and bottom surfaces of a rectangular-shaped solid, Fig. 9.7*b*—as you can confirm for yourself.

(*a*) *Floating bodies.* If a body floats partly immersed in a liquid (e.g. a ship), completely immersed in a liquid (e.g. a submarine) or a gas (e.g. a balloon), it appears to have zero weight and we can say

upthrust on body = weight of floating body

By Archimedes' principle

upthrust on body = weight of fluid displaced

Hence

weight of floating body = weight of fluid displaced

This result, sometimes called the 'principle of flotation', is a special case of Archimedes' principle and can be stated thus.

A floating body displaces its own weight of fluid.

If the body cannot do this, even when completely immersed, it sinks.

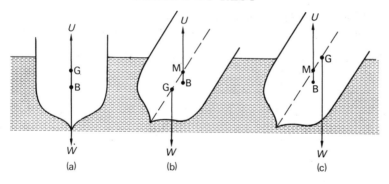

Fig. 9.8

The stability of a floating body such as a ship when it heels over depends on the relative positions of the ship's centre of gravity G, through which its weight W acts, and the centre of gravity of the displaced water, called the *centre of buoyancy* B, through which the upthrust U acts. In Fig. 9.8a the ship is on an even keel and B and G are on the same vertical line. If it heels over the shape of the displaced water changes, causing B to move and thereby setting up a couple which tends either to return the ship to its original position or to make it heel over more. The point of intersection of the vertical line from B with the central line of the ship is called the *metacentre* M. If M is above G as in Fig. 9.8b, the couple has an anticlockwise moment which acts to decrease the ship's heel and the equilibrium is stable. If M is below G as in Fig. 9.8c, equilibrium is unstable since the couple has a clockwise moment which causes further listing. For maximum stability G should be low and M high.

(*b*) *Hydrometer*. This is an instrument which uses the principle of flotation to give a rapid measurement of the relative density of a liquid, e.g. the acid in a lead accumulator. It consists of a narrow glass stem, a large buoyancy bulb and a smaller bulb loaded with lead shot to keep it upright when floating, Fig. 9.9. The relative density is found by floating the hydrometer

Fig. 9.9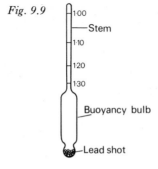

341

in the liquid and taking the reading on the scale inside the stem at the level of the liquid surface. The instrument shown is for use in the range 1.30 to 1.00, the numbers increase downwards (why?) and the scale is uneven.

Atmospheric pressure

The pressure due to a gas arises from the bombardment of the walls of the containing vessel by its molecules. In a small volume of gas the pressure is uniform throughout but in a large volume such as the atmosphere, gravity causes the density of the gas and therefore its pressure to be greater in the lower regions than in the upper regions. In fact atmospheric pressure at a height of about 6 km is half its sea-level value even though the atmosphere extends to a height of 150 km or so.

(a) *Value.* The statement that atmospheric pressure is '760 mm of mercury' means that it equals the pressure at the bottom of a column of mercury 760 mm high. Sometimes pressure is expressed in torrs (after Torricelli who made the first mercury barometer), 1 torr = 1 mmHg. To find the value of atmospheric pressure in S I units we use $p = h\rho g$ where $h = 0.760$ m, $\rho = 13.6 \times 10^3$ kg m^{-3}, $g = 9.81$ N kg^{-1}. Thus

$$p = 0.760 \times 13.6 \times 10^3 \times 9.81 \quad \text{m} \times \text{kg m}^{-3} \times \text{N kg}^{-1}$$

$$= 1.01 \times 10^5 \text{ Pa}$$

Standard pressure or 1 atmosphere is defined as the pressure at the foot of a column of mercury 760 mm high of specified density and subject to a particular value of g; it equals 1.01325×10^5 Pa.

(b) *Temperature corrections to the barometric height.* The temperatures at weather stations are generally different and before their pressure readings are compared the barometric height at each is often 'reduced' to the height it would have been at 0 °C. This may involve correcting for

(i) the change in density of mercury with temperature and
(ii) the metal scale on the barometer being used at a temperature other than its calibration temperature.

Consider (i). Suppose H_θ is the height recorded at the station at temperature θ and H_0 is the height of the mercury column to give the *same* pressure when the mercury is at 0 °C. We have

$$H_\theta \rho_\theta g = H_0 \rho_0 g$$

where ρ_θ and ρ_0 are the densities of mercury at θ and 0 °C respectively.

Hence

$$H_0 = H_\theta \times \frac{\rho_\theta}{\rho_0}$$

But
$$\rho_0 = \rho_\theta(1 + \gamma\theta)$$
(see p. 136)

where γ is the real cubic expansivity of mercury

$$\therefore \quad H_0 = \frac{H_\theta}{1 + \gamma\theta}$$

Consider (*ii*). If the scale is correct at 0 °C, the true reading at θ will be $H_\theta(1 + \alpha\theta)$ and not H_θ (see p. 146, question 9) where α is the linear expansivity of the metal scale. The final corrected height H_0 is given by

$$H_0 = \frac{H_\theta(1 + \alpha\theta)}{1 + \gamma\theta}$$

A numerical example will illustrate the use of this expression. Suppose the height of a mercury barometer read with a steel scale which is correct at 0 °C is 754 mm at 20 °C. What is the height reduced to 0 °C? Taking $\alpha_{steel} = 1.20 \times 10^{-5}$ °C^{-1} and $\gamma_{mercury} = 1.80 \times 10^{-4}$ °C^{-1} and substituting in the expression we get

$$H_0 = \frac{754(1 + 1.20 \times 10^{-5} \times 20)}{1 + 1.80 \times 10^{-4} \times 20}$$

$$= \frac{754(1 + 2.40 \times 10^{-4})}{1 + 3.60 \times 10^{-3}}$$

$$\simeq 754(1 + 2.40 \times 10^{-4})(1 - .3.60 \times 10^{-3})$$

since by the Binomial theorem $1/(1 + x) \simeq (1 - x)$ if x is small; therefore

$$H_0 \simeq 754(1 + 2.40 \times 10^{-4} - 3.60 \times 10^{-3})$$

neglecting $(2.40 \times 10^{-4}) \times (3.60 \times 10^{-3})$. Therefore

$$H_0 \simeq 754 \times 0.997$$

$$\simeq 752 \text{ mm}$$

Pressure gauges and vacuum pumps

The U-tube manometer can be used to measure moderate pressures, i.e. those in the region of atmospheric. High and low pressures require special gauges.

(*a*) *Bourdon gauge*. This measures pressures up to about 2000 atmospheres. It consists of a curved metal tube, sealed at one end, to which the pressure to be measured is applied, Fig. 9.10. As the pressure increases the tube uncurls and causes a rack and pinion to move a pointer over a scale.

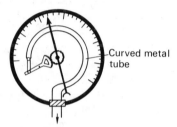

To unknown pressure

Fig. 9.10

(*b*) *McLeod gauge*. It is used to measure the very low pressures produced by vacuum pumps. The principle is to compress a sample of the gas whose pressure is required until its pressure is measurable on a mercury mano-meter and then, knowing the initial and final volumes of the gas, to apply Boyle's law.

A measurement is made by lowering the mercury reservoir on the gauge thus allowing gas from the evacuated vessel at unknown pressure p to enter bulb B and capillary tube C, Fig. 9.11. It is next raised until the mercury reaches a fixed mark M on C, Fig. 9.12. The pressure of the compressed

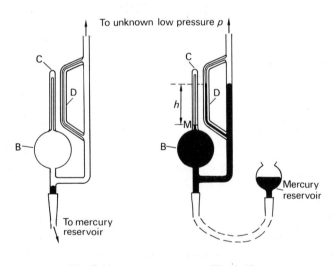

Fig. 9.11   *Fig. 9.12*

gas is then $p + h$. If V is the volume of B and all C and v is the volume of C above mark M, applying Boyle's law we have

$$pV = (p + h)v$$

Hence p can be found since the values of V and v are determined during manufacture of the gauge and h is read off directly. The capillary tube C has the same internal diameter (bore) as D thereby eliminating errors due to surface tension (see p. 351) when h is measured.

If V/v is of the order of 10^4, what will be the value of h when the pressure is 10^{-3} mmHg (torr)? The McLeod gauge can be used down to about 10^{-4} mmHg; for lower pressures (down to about 10^{-12} mmHg) an instrument called an *ion gauge* is used.

(*c*) *Rotary vacuum pump.* One rotary pump can produce pressures of about 10^{-2} mmHg; with two in series 10^{-4} mmHg is attainable. The pump comprises an eccentrically mounted cylindrical rotor inside and in contact at one point with a cylindrical stator, Fig. 9.13. Two spring-loaded vanes attached to the rotor press against the wall of the stator. The whole is im-

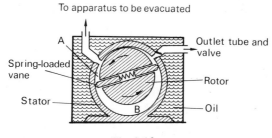

Fig. 9.13

mersed in a special low vapour pressure oil which both seals and lubricates the pump. As the rotor revolves, driven by an electric motor, each vane in turn draws gas into the increasing volume of space A on the intake side and then compresses it in space B where it is ejected from the outlet valve.

Lower pressures, down to 10^{-12} mmHg, require the use of a *diffusion pump*.

Surface tension

(*a*) *Some effects.* Various effects suggest that the surface of a liquid behaves like a stretched elastic skin, i.e. it is in a state of tension. For example, a steel needle will float if it is placed *gently* on the surface of a bowl of water, despite its greater density. (What else helps to support its weight?)

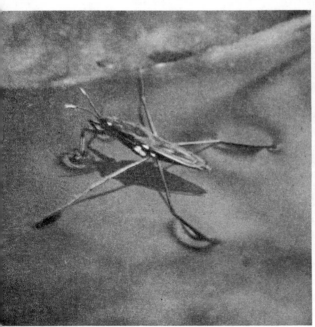

Fig. 9.14a   Fig. 9.14b

The effect, called *surface tension*, enables certain insects to run over the surface of a pond without getting wet, Fig. 9.14a.

Small liquid drops are nearly spherical, as can be seen when water drips from a tap, Fig. 9.14b; a sphere has the minimum surface area for a given volume. (What distorts larger drops?) This tendency of a liquid surface to shrink and have a minimum area can also be shown by the arrangement of Fig. 9.15a. When the soap film *inside* the loop of thread is punctured, the thread is pulled into the shape of a circle, Fig. 9.15b. Since a circle has the maximum area for a given perimeter, the area of the film outside the thread is a minimum.

Fig. 9.15

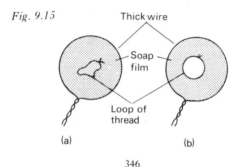

Thick wire

Soap film

Loop of thread

(a)   (b)

(b) *Definition and unit.* We can conclude from the circular shape of the thread in the previous demonstration that the liquid (soap solution) is pulling on the thread at right angles all along its circumference, Fig. 9.16a. This suggests that we might define the surface tension of a liquid in the following way.

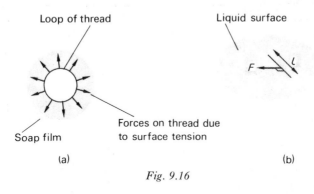

Loop of thread

Liquid surface

Soap film

Forces on thread due to surface tension

(a)

(b)

Fig. 9.16

Imagine a straight line of length l in the surface of a liquid. If the force acting at right angles to this line and in the surface is F, Fig. 9.16b, then the surface tension γ of the liquid is defined by

$$\gamma = \frac{F}{l}$$

In words, *γ is the force per unit length acting in the surface perpendicular to one side of a line in the surface.* The unit of γ is newton metre^{-1} (N m^{-1}). Its value depends on, among other things, the temperature of the liquid. At 20 °C, for water $\gamma = 72.6 \times 10^{-3}$ N m^{-1} and for mercury $\gamma = 465 \times 10^{-3}$ N m^{-1}.

It must be emphasized that normally surface tension acts equally on both sides of *any* line in the surface of a liquid; it creates a state of tension in the surface. The *effects* of surface tension are evident only when liquid is absent from one side of the line. For example in Fig. 9.17, to keep wire AB at rest an external force F has to be applied to the right to counteract the unbalanced surface tension forces acting to the left. A film has two surfaces and

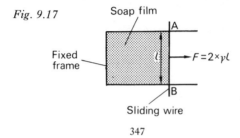

Fig. 9.17

Soap film

Fixed frame

Sliding wire

A

B

$F = 2 \times \gamma l$

so for a frame of width l, the surface tension force is $2\gamma l$ where γ is the surface tension of the liquid.

Or again, when a drop of methylated spirit or soap solution is dropped into the centre of a dish of water whose surface has been sprinkled with lycopodium powder, the powder rushes out to the sides leaving a clear patch. The effect is due to the surface tension of water being greater than that of meths or soap solution so that there is unbalance between the surface tension forces round the boundary of the two liquids. The powder is thus carried away from the centre by the water.

(c) *Molecular explanation.* Molecules in the surface of a liquid are farther apart than those in the body of the liquid, i.e. the surface layer has a lower density than the liquid in bulk. This follows because the increased separation of molecules which accompanies a change from liquid to vapour is not a sudden transition. The density of the liquid must therefore decrease *through* the surface.

The intermolecular forces in a liquid, like those in a solid, are both attractive and repelling and these balance when the spacing between molecules has its equilibrium value. However, from the intermolecular force–separation curve (Fig. 4.11a, p.132) we see that when the separation is greater than the equilibrium value (r_0), the attractive force between molecules exceeds the repelling force. This is the situation with the more widely spaced surface layer molecules of a liquid. They experience attractive forces on either side due to their neighbours which puts them in a state of tension, Fig. 9.18.

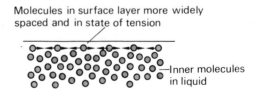

Fig. 9.18

The liquid surface thus behaves like a stretched elastic skin. If the tension-creating bonds between molecules are severed on one side by parting the liquid surface, then there is a resultant attractive force on the molecules due to the molecules on their other side. The effect of surface tension is then apparent.

The value of γ for a liquid does not increase when its surface area increases because more molecules enter the surface layer thereby keeping the molecular separation constant. Otherwise, any increase of separation would increase the attractive force between molecules and so also the surface tension.

Liquid surfaces

(a) *Shape of liquid surfaces.* The surface of a liquid must be at right angles to the resultant force acting on it, otherwise there would be a component of this force parallel to the surface which would cause motion. Normally a liquid surface is horizontal, i.e. at right angles to the force of gravity, but where it is in contact with a solid it is usually curved.

To explain the shape of the surface in Fig. 9.19 consider the liquid at B adjoining a vertical solid wall. It experiences an attractive force BC due to

Fig. 9.19

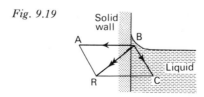

neighbouring liquid molecules; this is the *cohesive* force studied previously which binds liquid molecules together and makes them behave as a liquid. An attractive force BA is also exerted by neighbouring molecules of the solid; this is called the *adhesive* force and if it is greater than the cohesive force then the resultant force BR on the liquid at B will act to the left of the wall in the direction shown. The liquid surface at B has to be at right angles to this direction and so curves upwards. Since there is then equilibrium the resultant force must be balanced by appropriate intermolecular repulsive forces. At points on the liquid surface farther from the wall, the adhesive forces are smaller, the resultant force more nearly vertical and so the surface more nearly horizontal.

By contrast, when the cohesive force between molecules of the liquid is greater than the adhesive force between molecules of the liquid and molecules of the solid, the resultant force BR acts as in Fig. 9.20 and the surface curves downwards at the wall. This is the case with mercury against glass.

The *angle of contact θ* is defined as the angle between the solid surface and the tangent plane to the liquid surface, measured *through the liquid*. The

Fig. 9.20

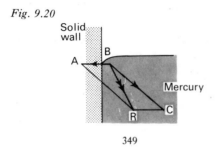

349

liquid in Fig. 9.21a has an acute angle of contact with this particular solid ($\theta < 90°$), while that in Fig. 9.21b has an obtuse angle of contact ($\theta > 90°$). Water, like many organic liquids, has zero angle of contact with a *clean* glass surface, i.e. the adhesive force is so much greater than the cohesive force that the water surface is parallel to the glass where it meets it, Fig. 9.21c. On

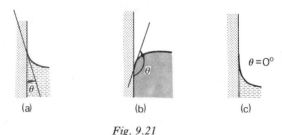

Fig. 9.21

a horizontal clean glass surface water tends to spread indefinitely and form a very thin film. Contamination of a surface affects the angle of contact appreciably; the value for water on greasy glass may be about 10° and causes it to form drops rather than spread. Mercury in contact with clean glass has an angle of contact of about 140° and tends to form drops instead of spreading over glass.

Liquids with acute angles of contact are said to 'wet' the surface, those with obtuse angles of contact do not 'wet' it. Fig. 9.22a shows a drop of a liquid which 'wets' the surface and b shows a drop on a surface which it does not 'wet'.

Fig. 9.22

(b) *Practical applications of spreading.* The behaviour of liquids in contact with solids is important practically. In soldering a good joint is formed only if the molten solder (a tin-lead alloy) 'wets' and spreads over the metal involved. Spreading occurs most readily if the liquid solder has a small surface tension. The use of a flux (e.g. resin) with the solder cleans the metal surface and acts as a 'wetting agent' which assists spreading. Metals like aluminium have an almost permanent oxide skin that resists the action of a flux and makes good soldered joints by normal methods very difficult.

'Wetting agents' play a key role in painting and spraying where the paint must not form drops but remain in a layer once spread out. The use of spreading agents (e.g. stearic acid) also assists lubricating oils to adhere to axles, bearings, etc.

If detergents are to remove the dirt particles that are held to fabrics usually by grease, they must be able to spread over the fabric before they can dislodge the grease. Detergent solutions should therefore, on this account, have low surface tensions and small angles of contact. By contrast, fabrics are weatherproofed by treatment with a silicone preparation which causes water to collect in drops and not to spread.

Capillarity

Surface tension causes a liquid with an angle of contact less than $90°$ to rise in a fine bore (capillary) tube above the level outside. The narrower the tube the greater the elevation, Fig. 9.23. The effect is called *capillarity* and is of practical importance.

Fig. 9.23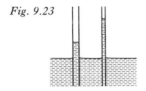

Why does the rise occur? In Fig. 9.24a, round the boundary where the liquid surface meets the tube, surface tension forces exert a *downwards* pull on the tube since they are not balanced by any other surface tension forces. The tube therefore exerts an equal but *upwards* force on the liquid, Fig. 9.24b, and causes it to rise (Newton's third law of motion). The liquid stops

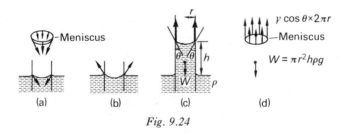

Fig. 9.24

rising when the weight of the raised column acting vertically downwards equals the *vertical component* of the upwards forces exerted by the tube on the liquid, Fig. 9.24c.

If the liquid has density ρ, surface tension γ and angle of contact θ and if the rise is h in a tube of radius r then neglecting the small amount of liquid in the meniscus,

$$\text{weight } W \text{ of liquid column} = \pi r^2 h \rho g$$

$$\text{vertical component of supporting forces} = \gamma \cos \theta \times 2\pi r$$

since these forces act round a circumference of $2\pi r$, Fig. 9.24*d*. Hence

$$\pi r^2 h \rho g = 2\pi r \, \gamma \cos \theta$$

$$\therefore \quad h = \frac{2\gamma \cos \theta}{r \rho g}$$

A more rigorous treatment shows that this equation only holds for very fine bore tubes in which the curvature of the meniscus is everywhere spherical; it also brings out the fact that r is the radius of the tube at the meniscus (p. 356).

An estimate of the rise h can be obtained by substituting values for γ, r, ρ and g in this expression. For example, for water in a very clean glass tube $\theta = 0°$, i.e. the water surface meets the tube vertically, and so $\cos \theta = 1$. Also $\rho = 1.0 \text{ g cm}^{-3} = 1.0 \times 10^3 \text{ kg m}^{-3}$ and $\gamma = 7.3 \times 10^{-2} \text{ N m}^{-1}$. If the capillary tube has radius 0.50×10^{-3} m then

$$h = \frac{2 \times 7.3 \times 10^{-2} \times 1}{0.50 \times 10^{-3} \times 1.0 \times 10^3 \times 9.8} \quad \frac{\text{N m}^{-1}}{\text{m} \times \text{kg m}^{-3} \times \text{N kg}^{-1}}$$

$$= 3.0 \times 10^{-2} \text{ m}$$

$$= 30 \text{ mm}$$

If θ is greater than 90°, the meniscus is convex upwards, $\cos \theta$ is negative and the expression shows that h will also be negative. This means the liquid falls in the capillary tube below the level of the surrounding liquid. Mercury in a glass capillary tube usually behaves in this way, Fig. 9.25. What happens when $\theta = 0°$?

Fig. 9.25

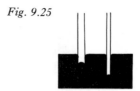

The drying action of blotting paper is due to the ink rising up the pores of the paper by capillarity. It also helps in soldering by causing the molten solder to penetrate any cracks. However, for this to happen the above expression for h indicates that the solder should have a high surface tension, a property which does not encourage spreading (see p. 350). Compromise is clearly necessary here as in the dyeing of fabrics where success depends largely on the dye penetrating into the fabric by capillarity.

Bubbles and drops

A study of bubbles and drops not only helps with the determination of the surface tension of liquids, as we shall see in a later section, but it also has practical relevance. Thus, the formation of gas bubbles plays an important part in the manufacture of expanded plastics such as polystyrene. In oil-fired boilers pressure burners depend on droplet formation for fast and efficient burning of the vapour. In steam heating systems the efficiency of heat transfer from the steam would be higher if instead of condensing as a film, which it does, it condensed in drops, and attempts are at present being made to achieve drop condensation.

A soap bubble blown on the end of a tube and then left open to the atmosphere gradually gets smaller, showing that the air is being forced out. Surface tension tries to make the film contract and thereby causes the pressure inside the bubble to exceed that outside. An expression can be obtained for the *excess pressure* inside a spherical soap bubble.

Consider a bubble of radius r, blown from a soap solution of surface tension γ. Let atmospheric pressure be P and suppose the pressure inside the bubble exceeds P by p, i.e. is $(P + p)$, Fig. 9.26a. Consider the equilibrium of *one half* of the bubble; there are two sets of opposing forces.

Fig. 9.26

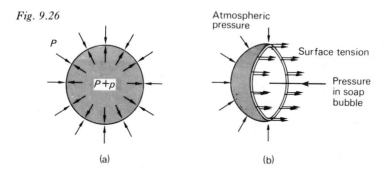

(a)   (b)

(*i*) Atmospheric pressure acts in different directions over the surface of the hemisphere but the resultant force in Fig. 9.26b acts *horizontally to the right* since the vertical components cancel (if any vertical variation of atmospheric pressure is neglected). It can be shown (see later *Note*) that the force exerted by a fluid in a certain direction on a curved surface equals the force on the projection of the surface on to a plane whose direction is perpendicular to the required direction. Here the projection of the hemisphere in a horizontal direction is a circle and so the horizontal force due to atmospheric pressure equals the product of the pressure and the area of projection i.e. $P \times \pi r^2$. Also, surface tension forces are exerted by the right-hand hemisphere (not shown in Fig. 9.26b) on the circular rim of the left-hand hemisphere along *both* its inside and outside surfaces. (Similar surface tension forces are exerted on the right-hand hemisphere by the left-hand one.) This force equals $2\gamma \times 2\pi r$ and so the total horizontal force to the right is $P\pi r^2 + 4\gamma \pi r$.

(*ii*) The pressure $(P + p)$ acts on the curved inside surface of the left-hand hemisphere and produces a *horizontal force to the left* equal to $(P + p)\pi r^2$.

Hence, if the horizontal forces balance, we have

$$P\pi r^2 + 4\gamma \pi r = (P + p)\pi r^2$$

$$\therefore \quad 4\gamma \pi r = p\pi r^2$$

The excess pressure p inside the bubble is then

$$p = \frac{4\gamma}{r}$$

Taking γ for a soap solution as 2.5×10^{-2} N m^{-1}, the excess pressure inside a bubble of radius 1.0 cm or 1.0×10^{-2} m is

$$p = \frac{4 \times 2.5 \times 10^{-2}}{1.0 \times 10^{-2}} \quad \frac{\text{N m}^{-1}}{\text{m}}$$

$$= 10 \text{ Pa}$$

If two soap bubbles of different radii are blown separately using the apparatus of Fig. 9.27, and then connected by opening taps T_1 and T_2 (T_3

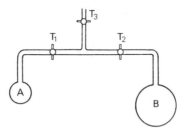

Fig. 9.27

being closed), the smaller bubble A gradually collapses while the larger bubble B expands. Why? Equilibrium is attained when A has become a small curved film of radius equal to that of bubble B.

A spherical drop of liquid in air or a bubble of gas in a liquid has only one surface and the excess pressure inside it is $2\gamma/r$, the proof being similar to that given above for a soap bubble.

Note. Force on a curved surface in a fluid exerting a uniform pressure. Consider a volume of fluid with end A hemispherical and end B plane, as in Fig. 9.28. The force on surface B is horizontal and must equal the horizontal

Fig. 9.28

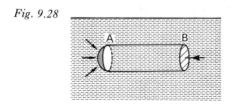

component of the resultant force on surface A (the vertical components at end A cancel) if the volume of fluid is in equilibrium. But surface B is the projection of surface A on a vertical plane, i.e. in a direction at right angles to the horizontal. Hence the force in, say, a horizontal direction on a curved surface equals the force on its projection on to a vertical plane.

Pressure difference across a spherical surface

It can be shown (in more advanced books) that due to surface tension, the pressure on the concave side of any spherical liquid surface of radius r exceeds that on the convex side by $2\gamma/r$ where γ is the surface tension of the liquid, Fig. 9.29. Bubbles and drops are special cases of this more general

Fig. 9.29

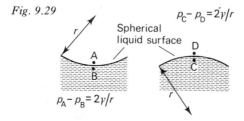

result which is useful when considering certain effects such as capillarity. Earlier we treated capillarity in terms of forces; the excess pressure method is more informative.

(a) *Capillary rise formula.* Consider a liquid of surface tension γ in a capillary tube. If the meniscus is everywhere spherical (as it will be in a very narrow tube) and if the angle of contact is zero, then the radius of the meniscus will equal the radius of the tube.

When the tube is first placed in the liquid, the curvature of the surface makes the pressure at D just below the meniscus less than that at E, which is atmospheric, Fig. 9.30a. Liquid will flow into the tube because of this pressure difference and capillary rise occurs.

Fig. 9.30

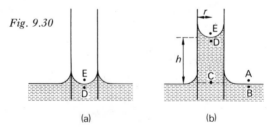

(a)　　　　　　　　　　　(b)

Let p_A, p_B, p_C, p_D and p_E be the pressures at A, B, C, D and E respectively, Fig. 9.30b. We have, once equilibrium is established,

$$p_A = p_B$$

(no pressure difference across a flat surface)

$$= p_C$$

$$= p_D + h\rho g$$

where ρ is the density of the liquid, r the radius of the tube and h the capillary rise. Also,

$$p_E = p_D + \frac{2\gamma}{r}$$

But $p_E = p_A$ (both are atmospheric pressure which is constant if we ignore the pressure due to the column AE of air)

$$\therefore \quad h\rho g = \frac{2\gamma}{r}$$

$$\therefore \quad h = \frac{2\gamma}{r\rho g}$$

If the angle of contact θ is zero this expression is the same as that obtained previously by the force method (p. 352). However, this derivation shows that the expression only holds strictly for tubes in which the meniscus is everywhere spherical, i.e. for fine bore tubes. The fact that r is the radius of the tube at the meniscus is also made clear.

(b) *Worked example. A* U-*tube with limbs of diameters 5.0 mm and 2.0 mm contains water of surface tension 7.0 $\times$ 10^{-2} N m^{-1}, angle of contact zero and density 1.0 $\times$ 10^3 kg m^{-3}. Find the difference in levels, Fig. 9.31 (g =* 10 m s^{-2}).

If the menisci are spherical they will be hemispheres since the angle of contact is zero; their radii will then equal the radii of the limbs. The pressure on the concave side of each surface exceeds that on the convex side by $2\gamma/r$ where γ is the surface tension and r is the radius of the limb concerned.

Fig. 9.31

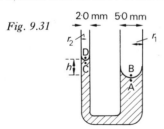

Now $r_1 = 2.5\,\text{mm} = 2.5 \times 10^{-3}\,\text{m}$ and $r_2 = 1.0\,\text{mm} = 1.0 \times 10^{-3}\,\text{m}$. Hence

$$p_B - p_A = \frac{2 \times 7.0 \times 10^{-2}}{2.5 \times 10^{-3}} = 56\,\text{Pa}$$

$$\therefore\ p_A = P - 56$$

where $p_B = P =$ atmospheric pressure. Also

$$p_D - p_C = \frac{2 \times 7.0 \times 10^{-2}}{1.0 \times 10^{-3}} = 140\,\text{Pa}$$

$$\therefore\ p_C = P - 140 \quad (\text{since } p_D = P)$$

$$\therefore\ p_A - p_C = (P - 56) - (P - 140)$$

$$= 84\,\text{Pa}$$

But
$$p_A = p_C + h\rho g$$

$$\therefore\ h\rho g = 84\,\text{Pa}$$

$$\therefore\ h = \frac{84}{10^3 \times 10}\,\text{m}$$

$$= 8.4\,\text{mm}$$

Methods of measuring γ

(a) *Capillary rise method.* The expression $\gamma = hr\rho g/(2 \cos \theta)$ is used (see p. 352) and so θ must be known. In the case of a liquid for which $\theta = 0°$ the expression becomes

$$\gamma = \frac{hr\rho g}{2}$$

Knowing the density ρ of the liquid and g, only h and r remain to be determined.

The apparatus is shown in Fig. 9.32, the capillary tube is previously cleaned thoroughly by immersing in caustic soda, dilute nitric acid and

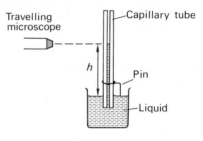

Fig. 9.32

distilled water in turn. A travelling microscope is focused first on the bottom of the meniscus in the tube and then, with the beaker removed, on the tip of the pin which previously just touched the surface of the liquid in the beaker. Hence h is obtained.

To find r the tube is broken at the meniscus level and the average reading of two diameters at right angles taken with the travelling microscope.

Surface tension decreases rapidly with temperature and so the temperature of the liquid should be stated.

(b) *Jaeger's method.* This method measures the excess pressure required to blow an air bubble in the liquid under investigation and then γ is calculated using $p = 2\gamma/r$.

The pressure inside the apparatus, Fig. 9.33a, is gradually increased by allowing water to enter the flask from the dropping funnel and the increase

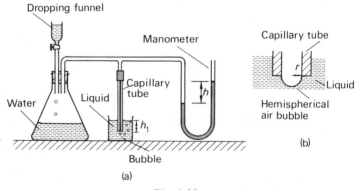

Fig. 9.33

is recorded on the manometer (containing a low density liquid such as xylol). An air bubble grows at the end of the capillary tube in the beaker of test liquid and as it does so the pressure rises to a maximum and then falls as the bubble breaks away. The maximum pressure will occur when the radius of the bubble is a minimum. Assuming that the bubble is then hemispherical with radius equal to that of the bore of the tube, Fig. 9.33b,

$$\text{pressure inside bubble} = P + h\rho g$$

where P is the atmospheric pressure, h the *maximum* manometer reading and ρ is the density of the liquid in the manometer. Also

$$\text{pressure in liquid outside bubble} = P + h_1\rho_1 g$$

where h_1 is the depth of the end of the capillary tube in the test liquid of density ρ_1. Hence

$$\text{excess pressure in bubble} = (P + h\rho g) - (P + h_1\rho_1 g)$$

$$= h\rho g - h_1\rho_1 g$$

But, $$\text{excess pressure} = \frac{2\gamma}{r}$$

where γ is the surface tension of the liquid and r is the radius of the tube at the end.

$$\therefore \quad \frac{2\gamma}{r} = (h\rho - h_1\rho_1)g$$

$$\therefore \quad \gamma = \frac{gr}{2}(h\rho - h_1\rho_1)$$

It is essential to measure h, h_1 and r carefully. One way of obtaining h is to arrange two pins so that their points mark the liquid levels in the manometer at the instant of maximum pressure and then to measure the distance between them afterwards with a travelling microscope. The same instrument should be used to find h_1 and r. Best results are achieved when a bubble is formed every few seconds.

Every bubble has a fresh surface and so the risk of contamination is small if the tube is clean. Also, measurements at different temperatures are easily made by changing and maintaining at any required value the temperature of the liquid in the beaker. The method is most suitable for accurate comparisons between different liquids and for one liquid at different temperatures. Absolute measurements are not reliable because the assumption that the minimum bubble radius equals the radius of the tube is not quite true.

FLUIDS AT REST

QUESTIONS

Pressure: Archimedes

1. Define *pressure at a point* in a fluid. In what unit is it measured?

State an expression for the pressure at a point at depth h in a liquid of density ρ. Does it also hold for a gas?

What force is exerted on the bottom of a tank of uniform cross-section area 2.0 m^2 by water which fills it to a depth of 0.50 m? (Density of water $= 1.0 \times 10^3$ kg m^{-3}; $g = 10$ N kg^{-1}.)

Find the extra force on the bottom of the tank when a block of wood of volume 1.0×10^{-1} m^3 and relative density 0.50 floats on the surface.

2. (a) State Archimedes' principle.

(b) A string supports a solid copper block of mass 1 kg (density 9×10^3 kg m^{-3}) which is completely immersed in water (density 1×10^3 kg m^{-3}). Calculate the tension in the string.

3. A specimen of an alloy of silver and gold, whose densities are 10.50 and 18.90 g cm^{-3} respectively, weighs 35.20 g in air and 33.13 g in water. Find the composition, by mass, of the alloy, assuming that there has been no volume change in the process of producing the alloy. (*W. part qn.*)

4. A simple hydrometer, consisting of a loaded glass bulb fixed at the bottom of a glass stem of uniform cross section, sinks in water of density 1.0 g cm^{-3} so that a certain mark X on the stem is 4.0 cm below the surface. It sinks in a liquid of density 0.90 g cm^{-3} until X is 6.0 cm below the surface. It is then placed in a liquid of density 1.1 g cm^{-3}. How far below the surface will X be? (Neglect surface tension effects.) (*S.*)

5. A barometer is exactly 760.0 mm high with the temperature at $20\,^\circ$C. What would be its height for the same pressure if the temperature were $0\,^\circ$C?

Calculate the expansion of a brass scale 760.0 mm long for a temperature rise of $20\,^\circ$C.

What height would have been measured for the barometer when it was exactly 760.0 mm high at $20\,^\circ$C if a brass scale had been used which was correct at $0\,^\circ$C?

(Linear expansivity of brass $1.800 \times 10^{-5}\,^\circ$C^{-1}; cubic expansivity of mercury $1.800 \times 10^{-4}\,^\circ$C^{-1}.) (*S.*)

6. A simple reciprocating exhaust pump has a piston area 0.002 m^2 and a stroke length 0.2 m. It is directly connected to a vessel of volume 0.01 m^3, containing air at atmospheric pressure.

Calculate the minimum number of strokes needed to reduce the pressure in the vessel to 0.01 atmosphere, assuming isothermal conditions.

Discuss briefly the validity of this assumption and state any further assumptions made. (*J.M.B. Eng. Sc.*)

Surface tension

7. Explain (a) in terms of molecular forces why the water is drawn up above the horizontal liquid level round a steel needle which is held vertically and partly immersed in water, (b) why, in certain circumstances, a steel needle will rest on a water surface. In each case show the relevant forces on a diagram. (*J.M.B.*)

8. Explain, using a simple molecular theory, why the surface of a liquid behaves in a different manner from the bulk of the liquid.

Giving the necessary theory, explain how the rise of water in a capillary tube may be used to determine the surface tension of water.

A microscope slide measures 6.0 cm × 1.5 cm × 0.20 cm. It is suspended with its face vertical and with its longest side horizontal and is lowered into water until it is half immersed. Its apparent weight is then found to be the same as its weight in air. Calculate the surface tension of water assuming the angle of contact to be zero. (*A.E.B.*)

9. A clean glass capillary tube of internal diameter 0.60 mm is held vertically with its lower end in water and with 80 mm of the tube above the surface. How high does the water rise in the tube?

If the tube is now lowered until only 30 mm of its length is above the surface, what happens? Surface tension of water is 7.2×10^{-2} N m^{-1}.

10. Describe and explain *two* experiments of a different nature to illustrate the phenomenon of surface tension.

Give a quantitative definition of *surface tension* and explain what is meant by *angle of contact*.

The internal diameter of the tube of a mercury barometer is 3.00 mm. Find the corrected reading of the barometer after allowing for the error due to surface tension, if the observed reading is 76.56 cm. (Surface tension of mercury = 4.80×10^{-1} N m^{-1}; angle of contact of mercury with glass = $140°$; density of mercury = 13.6×10^{3} kg m^{-3}.) (*L.*)

10 Fluids in motion

Introduction

The study of moving fluids is important in engineering. A large quantity of liquid may have to flow rapidly through a pipe from one location to another or air entering the inlet of a machine, e.g. a jet engine, may have to be transported to the outlet, undergoing changes of pressure, temperature and speed as it passes. In all such cases of *mass transport* a knowledge of the conditions existing at various points in the system is essential for efficient design.

Fluid dynamics is a complex subject which we shall only touch upon in this chapter.

Viscosity

If adjacent layers of a material are displaced laterally over each other as in Fig. 10.1a, the deformation of the material is called a *shear*. Basically the simplest type of fluid flow involves shear as we shall now see.

All liquids and gases (except *very* low density gases) stick to a solid surface so that when they flow the velocity must gradually decrease to zero as the wall of the pipe or containing vessel is approached. (The existence of a stationary layer may be inferred from the fact that whilst large particles of

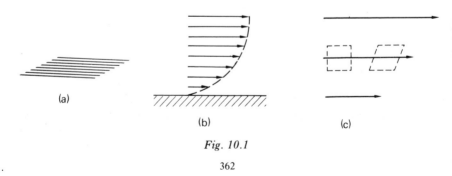

(a)

(b)

(c)

Fig. 10.1

dust can be blown off a shelf, small particles remain which can be wiped off subsequently with the finger.) A fluid is therefore sheared when it flows past a solid surface and the *opposition set up by the fluid to shear* is called its *viscosity*. Liquids such as syrup and engine oil which pour slowly are more viscous than water.

Viscosity is a kind of internal friction exhibited to some degree by all fluids. It arises in liquids because the forced movement of a molecule relative to its neighbours is opposed by the intermolecular forces between them.

When the fluid particles passing successively through a given point in a fluid always follow the same path afterwards, the flow is said to be *steady*. 'Streamlines' can be drawn to show the direction of motion of the particles and are shown in Fig. 10.1*b* for steady flow of the water at various depths near the centre of a wide river. The layer of water in contact with the bottom of the river must be at rest (or the river bed would be rapidly eroded) and the velocities of higher layers increase the nearer the layer is to the surface. The length of the streamlines represents the magnitude of the velocities. The water suffers shear, a cube becoming a rhombus, Fig. 10.1*c*, as if acted on by tangential forces at its upper and lower faces. Steady flow thus involves parallel layers of fluid sliding over each other with different velocities thereby creating viscous forces acting tangentially (as shear forces do) between the layers and impeding their motion.

Coefficient of viscosity

To obtain a definition of viscosity we consider two plane parallel layers of liquid separated by a very small distance δy and having velocities $v + \delta v$ and v, Fig. 10.2. The *velocity gradient* (i.e. change of velocity/distance) in a

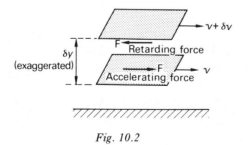

Fig. 10.2

direction perpendicular to the velocities is $\delta v/\delta y$. The slower, lower layer exerts a tangential retarding force F on the faster upper layer and experiences itself an equal and opposite tangential force F due to the upper layer (Newton's third law). The *tangential stress* between the layers is therefore

F/A where A is their area of contact. The *coefficient of viscosity* η is defined by the equation

$$\eta = \frac{tangential\ stress}{velocity\ gradient} = \frac{F/A}{\delta v/\delta y}$$

In words η is the *tangential force per unit area of fluid which resists the motion of one layer over another when the velocity gradient between the layers is unity.* If $\delta v/\delta y$ is small and F/A large, η is large and the fluid very viscous. For many pure liquids (e.g. water) and gases η is independent of the velocity gradient at a particular temperature, i.e. η is constant and so the *tangential stress is directly proportional to the velocity gradient.* Fluids for which this is true are called *Newtonian fluids* since Newton first suggested this relationship might hold. For some liquids such as paints, glues and liquid cements, η decreases as the tangential stress increases and these are said to be *thixotropic.*

The equation defining η shows that it can be measured in newton second metre^{-2} ($N\,s\,m^{-2}$) i.e. in pascal second ($Pa\,s$). Check this. At 20 °C, η for water is $1.0 \times 10^{-3}\,Pa\,s$ and for glycerine $8.3 \times 10^{-1}\,Pa\,s$. Experiment shows that the coefficient of viscosity of a liquid usually decreases rapidly with temperature rise.

Viscosity is an essential property of a lubricating oil if it is to keep apart two solid surfaces in relative motion. Too high viscosity on the other hand causes unnecessary resistance to motion. 'Viscostatic' oils have about the same value of η whether cold or hot.

It should be noted that viscous forces are called into play as soon as fluid flow starts. If the external forces causing the flow are constant, the rate of flow becomes constant and a steady state is attained with the resisting viscous forces equal to the applied force. The viscous forces stop the flow when the applied force is removed.

Poiseuille's formula: steady and turbulent flow

The streamlines for *steady* flow in a circular pipe are shown in Fig. 10.3. Everywhere they are parallel to the axis of the pipe and represent velocities varying from zero at the wall of the pipe to a maximum at its axis. The surfaces of equal velocity are the surfaces of concentric cylinders.

Fig. 10.3

An expression for the volume of liquid passing per second, V, through a pipe when the flow is *steady*, can be obtained by the method of dimensions (see p. 393). It is reasonable to assume that V depends on (*i*) the coefficient of viscosity η of the liquid, (*ii*) the radius r of the pipe and (*iii*) the pressure gradient p/l causing the flow, where p is the pressure difference between the ends of the pipe and l is its length, Fig. 10.4. We have

$$V = k\eta^x r^y (p/l)^z$$

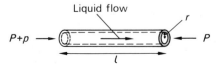

Liquid flow

$P+p \longrightarrow$ $\longleftarrow P$

l

$P = $ atmospheric pressure

Fig. 10.4

where x, y and z are the indices to be found and k is a dimensionless constant. The dimensions of V are $[L^3 T^{-1}]$, of η $[ML^{-1}T^{-1}]$, of r $[L]$, of p $[MLT^{-2}/L^2]$, i.e. $[ML^{-1}T^{-2}]$ (since pressure $=$ force/area), and of l $[L]$. Hence the dimensions of p/l are $[ML^{-2}T^{-2}]$.

Equating dimensions,

$$[L^3 T^{-1}] = [ML^{-1}T^{-1}]^x [L]^y [ML^{-2}T^{-2}]^z$$

Equating indices of M, L and T on both sides,

$$0 = x + z$$
$$3 = -x + y - 2z$$
$$-1 = -x - 2z$$

Solving, we get $x = -1$, $y = 4$ and $z = 1$. Hence

$$V = \frac{kpr^4}{\eta l}$$

The value of k cannot be obtained by the method of dimensions but a fuller analysis shows that it equals $\pi/8$ and so the complete expression is

$$V = \frac{\pi p r^4}{8\eta l}$$

This is known as *Poiseuille's formula* since he made the first thorough experimental investigation of the *steady* flow of liquid through a pipe in 1844.

So far we have considered only steady flow. When the velocity of flow exceeds a certain critical value the motion becomes *turbulent*, the liquid is churned up and the streamlines are no longer parallel and straight. The change from steady to turbulent flow can be studied with the apparatus of Fig. 10.5. The flow of water along the tube T is controlled by the clip C. Potassium permanganate solution from the reservoir R is fed into the water flowing through T by a fine jet J. At low flow velocities a fine coloured stream is observed along the centre of T, but as the rate of flow increases it starts to break up and the colour rapidly spreads throughout T indicating the onset of turbulence.

Reynolds's number (*Re*) is useful in the study of the stability of fluid flow.

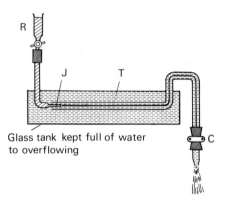

Fig. 10.5

It is defined by the equation

$$Re = \frac{vl\rho}{\eta}$$

where η and ρ are the viscosity and density respectively of the fluid, v is the speed of the bulk of the fluid and l is a characteristic dimension of the solid body concerned. For a cylindrical pipe l is usually the diameter ($2r$) of the pipe. Experiment shows that for cylindrical pipes, when

$Re < 2200$, flow is steady

$Re \simeq 2200$, flow is unstable (critical velocity v_c)

$Re > 2200$, flow is usually turbulent

Hence large η and small v, r and ρ promote steady flow.

Poiseuille's formula holds for velocities of flow below v_c.

Motion in a fluid: Stokes' law

The streamlines for a fluid flowing *slowly* past a stationary solid sphere are shown in Fig. 10.6. When the sphere moves slowly rather than the fluid, the pattern is similar but the streamlines then show the apparent motion of the fluid particles as seen by someone on the moving sphere. In this latter

Fig. 10.6   Viscous fluid

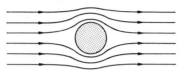

case it is known that the layer of fluid in contact with the sphere moves with it, thus creating a velocity gradient between this layer and other layers of the fluid. Viscous forces are thereby brought into play and constitute the resistance experienced by the moving sphere.

If we make the plausible assumption that the viscous retarding force F depends on (*i*) the viscosity η of the fluid, (*ii*) the velocity v and radius r of the sphere, then an expression can be derived for F by the method of dimensions. Thus

$$F = k\eta^x v^y r^z$$

where x, y and z are the indices to be found and k is a dimensionless constant. The dimensional equation is

$$[MLT^{-2}] = [ML^{-1}T^{-1}]^x [LT^{-1}]^y [L]^z$$

Equating indices of M, L and T on both sides,

$$1 = x$$
$$1 = -x + y + z$$
$$-2 = -x - y$$

Solving, we get $x = 1$, $y = 1$ and $z = 1$. Hence

$$F = k\eta v r$$

A detailed treatment, first done by Stokes, gives $k = 6\pi$ and so

$$F = 6\pi\eta v r$$

This expression, called *Stokes' law*, only holds for *steady motion* in a fluid of *infinite extent* (otherwise the walls and bottom of the vessel affect the resisting force).

Now consider the sphere falling vertically under gravity in a viscous fluid. Three forces act on it, Fig. 10.7,

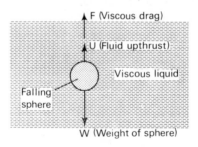

Fig. 10.7

(i) its weight W, acting downwards,

(ii) the upthrust U due to the weight of fluid displaced, acting upwards, and

(iii) the viscous drag F, acting upwards.

The resultant downward force is $(W - U - F)$ and causes the sphere to accelerate until its velocity, and so the viscous drag, reach values such that

$$W - U - F = 0$$

The sphere then continues to fall with a constant velocity, known as its *terminal velocity*, of say v_t. Now

$$W = \tfrac{4}{3}\pi r^3 \rho g \text{ where } \rho \text{ is the density of the sphere}$$

$$U = \tfrac{4}{3}\pi r^3 \sigma g \text{ where } \sigma \text{ is the density of the fluid}$$

Also, *if steady conditions still hold* when velocity v_t is reached then by Stokes' law

$$F = 6\pi\eta r v_t$$

Hence

$$\tfrac{4}{3}\pi r^3 \rho g - \tfrac{4}{3}\pi r^3 \sigma g - 6\pi\eta r v_t = 0$$

$$\therefore \quad v_t = \frac{2r^2(\rho - \sigma)g}{9\eta}$$

All that has been said so far applies to steady flow. As the velocity of the sphere increases, a critical velocity v_c is reached when the flow breaks up, eddies are formed as in Fig. 10.8a and the motion becomes *turbulent*. At velocities greater than v_c the resistance to motion, called the *drag*, increases sharply and is roughly proportional to the square of the velocity.

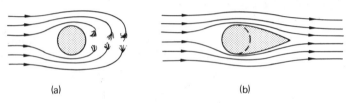

(a)   (b)

Fig. 10.8

(Below v_c Stokes' law indicates that the resistance is proportional to the velocity.) For highly turbulent flow resistance is dependent on density, not viscosity; this is the ordinary case of air resistance to vehicles.

By modifying the shape of a body the critical velocity can be raised and the drag thereby reduced at a particular speed if steady flow replaces turbulent flow. This is called *streamlining* the body and Fig. 10.8*b* shows how it is done for a sphere. The pointed tail can be regarded as filling the region where eddies occur in turbulent motion, thus ensuring that the streamlines merge again behind the sphere. Streamlining is particularly important in the design of high-speed aircraft.

The air flow past a model 'Mini' car in a wind tunnel is shown in Fig. 10.9. The flow is visualized by streams of finely condensed paraffin vapour giving the appearance of smoke trails. Features illustrated are flow separation part-way along the bonnet, vortex ('whirlwind') flow across the foot of the windscreen, flow separation just off the top of the windscreen roof joint and unsteady flow behind the car (as shown by the dispersion of the smoke trails).

Fig. 10.9

Methods of measuring η

(a) *Using Poiseuille's formula.* The method is suitable for a liquid obtainable in large quantities and which flows easily, e.g. water. The liquid passes *slowly* from a constant head apparatus through a capillary tube and the volume collected in a certain time is found, Fig. 10.10a.

Fig. 10.10

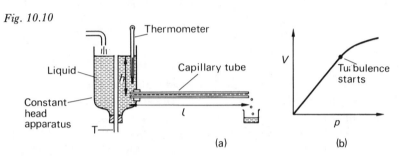

(a)

(b)

By altering the position of tube T, the rates of flow for different pressure differences can be measured. If a graph of volume delivered per second V against pressure difference p is plotted, the onset of turbulence, to which Poiseuille's formula does not apply, will be shown by non-linearity, Fig. 10.10b. The slope of the linear part of the graph gives an average value for V/p where $p = h\rho g$, ρ being the density of the liquid and h the pressure head. Knowing the length l of the capillary tube and its radius r, the viscosity η of the liquid at the particular temperature is calculated from

$$\eta = \frac{\pi p r^4}{8Vl}$$

To measure r with the care required (since it is small and appears to the fourth power) a long thread of mercury is introduced into the tube and its length and mass found. A narrow bore capillary tube is used so that steady flow is obtained with pressure differences that are not so small as to be difficult to measure accurately.

(b) *Ostwald's viscometer.* Viscosities can be easily and rapidly *compared* using this instrument, Fig. 10.11. A certain volume of liquid is introduced via E and sucked up into bulb B until its upper level is above mark A. It is then allowed to flow under its own weight through capillary CD and the time t_1 found for the upper level to fall between marks A and C. This is also the time for a volume of liquid equal to the volume V of the viscometer between A and C to flow through CD. The experiment is repeated and t_2 found with the *same volume* of another liquid (or with the same liquid at a different temperature if the variation of η with temperature of a given liquid is being studied).

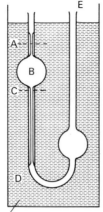

Fig. 10.11

Constant temperature bath

The pressure difference causing the flow decreases during the flow but since the viscometer always contains the same volume of liquid, the average difference of level, say h, is always the same. Hence for liquids of densities ρ_1 and ρ_2 the average pressure differences are $h\rho_1 g$ and $h\rho_2 g$ respectively. Thus for steady flow of the first liquid of viscosity η_1

$$\frac{V}{t_1} = \frac{\pi h \rho_1 g r^4}{8\eta_1 l}$$

and for the second of viscosity η_2

$$\frac{V}{t_2} = \frac{\pi h \rho_2 g r^4}{8\eta_2 l}$$

where r and l are the radius and length of the capillary respectively. Therefore

$$\frac{\eta_1}{\eta_2} = \frac{\rho_1 t_1}{\rho_2 t_2}$$

The method is widely used in practice because of its simplicity and accuracy.

(c) *Using Stokes' law.* The viscosity of a liquid such as glycerine or a heavy oil, whose high viscosity makes the previous methods unsuitable, may be found by timing a small ball-bearing falling with its terminal velocity through the liquid. So long as the terminal velocity does not exceed the critical velocity, i.e. the flow is steady, Stokes' law applies and we can therefore say

$$v_t = \frac{2r^2(\rho - \sigma)g}{9\eta} \qquad \text{(p. 368)}$$

where v_t is the terminal velocity, r and ρ the radius and density of the ball-bearing and σ the density of the liquid. The viscosity η can then be calculated.

To satisfy as far as possible the assumption in Stokes' law that the liquid is infinite in extent, the vessel of liquid must be wide compared with the diameter of the ball-bearing (less than 2 mm for a wide measuring cylinder); it should also be deep, Fig. 10.12. The terminal velocity v_t is obtained by

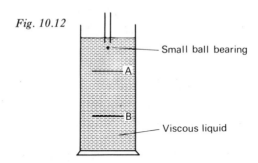

Fig. 10.12

Small ball bearing

A

B

Viscous liquid

finding the average time t taken by balls of the same size to fall from mark A (which is far enough below the surface for the ball to have its terminal velocity at A) to mark B (which is not too near the bottom of the vessel). Then $v_t = AB/t$.

To reduce the chance of air bubbles adhering to the falling ball it should be dipped in some of the liquid and thereby coated, before dropping. The temperature of the liquid should also be kept constant.

Bernoulli's equation

The pressure is the same at all points on the same horizontal level in a fluid at rest; this is not so when the fluid is in motion. The pressure at different points in a liquid flowing through (a) a uniform tube and (b) a tube with a narrow part, is shown by the height of liquid in the vertical manometers in Figs. 10.13a and b. In (a) the pressure drop along the tube is steady and maintains the flow against the viscosity of the liquid. In (b) the pressure falls in the narrow part B but rises again in the wider part C. If the liquid can be

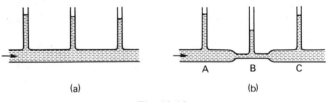

(a)   A   B   C
                                          (b)

Fig. 10.13

assumed to be incompressible, the same volume of liquid passes through B in a given time as enters A and so the velocity of the liquid must be greater in B than in A or C. Therefore a decrease of pressure accompanies an increase of velocity. This may be shown by blowing into a 'tunnel' made from a sheet of paper, Fig. 10.13c. The faster one blows the more does the tunnel collapse.

Fig. 10.13c

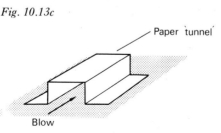

Paper tunnel

Blow

A useful relation can be obtained between the pressure and the velocity at different parts of a fluid in motion.

Suppose a fluid flows through a non-uniform tube from X to Y, Fig. 10.14, and its velocity changes from v_1 at X where the cross-section area is A_1 to

Fig. 10.14

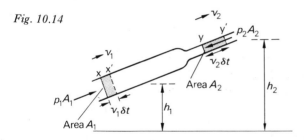

v_2 at Y where the cross-section is A_2. The flow of fluid between X and Y is caused by the forces acting on its ends which arise from the pressure exerted on it by the fluid on either side of it. At X, if the fluid pressure is p_1, there is a force p_1A_1 acting in the direction of flow and at Y, if the fluid pressure is p_2, a force p_2A_2 opposes the flow. Consider a small time interval δt in which the fluid at X has moved to X' and that at Y to Y'.

At X, work done during δt *on* the fluid XY by p_1A_1 pushing it into the tube

$$= \text{force} \times \text{distance moved} = \text{force} \times \text{velocity} \times \text{time}$$

$$= p_1A_1 \times v_1 \times \delta t$$

At Y, work done during δt *by* the fluid XY emerging from the tube against $p_2A_2 = p_2A_2 \times v_2 \times \delta t$. Therefore

$$\text{net work } W \text{ done } on \text{ the fluid} = (p_1A_1v_1 - p_2A_2v_2)\,\delta t$$

373

If the fluid is incompressible, volume between X and X' equals volume between Y and Y', i.e.

$$A_1 \times v_1 \, \delta t = A_2 \times v_2 \, \delta t$$

$$\therefore \quad W = (p_1 - p_2)A_1 v_1 \, \delta t$$

As a result of the work done on it, the fluid gains p.e. and k.e. when XY moves to X'Y'.

Gain of p.e. = p.e. of X'Y' − p.e. of XY

= p.e. of X'Y + p.e. of YY' − p.e. of XX' − p.e. of X'Y

= p.e. of YY' − p.e. of XX'

= $(A_2 v_2 \, \delta t \rho)gh_2 - (A_1 v_1 \, \delta t \rho)gh_1$ (since p.e. = mgh)

= $A_1 v_1 \, \delta t \rho g(h_2 - h_1)$ (since $A_1 v_1 \, \delta t = A_2 v_2 \, \delta t$)

where h_1 and h_2 are the heights of XX' and YY' above an arbitrary horizontal reference level and ρ is the density of the fluid. Similarly,

gain of k.e. = k.e. of YY' − k.e. of XX'

= $\frac{1}{2}(A_2 v_2 \, \delta t \rho)v_2^2 - \frac{1}{2}(A_1 v_1 \, \delta t \rho)v_1^2$ (since k.e. = $\frac{1}{2}mv^2$)

= $\frac{1}{2}A_1 v_1 \, \delta t \rho(v_2^2 - v_1^2)$

If the fluid is non-viscous (i.e. inviscid) no work is done against viscous forces to maintain the flow, no change of internal energy of the fluid occurs and by the principle of conservation of energy we have

net work done *on* fluid = gain of p.e. + gain of k.e.

$$\therefore \quad (p_1 - p_2)A_1 v_1 \, \delta t = A_1 v_1 \, \delta t \rho g(h_2 - h_1) + \frac{1}{2}A_1 v_1 \, \delta t \rho(v_2^2 - v_1^2)$$

$$p_1 - p_2 = \rho g(h_2 - h_1) + \frac{1}{2}\rho(v_2^2 - v_1^2)$$

or $\qquad p_1 + h_1 \rho g + \frac{1}{2}\rho v_1^2 = p_2 + h_2 \rho g + \frac{1}{2}\rho v_2^2$

This is *Bernouilli's equation* and it is usually stated by saying that *along a streamline in an incompressible, inviscid fluid*

$$p + h\rho g + \frac{1}{2}\rho v^2 = constant$$

In deriving the equation we have in effect assumed that the pressure and velocity are uniform over any cross-section of the tube. This is not so for a real (viscous) fluid and so it only applies strictly to a single streamline in the fluid. In addition, actual fluids, especially gases, are compressible. The equation has therefore to be applied with care or the results will be misleading.

Applications of Bernoulli

(a) *Jets and nozzles.* Bernoulli's equation suggests that for fluid flow where the potential energy change $h\rho g$ is very small or zero, as in a horizontal pipe, *the pressure falls when the velocity rises.* The velocity increases at a constriction—a slow stream of water from a tap can be converted into a fast jet by narrowing the exit with a finger—and the greater the change in cross-sectional area, the greater is the increase of velocity and so the greater is the pressure drop. Several devices with jets and nozzles use this effect; Fig. 10.15 shows the action of a Bunsen burner, a filter pump and a paint spray.

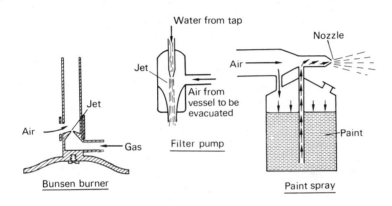

Fig. 10.15

(b) *Spinning ball.* If a tennis ball is 'cut' or a golf ball 'sliced' it spins as it travels through the air and experiences a sideways force which causes it to curve in flight. This is due to air being dragged round by the spinning ball, thereby increasing the air flow on one side and decreasing it on the other. A pressure difference is thus created, Fig. 10.16. The swing of a spinning cricket ball is complicated by its raised seam.

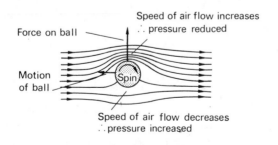

Fig. 10.16

(c) *Aerofoil*. This is a device which is shaped so that relative motion between it and a fluid produces a force perpendicular to the flow. Examples of aerofoils are aircraft wings, turbine blades and propellors.

The shape of the aerofoil section in Fig. 10.17 is such that fluid flows faster over the top surface than over the bottom, i.e. the streamlines are

Fig. 10.17

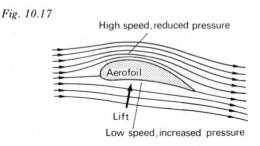

High speed, reduced pressure

Aerofoil

Lift

Low speed, increased pressure

closer above than below the aerofoil. By Bernoulli, it follows that the pressure underneath is increased and that above reduced. A resultant upwards force is thus created, normal to the flow and it is this force which provides most of the 'lift' for an aircraft. Its value increases with the angle between the wing and the air flow (called the 'angle of attack') until at a certain angle the flow separates from the upper surface, lift is lost almost completely, drag increases sharply, the flow downstream becomes very turbulent and the aircraft stalls.

The sail of a yacht 'tacking' into the wind is another example of an aerofoil. The air flow over the sail produces a pressure increase on the windward side and a decrease on the leeward side. The resultant force is roughly

Fig. 10.18

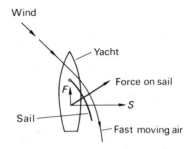

Wind

Yacht

Force on sail

F

S

Sail

Fast moving air

normal to the sail and can be resolved into a component F producing forward motion and a greater component S acting sideways, Fig. 10.18. The keel produces a lateral force to balance S.

FLUIDS IN MOTION

Flowmeters

These measure the rate of flow of a fluid through a pipe. Two types will be considered.

(a) *Venturi meter*. This consists of a horizontal tube with a constriction and replaces part of the piping of a system, Fig. 10.19. The two vertical

Fig. 10.19

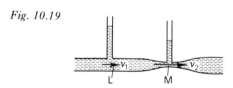

tubes record the pressures (above atmospheric) in the fluid flowing in the normal part of the tube and in the constriction.

If p_1 and p_2 are the pressures and v_1 and v_2 the velocities of the fluid (density ρ) at L and M on the same horizontal level, then assuming Bernoulli's equation holds

$$p_1 + \tfrac{1}{2}\rho v_1{}^2 = p_2 + \tfrac{1}{2}\rho v_2{}^2 \qquad (\text{since } h_1 = h_2)$$

$$\therefore \quad p_1 - p_2 = \tfrac{1}{2}\rho(v_2{}^2 - v_1{}^2)$$

If A_1 and A_2 are the cross-sectional areas at L and M and if the fluid is incompressible, the same volume passes each section of the tube per second

$$\therefore \quad A_1 v_1 = A_2 v_2$$

Hence
$$p_1 - p_2 = \tfrac{1}{2}\rho v_1{}^2 \left(\frac{A_1{}^2}{A_2{}^2} - 1 \right)$$

Knowing A_1, A_2, ρ and $(p_1 - p_2)$, v_1 can be found and so also the rate of flow $A_1 v_1$. Why is the above equation not valid for (*i*) a gas, (*ii*) a heavy oil, (*iii*) very rapid flow?

(b) *Pitot tube*. The pressure exerted by a moving fluid, called the *total pressure*, can be regarded as having two components: the *static pressure* which it would have if it were at rest and the *dynamic pressure* which is the pressure equivalent of its velocity. The Pitot tube measures total pressure and in essence is a manometer with one limb parallel to the flow and open to the oncoming fluid, Fig. 10.20. The fluid at the open end is at rest and a

Fig. 10.20

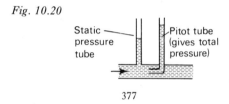

FLUIDS IN MOTION

'stagnant' region exists there. The total pressure is also called the *stagnation pressure*. The static pressure is measured by a manometer connected at right angles to the pipe or surface over which the fluid is passing.

In Bernoulli's equation

$$p + h\rho g + \tfrac{1}{2}\rho v^2 = constant$$

the static pressure is given by $p + h\rho g$ or by p if the flow is horizontal, the dynamic pressure by $\tfrac{1}{2}\rho v^2$ and the total pressure by $p + \tfrac{1}{2}\rho v^2$. Hence

$$total\ pressure\ -\ static\ pressure = p + \tfrac{1}{2}\rho v^2 - p = \tfrac{1}{2}\rho v^2$$

$$\therefore\quad v = \sqrt{\frac{2}{\rho}\ (total\ pressure\ -\ static\ pressure)}$$

This expression enables a value for the velocity of flow v of an incompressible, inviscid fluid to be calculated from the readings of Pitot-static tubes. In real cases v varies across the diameter of the pipe carrying the fluid (because of its viscosity) but it can be shown that if the open end of the Pitot tube is offset from the axis of the pipe by $0.7 \times$ radius of the pipe, then v is the *average* flow velocity.

Fluid flow calculations

1. *A garden sprinkler has 150 small holes each 2.0 mm^2 in area, Fig. 10.21. If water is supplied at the rate of 3.0 $\times$ 10^{-3} m^3 s^{-1}, what is the average velocity of the spray?*

Fig. 10.21

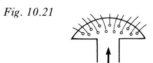

Volume of water per second from sprinkler

     = volume supplied per second

     = 3 $\times$ 10^{-3} m^3 s^{-1}

     = total area of sprinkler holes $\times$ average velocity of spray

     = 300 $\times$ 10^{-6} m^2 $\times$ average velocity of spray

Therefore average velocity of spray

$$= \frac{3 \times 10^{-3}}{300 \times 10^{-6}} \frac{m^3\ s^{-1}}{m^2}$$

$$= 10\ m\ s^{-1}$$

FLUIDS IN MOTION

2. Obtain an estimate for the velocity of emergence of a liquid from a hole in the side of a wide vessel 10 cm below the liquid surface.

Consider the general case in which the hole is at depth h below the surface of the liquid of density ρ, Fig. 10.22. If the liquid is incompressible and inviscid and the motion is steady we can apply Bernoulli's equation to points A and B on the streamline AB.

Fig. 10.22

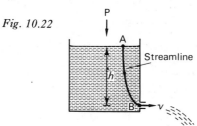

At A
$$p_1 = \text{atmospheric pressure} = P$$
$$h_1 = h$$
$$v_1 = 0$$

(assuming the rate of fall of the surface can be neglected compared with the speed of emergence since the vessel is wide).

At B $\quad p_2$ = pressure of air into which the liquid emerges = P
$$h_2 = 0$$
$$v_2 = v$$

Substituting in Bernoulli's equation

$$P + h\rho g + 0 = P + 0 + \tfrac{1}{2}\rho v^2$$
$$\therefore \quad h\rho g = \tfrac{1}{2}\rho v^2$$

From which we see that the potential energy lost by unit volume of liquid (mass ρ) in falling from the surface to depth h is changed to kinetic energy. The velocity of emergence is given by

$$v^2 = 2gh$$

and is the same as the vertical velocity which would be acquired in free fall —a statement known as *Torricelli's theorem*. In fact, v is always less than $\sqrt{2gh}$ due to the viscosity of the liquid.

If $h = 10$ cm $= 0.1$ m and $g = 9.8$ m s^{-2} then

$$v = \sqrt{2 \times 9.8 \times 0.1} \quad \text{m s}^{-1}$$
$$= 1.4 \text{ m s}^{-1}$$

Flow of mass, energy and charge

The transfer of mass, energy or electrical charge from one place to another is an important engineering problem. For example, the transportation of all three occurs in the generation and distribution of electricity. The flow of each quantity is expressed by the same general expression

$$\text{flow rate} \propto \frac{\text{'pressure' causing flow}}{\text{resistance to flow}}$$

(a) *Mass*. Mass transport in the form of *steady* fluid flow along a pipe is given by Poiseuille's formula (p. 365)

$$V = \frac{\pi p r^4}{8 \eta l}$$

where V is the volume of fluid passing per second, p the pressure difference between the ends of the pipe of length l and radius r and η is the coefficient of viscosity of the fluid.

If ρ_m is the density of the fluid then the mass passing per second is $\rho_m V$ and we can say

$$\text{mass flow rate} = p \frac{\rho_m \pi r^4}{8 \eta l} = \frac{p}{R_m}$$

where R_m is a constant which incorporates the resistance to flow, i.e. η, and equals $8 \eta l / (\rho_m \pi r^4)$. Hence

$$\text{fluid mass flow rate} \propto \frac{\text{fluid pressure difference}}{\text{resistance to flow}}$$

(b) *Energy* (*heat*). The transport of heat by conduction is expressed by Fourier's law (p. 139)

$$\frac{Q}{t} = \frac{kA(\theta_2 - \theta_1)}{x}$$

where Q is the quantity of heat passing in time t down a lagged bar of cross-sectional area A, length x and thermal conductivity k when its opposite ends are at steady temperatures θ_2 and θ_1. Hence

$$\text{heat flow rate} = (\theta_2 - \theta_1) \frac{kA}{x} = \frac{\theta_2 - \theta_1}{R_e}$$

where R_e is a constant incorporating the resistance to flow, i.e. $1/k$, and equals $x/(kA)$. Hence

$$heat\ flow\ rate \propto \frac{temperature\ difference}{resistance\ to\ flow}$$

(c) *Charge.* The rate of electric charge flow, i.e. current, in metallic conductors obeys Ohm's law (p. 61)

$$I = \frac{Q}{t} = \frac{V}{R}$$

where I is the current and V the potential difference across the ends of a conductor of resistance R. Since $R = \rho l/A$ where ρ is the resistivity of the material of the conductor of length l and cross-sectional area A we can also write

$$I = V\frac{A}{\rho l}$$

. Hence

$$electric\ charge\ flow\ rate \propto \frac{potential\ difference}{resistance\ to\ flow}$$

Further comparison of these transport effect expressions shows that for all three we can also say

$$flow\ rate = \frac{driving}{factor} \times \frac{physical}{factor} \times \frac{conductivity}{factor}$$

In particular,

$$mass\ flow\ rate = p \times \frac{\pi r^2}{l} \times \frac{\rho_m r^2}{8\eta}$$

$$heat\ flow\ rate = (\theta_2 - \theta_1) \times \frac{A}{x} \times k$$

$$electric\ charge\ flow\ rate = V \times \frac{A}{l} \times \frac{1}{\rho}$$

In each case the physical factor equals (cross-sectional area of flow path)/(length of flow path).

Film (16 mm): *Fluid flow*—Unilever Film Library, Unilever House, London, E.C.4.

381

FLUIDS IN MOTION

QUESTIONS

1. What do you understand by the *dimensions* of a physical quantity? Derive the dimensions of coefficient of viscosity.

Explain the value, and the limitations, of the method of dimensions as a means of checking, and sometimes deriving, the form of equations involving physical quantities.

Confirm the dimensional consistency of the following statements in which η represents the coefficient of viscosity of a liquid, a and l are lengths, ρ a density, p a pressure-difference and v a speed:

(a) The product $lv\rho/\eta$, known as Reynolds's number, is dimensionless.

(b) According to Poiseuille, the volume of liquid flowing per second steadily through a capillary tube is $\pi pa^4/8l\eta$.

(c) For a sphere of density ρ falling steadily under gravity through an expanse of liquid of density ρ'

$$6\pi\eta av = \tfrac{4}{3}\pi a^3(\rho - \rho')g \qquad (O.)$$

2. An aluminium sphere is suspended by a thread below the surface of a liquid. Show on a sketch the forces acting on the sphere, and explain its equilibrium. (No formal proof is required.)

The thread is now cut. Show on a second sketch, or explain in words, the forces which act on the sphere when it is in motion.

The following figures for x, the total distance travelled by the sphere in the liquid at time t, were obtained:

t (s)	1.0	2.0	3.0	4.0	5.0
x (cm)	3.60	10.3	18.6	27.9	37.4

Draw a graph to display the relation between x and t, and explain its form. Find the terminal velocity of the sphere. Give a qualitative account of how you would expect the graph to be modified if the temperature of the liquid were increased. (C.)

3. Define *coefficient of viscosity* η and obtain its dimensions in terms of M, L and T.

Stokes's law for the viscous force F acting on a sphere of radius a falling with velocity v through a large expanse of fluid of coefficient of viscosity η is expressed by the equation

$$F = 6\pi a\eta v$$

Show that this equation is correct dimensionally and state why it is true only for sufficiently low velocities.

Explain why a sphere released in a fluid will fall with diminishing acceleration until it attains a constant terminal velocity.

Calculate this velocity for an oil drop of radius 3.0×10^{-6} m falling through air of coefficient of viscosity 1.8×10^{-5} Pa s, given that the density of the oil is 8.0×10^2 kg m^{-3} and that the density of air may be neglected. (L.)

4. In the simplified petrol engine carburettor shown in Fig. 10.23, air is drawn into the carburettor by the action of the engine piston, and the petrol enters at the point of

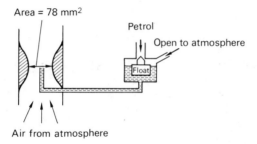

Area = 78 mm²

Petrol

Open to atmosphere

Float

Air from atmosphere

Fig. 10.23

minimum cross-sectional area. If the throat of the venturi section has an area of 78 mm², calculate the area of the fuel jet required to produce an air–fuel mass ratio of 12:1.

The density of air is 1.2 kg m⁻³ and that of petrol is 7.8 × 10² kg m⁻³.

(J.M.B. Eng. Sc.)

5. A toy designer has submitted a design for a water pistol with a barrel area 75 mm² and jet area 1.0 mm². The manufacturer required that when the pistol was fired horizontally, the jet should be able to hit a target 3.5 m away not more than 1.0 m below the firing line. Given that the average child is able to exert a force of 10 N on the plunger, has the designer satisfied the requirements? You may neglect barrel friction and energy loss at the exit jet.

Atmospheric pressure = 1.0 × 10⁵ N m⁻², density of water = 1.0 × 10³ kg m⁻³, g = 9.8 m s⁻².

(J.M.B. Eng. Sc.)

Objective-type revision questions

The first figure of a question number gives the relevant chapter, e.g. **2.3** is the third question for chapter 2.

Multiple choice

Select the response which you think is correct.

1.1. The density of aluminium is 2.7 g cm^{-3}, its atomic mass is 27 and the Avogadro constant is 6.0×10^{23} atoms per mole. If aluminium atoms are assumed to be spheres, packed so that they occupy three-quarters of the total volume, the volume of an aluminium atom in cm^3 is

A $4 \times 27/(3 \times 2.7 \times 6.0 \times 10^{23})$ **B** $3 \times 27/(4 \times 2.7 \times 6.0 \times 10^{23})$
C $3 \times 2.7/(4 \times 27 \times 6.0 \times 10^{23})$ **D** $4 \times 27 \times 6.0 \times 10^{23}/(3 \times 2.7)$
E $3 \times 2.7 \times 6.0 \times 10^{23}/(4 \times 2.7)$

1.2. The weakest form of bonding in materials is

A van der Waals   **B** ionic   **C** covalent   **D** metallic

2.1. Which one of the following is Young's modulus (in Pa) for the wire having the stress-strain curve of Fig. 1 ?

A 36×10^{11} **B** 8.0×10^{11} **C** $2.0 \times 10^{11 \cdot}$ **D** 0.50×10^{11}
E 16×10^{11}

(J.M.B.)

Fig. 1

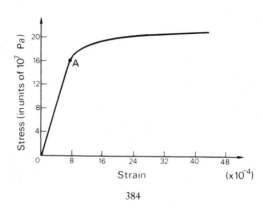

2.2. Which one of the following statements is a correct statement about the evidence provided by Fig. 1 ?

A The wire only obeys Hooke's law between O and A and after A it becomes much more difficult to stretch it.

B The wire does not obey Hooke's law between O and A and after A it becomes much more difficult to stretch it.

C The wire only obeys Hooke's law between O and A and after A it becomes much easier to stretch it.

D The wire does not obey Hooke's law between O and A and after A it becomes much easier to stretch it.

E The wire does not obey Hooke's law at all and is no harder or easier to stretch before A than after A.

(J.M.B.)

2.3. Wires X and Y are made from the same material. X has twice the diameter and three times the length of Y. If the elastic limits are not reached when each is stretched by the same tension, the ratio of energy stored in X to that in Y is

A 2:3   **B** 3:4   **C** 3:2   **D** 6:1   **E** 12:1

2.4. Reasons for the good stiffness and strength of three different materials are given below. Select from the list of five materials the *one* to which *each* statement best applies (three answers).

  (*i*) It has 'foreign' atoms in the lattice which oppose dislocation movement.

  (*ii*) It has high covalent bond density.

  (*iii*) It has long-chain molecules lying more or less parallel along their length.

A steel   **B** rubber   **C** copper   **D** polythene   **E** glass

3.1. The drift velocity of the free electrons in a conductor is independent of one of the following. Which is it?

A The length of the conductor.

B The number of free electrons per unit volume.

C The cross-sectional area of the conductor.

D The electronic charge.

E The current.

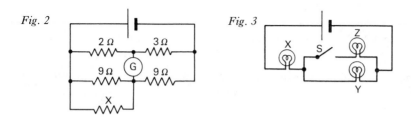

Fig. 2   *Fig. 3*

3.2. The value of X in ohms which gives zero deflection on the galvanometer in Fig. 2 is

A 3   **B** 6   **C** 15   **D** 18   **E** 27

3.3. If X, Y and Z in Fig. 3 are identical lamps, which of the following changes to the brightnesses of the lamps occur when switch S is closed?

A X stays the same   Y decreases   **B** X increases   Y stays the same
C X increases   Y decreases   **D** X decreases   Y increases
E X decreases   Y decreases

3.4. A moving coil galvanometer has a resistance of 10 Ω and gives a full scale deflection for a current of 0.01 A. It could be converted into a voltmeter reading up to 10 V by connecting a resistor of value

A 0.10 Ω in parallel with it   **B** 90 Ω in series with it
C 0.10 Ω in series with it   **D** 990 Ω in parallel with it
E 990 Ω in series with it   (*J.M.B.*)

3.5. Which of the graphs in Fig. 4 best shows the variation of current with time in a tungsten filament lamp, from the moment current flows?

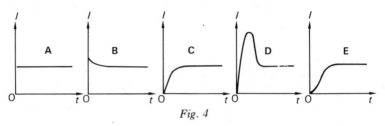

Fig. 4

3.6. Two resistors are connected in parallel as shown in Fig. 5. A current passes through the parallel combination. The power dissipated in the 5.0 Ω resistor is 40 W. Which one of the following is the power dissipated in watts in the 10 Ω resistor?

A 10   **B** 20   **C** 40   **D** 80

(*J.M.B. Eng. Sc.*)

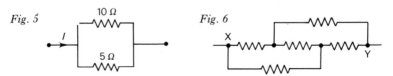

3.7. If each resistor in Fig. 6 is 2 Ω, the effective resistance in ohms between X and Y is

A 2/5   **B** 1   **C** 2   **D** $2\frac{2}{3}$ **E** $3\frac{1}{2}$

3.8. A cylindrical copper rod is re-formed to twice its original length. Which one of the following statements describes the way in which the resistance is changed?

A The resistance remains constant.
B The resistance increases by a factor of two.
C The resistance increases by a factor of four.
D The resistance increases by a factor of eight.

(*J.M.B. Eng. Sc.*)

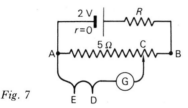

Fig. 7

3.9. A thermocouple thermometer is to be designed using the circuit of Fig. 7. AB is a potentiometer wire of resistance 5.0 Ω and ED is a thermocouple whose e.m.f. is 20 mV at 400 °C and zero at 0 °C. For a temperature measurement range from 0 °C to 400 °C, the required value for resistor R in ohms is

A 195   **B** 295   **C** 395   **D** 495

(J.M.B. Eng. Sc.)

4.1. Choose from the following statements one which does *not* apply to the platinum resistance thermometer.

  A It can give high accuracy.
  B It is suitable for measuring the temperature in a small object.
  C It has a high heat capacity.
  D It can cover a wide range of temperature.
  E It can only be used for steady temperatures.

4.2. Spheres P and Q are uniformly constructed from the same material which is a good conductor of heat and the radius of Q is twice the radius of P. The rate of fall of temperature of P is x times that of Q when both are at the same surface temperature. The value of x is

A $\frac{1}{4}$ **B** $\frac{1}{2}$ **C** 2   **D** 4   **E** 8

Fig. 8a

4.3. Heat flows through the bar XYZ in Fig. 8a, the ends X and Z being maintained at fixed temperatures (temperature at X > temperature at Z). If only the part YZ is lagged, which graph in Fig. 8b shows the variation of temperature (θ) with distance along XZ for steady state conditions?

Fig. 8b

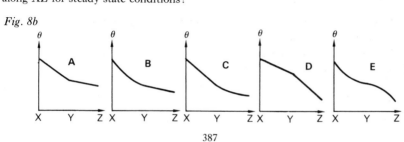

4.4. The dimensions of specific heat capacity are

A $ML^2T\theta^{-1}$ **B** $L^2T^{-2}\theta^{-1}$ **C** $L^2T^2\theta^{-1}$ **D** $M^{-1}LT^{-1}\theta^{-1}$
E $ML^{-1}T^{-2}\theta^{-1}$

5.1. A lens of focal length 12 cm forms an upright image three times the size of a real object. The distance in cm between the object and image is

   A 8.0   **B** 16   **C** 24   **D** 32   **E** 48

5.2. When a lens is inserted between an object and a screen which are a fixed distance apart the size of the image is either 6 cm or $\frac{2}{3}$ cm. The size of the object in cm is

   A 2   **B** 3   **C** 4   **D** $4\frac{1}{3}$ **E** 9

5.3. Which *one* of the following combinations of lenses is used as a compound microscope? (The objective is listed first.)

   A long focus converging and shorter focus converging
   B long focus converging and shorter focus diverging
   C long focus converging and long focus converging
   D short focus converging and longer focus converging
   E short focus converging and longer focus diverging

(J.M.B.)

6.1. A pendulum bob suspended by a string from the point P, Fig. 9, is in equilibrium under the action of three forces: W, the weight of the bob; T, the tension in the string; and F, a horizontally applied force. Which one of the following statements is untrue?

   A $F^2 + W^2 = T^2$ **B** F and W are the components of T
   C $W = T\cos\theta$ **D** $F = W\tan\theta$

(J.M.B. Eng. Sc.)

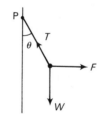

Fig. 9

6.2. Forces of 3 N, 4 N and 12 N act at a point in mutually perpendicular directions. The magnitude of the resultant force in newtons is

   A 5   **B** 11   **C** 13   **D** 19   **E** indeterminate from information given

6.3. Which graph in Fig. 10 best represents the variation of velocity with time of a ball which bounces vertically on a hard surface, from the moment when it rebounds from the surface?

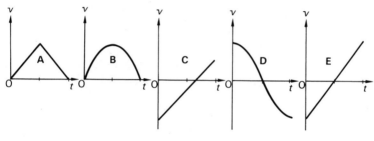

Fig. 10

6.4. A ball is projected horizontally at 15 m s^{-1} from a point 20 m above a horizontal surface ($g = 10$ m s^{-2}). The magnitude of its velocity in m s^{-1} when it hits the surface is

A 10　　**B** 15　　**C** 20　　**D** 25　　**E** 35

6.5. A trolley of mass 60 kg moves on a frictionless horizontal surface and has kinetic energy 120 J. A mass of 40 kg is lowered vertically on to the trolley. The total kinetic energy of the system is now

A 60 J　　**B** 72 J　　**C** 120 J　　**D** 144 J　　**E** another answer

7.1. A mass of 2.0 kg describes a circle of radius 1.0 m on a smooth horizontal table at a uniform speed. It is joined to the centre of the circle by a string which can just withstand 32 N. The greatest number of revolutions per minute the mass can make is

A 38　　**B** 4　　**C** 76　　**D** 240　　**E** 16

7.2. In order to turn in a horizontal circle an aircraft banks so that

A there is a resultant force on the wings from the centre of the circle
B the weight of the aircraft has a component towards the centre of the circle
C the drag on the plane is reduced
D the lifting force on the wings has a component towards the centre of the circle

(*J.M.B. Eng. Sc.*)

7.3. If a small body of mass m is moving with angular velocity ω in a circle of radius r, what is its kinetic energy?

A $m\omega r/2$　　**B** $m\omega^2 r/2$　　**C** $m\omega r^2/2$　　**D** $m\omega^2 r^2/2$

(*J.M.B. Eng. Sc.*)

7.4. Planet X is twice the radius of planet Y and is of material of the same density. The ratio of the acceleration due to gravity at the surface of X to that at the surface of Y is

A 1:4　　**B** 1:2　　**C** 2:1　　**D** 4:1　　**E** 8:1

8.1. The frequency of oscillation of a mass m suspended at the end of a vertical spring having a spring constant k is directly proportional to

A mk　　**B** m/k　　**C** $m^2 k$　　**D** $1/(mk)^{1/2}$　　**E** $(k/m)^{1/2}$

389

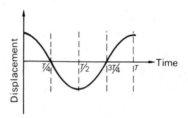

Fig. 11

8.2. The graph of Fig. 11 shows how the displacement of a particle describing s.h.m. varies with time. Which one of the following statements is, from the graph, false?

  A The restoring force is zero at time $T/4$.
  B The velocity is a maximum at time $T/2$.
  C The acceleration is a maximum at time T.
  D The displacement is a maximum at time T.
  E The kinetic energy is zero at time $T/2$.

9.1. When a capillary tube of uniform bore is dipped in water the water level in the tube rises 10 cm higher than in the vessel. If the tube is lowered until its open end is 5.0 cm above the level in the vessel, the water in the tube appears as in (Fig. 12)

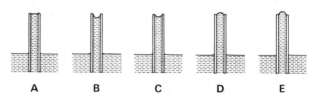

Fig. 12

10.1. Spheres X and Y of the same material fall at their terminal velocities through a liquid without causing turbulence. If Y has twice the radius of X, the ratio of the terminal velocity of Y to that of X is

  A 1:4   **B** 1:2   **C** 1:1   **D** 2:1   **E** 4:1

OBJECTIVE-TYPE REVISION QUESTIONS

Multiple selection

In each question one or more of the responses may be correct. Choose one letter from the answer code given.

Answer **A** *if* (*i*), (*ii*) *and* (*iii*) *are correct*
Answer **B** *if only* (*i*) *and* (*ii*) *are correct*
Answer **C** *if only* (*ii*) *and* (*iii*) *are correct*
Answer **D** *if* (*i*) *only is correct*
Answer **E** *if* (*iii*) *only is correct*

3.10. In the potentiometer circuit of Fig. 13 the galvanometer reveals a current in the direction shown wherever the sliding contact touches the wire. This could be caused by

(*i*) E_1 being too low
(*ii*) 3.0 Ω being too high
(*iii*) a break in PQ

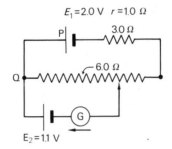

$E_1 = 2.0$ V $\quad r = 1.0$ Ω

Fig. 13

5.4. A microscope with a short focus objective

(*i*) allows more light to be collected
(*ii*) keeps the distance between objective and eyepiece small
(*iii*) gives high magnifying power

6.6. In the equation $Ft = mv - mu$

(*i*) the dimensions of F are MLT^2
(*ii*) the dimensions of mv are MLT^{-1}
(*iii*) the dimensions of all three terms are the same

8.3. The period of a simple pendulum oscillating in a vacuum depends on

(*i*) the mass of the pendulum bob
(*ii*) the length of the pendulum
(*iii*) the acceleration due to gravity

OBJECTIVE-TYPE REVISION QUESTIONS

Assertion-reason

*Each question consists of an assertion (statement) followed by a reason. Only if you decide that **both** are true do you need to consider if the reason is a valid explanation of the assertion. Select your answer according to the code below*

Answer	Assertion	Reason	
A	True	True	Reason correct
B	True	True	Reason incorrect
C	True	False	—
D	False	True	—
E	False	False	—

5.5. The deviation of blue light in a glass prism is greater than the deviation of red light *because* blue light travels faster in glass than red light.

7.5. An astronaut inside a space capsule in earth orbit feels weightless *because* he is where the gravitational fields of the earth and moon counteract each other.

8.4. The time period of the same object oscillating on a certain spring is the same on the moon as it is on the earth *because* the static extension is less on the moon than on the earth.

9.2. An object immersed in a liquid experiences an upthrust *because* the pressure in a liquid increases with depth.

Appendix 1

Method of dimensions

(a) *Physical quantities.* These can be classified as *basic* quantities and *derived* quantities. Seven basic quantities are chosen for their convenience and are: *mass, length, time, electric current, temperature, luminous intensity* and *amount of substance.* All other quantities are derived from one or more of the basic quantities.

(b) *Dimensions.* The dimensions of a quantity show how it is related to the basic quantities. For example, the derived quantity *volume* has three dimensions in length; it is measured basically by multiplying three lengths together and this is shown by writing $[V] = [L^3]$. The square brackets round V indicate that we are dealing with the dimensions of V. *Density* is measured by dividing a mass by a volume and has dimensions $[ML^{-3}]$. The dimensions of *velocity* are $[LT^{-1}]$ and of *acceleration* $[LT^{-2}]$. Fundamentally *force* is measured by multiplying a mass by an acceleration $(F = ma)$ and so has dimensions $[MLT^{-2}]$. Every derived quantity has dimensions. What are they for *pressure*?

Some quantities are partly dimensional. For example, *frequency* is a number of oscillations per unit time and has dimension $[T^{-1}]$ since the number of oscillations part is dimensionless. Some quantities such as refractive index are dimensionless.

(c) *Dimensional analysis.* If an equation is correct the dimensions of the quantities on either side must be identical. This fact is used in the method of Dimensional analysis which enables predictions to be made about how quantities may be related. The method is particularly helpful when dealing with viscosity problems and is used to derive Poiseuille's formula on p. 365 and Stokes' law on p. 367.

Assumptions based on experiment or intuition have first to be made about what quantities could be involved and in general no more than *three* such dependent quantities can be considered. Neither does the method yield the value of any dimensionless constants, e.g. π.

393

Appendix 2

SI units

S I units (standing for *Système International d'Unités*) were adopted internationally in 1960.

(*a*) *Basic units.* The system has seven basic units, one for each of the basic quantities.

Basic quantity	Unit	
	Name	Symbol
mass	kilogram	kg
length	metre	m
time	second	s
electric current	ampere	A
temperature	kelvin	K
luminous intensity	candela	cd
amount of substance	mole	mol

Units must be easily reproducible and unvarying with time and so are often based on the properties of atoms. Thus the *metre* is now the length which equals 1 650 763.73 wavelengths in a vacuum of a specified radiation from a krypton-86 atom. Definitions of some of the other units are given when the quantity arises in the text.

(*b*) *Derived units.* These are obtained from the basic units by multiplication or division; no numerical factors are involved. Some derived units with complex names:

Derived quantity	Unit	
	Name	Symbol
volume	cubic metre	m^3
density	kilogram per cubic metre	$kg\ m^{-3}$
velocity	metre per second	$m\ s^{-1}$
acceleration	metre per second squared	$m\ s^{-2}$
momentum	kilogram metre per second	$kg\ m\ s^{-1}$

394

APPENDIX 2

Some derived units are given special names due to their complexity when expressed in terms of the basic units.

Derived quantity	Unit	
	Name	Symbol
force	newton	N
pressure	pascal	Pa
energy, work	joule	J
power	watt	W
frequency	hertz	Hz
electric charge	coulomb	C
electric resistance	ohm	Ω
electromotive force	volt	V

When the unit is named after a person the *symbol* has a capital letter.

(c) *Standard prefixes.* Decimal multiples and submultiples are attached to units when appropriate. In general, prefixes involving powers which are multiples of three are preferred but others are used, e.g. 10^{-2} (centi).

Multiple	Prefix	Symbol	Submultiple	Prefix	Symbol
10^3	kilo	k	10^{-3}	milli	m
10^6	mega	M	10^{-6}	micro	μ
10^9	giga	G	10^{-9}	nano	n
10^{12}	tera	T	10^{-12}	pico	p

(d) *Coherence.* S I units are coherent. This means that there is only *one* unit for each quantity (ignoring multiples and submultiples) and if these are used for the quantities in an expression, the answer is obtained in the correct S I unit. For example if in $F = ma$, m is expressed in kg and a in m s^{-2} then F will be automatically in newtons.

Appendix 3

Measurement of length

(a) *Vernier scale*. The simplest type enables a length to be measured to 0.01 cm. It comprises a small sliding scale which is 9 mm long and is divided into 10 equal divisions, Fig. A3.1a. Hence

1 vernier division = 9/10 mm = 0·9 mm = 0·09 cm

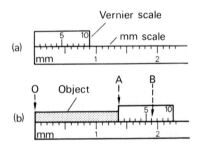

Fig. A3.1

One end of the length to be measured is made to coincide with the zero of the milli-metre scale and the other end with the zero of the vernier scale. The length of the object in Fig. A3.1b is between 1.3 cm and 1.4 cm. The reading to the second place of decimals is obtained by finding the *vernier mark* which is exactly opposite (or nearest to) a mark on the millimetre scale. In this case it is the 6th mark and the length is 1.36 cm since

$$OA = OB - AB$$

$$= (1.90 \text{ cm}) - (6 \text{ vernier divisions})$$

$$= 1.90 - 6(0.09) = 1.90 - 0.54 \text{ cm}$$

$$= 1.36 \text{ cm}$$

Vernier scales are often used on calipers, barometers, travelling microscopes and spectrometers.

(*b*) *Micrometer screw gauge.* This measures very small objects to 0.001 cm. One revolution of the drum opens the accurately plane, parallel jaws by 1 division on the scale on the *shaft* of the gauge; this is usually $\frac{1}{2}$ mm, i.e. 0.05 cm. If the *drum* has a scale of 50 divisions round it, then rotation of the drum by 1 division opens the jaws

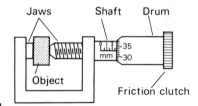

Fig. A3.2

by $0.05/50 = 0.001$ cm, Fig. A3.2. A friction clutch ensures that the jaws always exert the same forces when the object is gripped. The object shown has a length

$= 2.5$ mm on the shaft scale $+ 33$ divisions on the drum scale

$= 0.25$ cm $+ 33(0.001)$ cm

$= 0.283$ cm.

Appendix 4

Graphs

When plotting a graph from experimental results, as much of the paper as possible should be used, points should be marked O or X and a smooth curve or straight line drawn so that the points are distributed equally on either side of it.

(a) *Straight line graph.* If a straight line is obtained its equation is of the form

$$y = mx + c$$

where m is the slope of the line (QR/PQ in Fig. A4.1a) and c is the intercept (OS) on the y-axis.

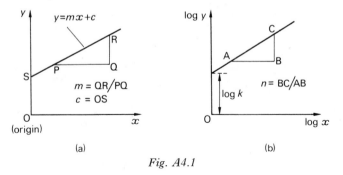

Fig. A4.1

In the magnification method for finding the focal length of a converging lens (p. 192), if a graph is plotted of magnification m against image distance v, a straight line should be obtained of equation $m = v/f - 1$. The slope of the graph is $1/f$ and the intercept on the m-axis is -1.

If a graph passes through the origin O then $c = 0$ and $y = mx$, i.e. $y \propto x$ and so y is directly proportional to x. A straight line graph *not* passing through the origin only indicates that equal increases of the independent variable x cause equal increases of the dependent variable y but $y \not\propto x$.

If it is not possible to include on the graph the origins for one or both axes, c can be calculated by substituting the co-ordinates of a point on the graph, say (x_1, y_1) and the value of m, in the equation $y = mx + c$ and solving for c. Thus if $x = 2$, $y = 3$ and $m = +1$ then $3 = 1 \times 2 + c$, therefore $c = 1$ and the graph does not pass through the origin of the y-axis.

When possible, quantities are usually plotted which give a straight-line graph since this is easier to interpret. Thus if quantities T and l are related by the equation $aT = bl^2 + c$, (*i*) what would you plot to obtain a straight line and (*ii*) what would be the slope?

(*b*) *Log graph.* Sometimes we wish to find experimentally the relationship between two quantities x and y. If we assume that

$$y = kx^n$$

where k and n are constants, then taking logs

$$\log y = \log k + n \log x$$

This is of the form $y = c + mx$ and so a graph of $\log y$ against $\log x$ should be a straight line of slope n and intercept $\log k$ on the $\log y$ axis, Fig. A4.1*b*. Hence k and n can be found.

Appendix 5

Treatment of errors

(*a*) *Types of error.* Experimental errors cause a measurement to differ from its true value and are of two main types.

(*i*) A *systematic error* may be due to an incorrectly calibrated scale on, for example, a ruler or ammeter. Repeating the observation does not help and the existence of the error may not be suspected until the final result is calculated and checked, say by a different experimental method. If the systematic error is small a measurement is accurate.

(*ii*) A *random error* arises in any measurement, usually when the observer has to estimate the last figure, possibly in an instrument which lacks sensitivity. Random errors are small for a good experimenter and taking the mean of a number of separate measurements reduces them in all cases. A measurement with a small random error is precise but it may not be accurate.

Fig. A5.1*a* shows random errors only in a meter reading, whilst in Fig. A5.1*b* there is a systematic error as well.

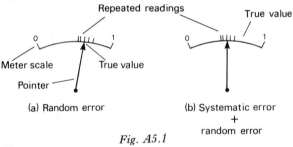

(a) Random error   (b) Systematic error
+
random error

Fig. A5.1

(*b*) *Estimating errors in single measurements.* In more advanced work, if systematic errors are not eliminated they can be corrected from the observations made. Here we shall assume they do not exist and then make a reasonable estimate of the likely random error. Two examples follow.

Using a metre rule the length of an object is measured as 2.3 cm. At very worst the answer might be 2.2 or 2.4 cm, i.e. the maximum error it is possible to make using a ruler marked in mm is ±0.1 cm. The *possible error* (p.e.), is said to be ±0.1 cm and the length is written (2.3 ± 0.1) cm. The *percentage possible error* (p.p.e.) is (±0.1 × 100)/2.3 ≃ ±4%.

Using vernier calipers capable of measuring to 0.01 cm, the length of the same object might be read as (2.36 ± 0.01) cm. In this case the p.e. is ±0.01 cm and the p.p.e. is (±0.01 × 100)/2.36 ≃ ±0.4%.

If a large number of readings of one quantity are taken the mean value is likely to be close to the true value and statistical methods enable a *probable error* to be estimated. Here, we adopt the simpler procedure of estimating the maximum error likely, i.e. the *possible error*.

(c) *Combining errors*. The result of an experiment is usually calculated from an expression containing the different quantities measured. The combined effect of the errors in the various measurements has to be estimated. Three simple cases will be considered.

(i) *Sum*. Suppose the quantity Q we require, is related to quantities a and b which we have measured, by the equation

$$Q = a + b$$

Then total p.e. in Q = p.e. in a + p.e. in b. Thus if a = 5.1 ± 0.1 cm and b = 3.2 ± 0.1 cm then Q = 8.3 ± 0.2 cm. That is, in the worst cases, if both a and b are read 0.1 cm too high Q = (5.2 + 3.3) = 8.5 cm, but if both are 0.1 cm too low then Q = (5.0 + 3.1) = 8.1 cm.

(ii) *Difference*. If $Q = a - b$, the same rule applies, i.e. the total p.e. in Q = p.e. in a + p.e. in b.

(iii) *Product and quotient*. If the individual measurements have to be multiplied or divided it can be shown that the *total percentage possible error equals the sum of the separate percentage possible errors*. For example, if a, b and c are measurements made and

$$Q = \frac{ab^2}{c^{1/2}}$$

then if the p.p.e. in a is ±2%, that in b is ±1% and that in c is ±2%, then the p.p.e. in b^2 is 2(±1)% and in $c^{1/2}$ is $\frac{1}{2}$(±2)% = ±1%. Hence

total p.p.e. in Q = ±(p.p.e. in a + p.p.e. in b^2 + p.p.e. in $c^{1/2}$)

= ±(2 + 2 + 1) = ±5%

The answer for Q will therefore be accurate to 1 part in 20 and if the numerical result for Q is 1.8 then it is written

$$Q = 1.8 \pm \tfrac{1}{20} \times 1.8 = 1.8 \pm 0.1$$

It would not be justifiable to write Q = 1.852 since this would be claiming an accuracy of a few units in 1852. According to our estimate this accuracy is not possible with the apparatus used.

It is instructive to estimate whenever possible the total p.p.e. for an experiment; it indicates (i) the number of significant figures that can be given in the result, (ii) the limits within which the result lies and (iii) the measurements requiring particular care. There is little point in making one measurement to a very high degree of accuracy if it is not possible with the others; a chain is only as strong as the weakest link.

Appendix 6

Construction of model crystal for microwave analogue demonstration
(p. 22)

The model has a face-centred cubic structure and is made from 190 5-cm-diameter polystyrene balls glued together with Durofix in seven hexagonal close-packed layers, Fig. A6.1.

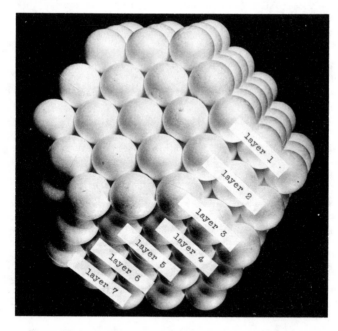

Fig. A6.1

Start with *layer 4* which has 37 balls, Fig. A6.2*a* (the black dots represent the centres of the balls).

The black dots in Fig. A6.2*b* show the 36 balls in *layer 3* and their positions in the hollows of layer 4 which is shown by circles. Each ball should be glued to all those it touches.

APPENDIX 6

The 27 balls of *layer 2* are the black dots in Fig. A6.2c (the circles are layer 3) and the layer 2 balls should be placed *over the hollows in layer 4* (not over the balls in layer 4) to give the ABCABC stacking of an FCC crystal.

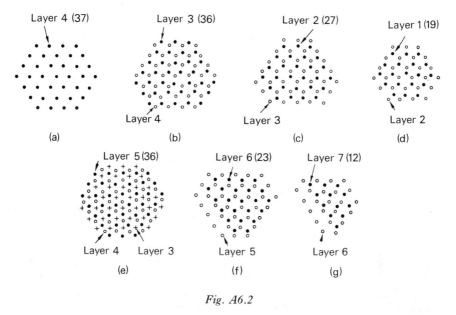

Fig. A6.2

Fig. A6.2d shows the 19 balls of *layer 1* as black dots and layer 2 as circles.

Half the model is now made and when set, it can be turned over to build the other half.

In Fig. A6.2e the circles are layer 4 (now uppermost), the crosses represent layer 3. *Layer 5* is shown by 36 black dots.

The 23 black dots in Fig. A6.2f show *layer 6* and the circles layer 5. Care should again be taken to ensure that the balls in layer 6 are over hollows in layer 4.

Fig. A6.2g gives the position of the 12 balls in *layer 7*.

Appendix 7

Speed of an electrical pulse along a cable (p. 59)

The drift speed of the electrons forming the current in a conductor is about 1 mm s^{-1} but the electric and magnetic fields which constitute the signal that sets them in motion almost simultaneously round the circuit, travel at very great speeds. Basically these fields are the same as those of an electromagnetic wave in free space.

Using a double beam C.R.O. on a fast time base speed (1 μs cm^{-1}), the time taken by an electrical pulse from a 200 kHz pulse generator to travel along a 200 m length of coaxial cable can be found, Fig. A7.1. The in-going pulse is applied to the Y_1 input (set at 0.2 V cm^{-1}) as well as to the near end of the cable and the out-coming pulse from the far end of the cable is applied to the Y_2 input (also on

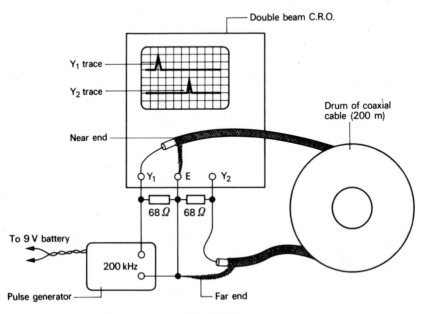

Fig. A7.1

404

0.2 V cm^{-1}). The distance between the pulses is measured and the time it represents calculated. The speed is roughly 2×10^8 m s^{-1}. If air or a vacuum replaced the polythene insulation between the central and outer conductors of the cable and through which the fields travel, the speed would be that of light (3×10^8 m s^{-1}). Fig. A7.2 shows the shapes of the electric and magnetic fields travelling along the cable; they are at right angles to each other and to the direction of the current.

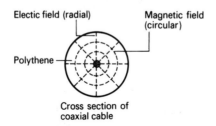

Fig. A7.2

Coaxial cable is used because it does not pick up unwanted interference if the outer conductor is connected to E on the C.R.O. at both ends. The 68 Ω resistors should be connected to the C.R.O. terminals directly. A full explanation of their action requires a more advanced treatment but without them the pulse would be reflected backwards and forwards along the cable, setting up a standing wave system on it. Instead, the resistors 'absorb' the pulses and ensure they are applied to the C.R.O.

Answers

Chapter 1. Structure of materials

1. (a) $4:3$ (b) 16 (c) 16 g
2. (a) 14 g (b) 70 g
3. (a) 1.5×10^{23} (b) 18×10^{23} (c) 3.0×10^{23}
4. (a) 4.0×10^{-23} g (b) 1.2×10^{-21} g (c) 3.0×10^{-23} g
5. (a) 6.0×10^{23} (b) 2.0 g (c) 2.2×10^4 cm^3 (d) 2.7×10^{17}
6. 6.1×10^{23} mol^{-1}
7. (a) 3.0×10^{-23} g (b) 3.9×10^{-10} m (0.39 nm)
8. (a) 5.0×10^3 J g^{-1} (b) 5.3×10^{-19} J/atom
9. (a) (i) 4 (ii) 6. Closest in hexagonal packing
   (b) (i) hexagonal (ii) square
10. (a) $2\sqrt{2}\,r$ cm (b) $16\sqrt{2}\,r^3$ cm^3 (c) $1/(16\sqrt{2}\,r^3)$
11. (a) 8 (b) 8 (c) $8/8 = 1$ (d) 6 (e) 2 (f) $6/2 = 3$ (g) $1 + 3 = 4$ (h) $1/(4\sqrt{2}\,r^3)$
12. 8.4×10^{22}
13. 1.3×10^{-10} m (0.13 nm)
14. If a is the separation between adjacent Na and Cl ions, a cube of side a and volume a^3 will contain 4 Na ions and 4 Cl ions, i.e. 4 NaCl molecules. But each corner (ion) of the cube is shared by 8 neighbouring cubes and so each volume a^3 contains $\frac{1}{8}$ of 4 molecules of NaCl, i.e. $\frac{1}{2}$ molecule of NaCl.

$$\therefore \quad \text{Volume occupied by } \tfrac{1}{2} \text{ molecule} = a^3 = \frac{1}{2} \cdot \frac{M}{N_A \rho}$$

where M = formula weight (molecular mass) and ρ = density of NaCl and N_A = Avogadro's constant.
$$a = 2.82 \times 10^{-8} \text{ cm (0.282 nm)}$$

Chapter 2. Mechanical properties

1. (b) 4.0×10^7 Pa
   (c) Yes
   (d) (i) $\frac{1}{2}$ (ii) 50%
   (e) 2.002 m
2. (c) 1.0×10^{11} Pa; 0.040%; 4.0×10^2 N; no; smaller.
4. (a) 1.1 (1.06) mm (b) 1.0×10^2 (101) N
6. (a) 1.86×10^{-3} m (b) 7.75×10^{-2} J

Chapter 3. Electrical properties

1. 2.4×10^2 C; 1.5×10^{21}
2. (a) 8.5×10^{28} (b) 8.5×10^{22} (c) 1.4×10^4 C
   (d) 6.8×10^3 s (e) 0.15 mm s^{-1}

3. 8.0 A
4. **(a)** 8.0 V; 2.5 A   **(b)** 240 V
5. **(a)** 4.0 V   **(b)** 240 J
6. **(a)** 2 A   **(b)** 2 A   **(c)** 6 V   **(d)** 10 V   **(e)** 5 Ω
7. **(a)** $I = I_1 + I_2$ **(b)** 6 V   **(c)** 3 A, 2 A, 1 A   **(d)** 2 Ω
8. **(a)** 4 V; 2 V   **(b)** 4.8 V; 1.2 V
9. 150 kΩ, 25 kΩ
10. **(a)** 10 V   **(b)** 40 V
12. **(i)** 95 Ω in series   **(ii)** 0.050 Ω in parallel
13. 3:2
14. $4.20 \times 10^{-3} \, °C^{-1}$
15. **(a)** $1.00 \times 10^3 \, J \, s^{-1}$ **(b)** 798 °C
16. **(a)** 12 J   **(b)** 36 J   **(c)** 240 J
17. **(a)** 6 V   **(b) (i)** 2 J   **(ii)** 6 J   **(c)** 0.03 Ω   **(d)** 3 A
18. **(b)** 3×10^4 J   **(c)** 960 Ω
19. Internal resistance 5 Ω
20. 2.0 (4) Ω; 0.50 A; 0.75 W
22. 4.6 Ω; 8.6 Ω
23. 0.0020 °C⁻¹; 50.2 cm
24. **(i)** 2.0 V   **(ii)** 2.7 Ω in parallel
25. 996 Ω; 0.2 cm
26. 2.00 Ω
27. 1.8×10^5 s
28. 1.5×10^2 A; 120 times

Chapter 4. Thermal properties

2. **(a)** 63.16 °C   **(b)** 47.7 °C
3. 16.0 V
4. $1.70 \, J \, g^{-1} \, °C^{-1}$
5. $4.4 \, J \, g^{-1} \, °C^{-1}$; ±10%
6. 0.050 °C s⁻¹; 6.8 revolutions per second
7. 26.0%
8. 3.3×10^2 °C
9. True height is 76.46 *divisions* of scale. A division marked as 1 cm is only 1 cm at 0 °C. At 15 °C 1 division has length $1(1 + 1.9 \times 10^{-5} \times 15)$ cm, therefore 76.46 divisions at 15 °C have a true length of $76.46(1 + 1 \times 1.9 \times 10^{-5} \times 15)$, i.e. 76.48 cm
10. 1.2 N
11. **(a)** 1:15 (14.5)   **(b)** 66.7 °C
12. 1.9×10^3 W
13. Copper 3.0 (2.98) °C cm⁻¹; aluminium 5.5 °C cm⁻¹
14. 4.4×10^{-3} W cm⁻²
15. 18 °C

Chapter 5. Optical properties

1. 15 cm; 5.0 cm
2. 20 cm; 10 cm behind mirror
3. Virtual, 13 (12.9) cm behind convex mirror
4. **(a)** 2.4 m   **(b)** 1.3×10^{-2} m
5. **(a)** 40 (.5)°   **(b)** 40 (.5)°   **(c)** 35 (.4)°
6. **(i)** ray refracted in water at an angle of 34 (.2)°
   (ii) critical angle for glass–water boundary = 63 (62.7)°, therefore ray totally internally reflected
7. Angle in liquid must exceed 67.9°
8. **(a)** 37 (.4)°
   (b) 28 (27.9)°
9. **(a)** 55.6°; 39.6° to 90.0°   **(b)** 60.3°

ANSWERS

10. 48.8°

11. (a) 20 cm; +15 cm   **(b)** +4.0 cm

13. 20 cm from first position of screen

14. (a) +60 cm   **(b)** +2.4 m

15. +20 cm, +20 cm, −20 cm, −60 cm

16. Virtual, 39 cm from lens on same side as object

17. $f = -500$ cm gives far point at infinity.

The nearest distance an object can now be brought up to his eye is that object distance which has its virtual image at 60.0 cm, i.e. we have to find u when $v = -60.0$ cm and $f = -500$ cm. Using $1/f = 1/u + 1/v$ we get $u = 68.2$ cm. His range of vision is now infinity to 68.2 cm, i.e. when wearing concave spectacle lenses of $f = -500$ cm objects within this range can be seen.

18. $f = 66.7$ cm gives near point at 25.0 cm.

The new far point will be the object distance which gives a virtual image 200 cm from the spectacle lens, i.e. we have to find u when $v = -200$ cm and $f = 200/3$ cm. From $1/f = 1/v + 1/u$ we get $u = 50.0$ cm. The new range of vision is therefore 25.0 to 50.0 cm.

19. magnifying power = 5.0; magnification = 6.0

20. (b) 58 (.3)

21. Separation of lenses 25 cm; $f = 6.0$ cm

22. 107 (.2) cm; 14.0

23. 80 cm; 16; 5.3 cm; 13 (12.8) cm

24. Reduced $\frac{1}{8}$; 5 times larger

Chapter 6. Statics and dynamics

2. (a) 1.2×10^3 N   **(b)** 1.1×10^3 N; 74° to horizontal.

3. 4.3×10^2 N; 69° to the horizontal.

4. AB $\sqrt{3} \times 10^6$ N; BC 2×10^6 N; AC 10^6 N; CD $\sqrt{3} \times 10^6$ N. Minimum cross-section area for BC = $2(.5) \times 10^{-2}$ m^2

5. (i) 0.245 s   **(ii)** 12.2 m s^{-1} **(iii)** 12.5 m s^{-1}

6. 2.17×10^4 m; 3.13×10^3 m; 25.0 s.

7. $\sqrt{3}/3$; $\sqrt{3}\,g/3$ m s^{-2}; $2\sqrt{3}\,mg/3$(m = mass of body)

9. $P/(5\ m)$; $P/5$; $2\ P/5$

10. 3.6×10^3 N; 5.4×10^4 W

11. $mg/2$

12. (a) 10/3 N   **(b)** 5/9 W   **(c)** 5/18 W

13. 8.00×10^3 N; 1.20×10^4 W

14. $2E/103$

Chapter 7. Circular motion and gravitation

1. (a) time $= \dfrac{\text{arc AB}}{\text{speed}} = \dfrac{\pi \times 5\,\text{m}}{11\,\text{m s}^{-1}} = \dfrac{22 \times 5\,\text{s}}{7 \times 11} = \dfrac{10\,\text{s}}{7}$

(b) average velocity $= \dfrac{\text{displacement}}{\text{time}} = \dfrac{10\,\text{m}}{10/7\,\text{s}}$

$= 7.0$ m s^{-1} *to the right*

The displacement is diameter AB to the right.

(c) average acceleration $= \dfrac{\text{change of velocity}}{\text{time}} = \dfrac{22\,\text{m s}^{-1}\ \text{downward}}{10/7\,\text{s}}$

$= 154/10$ m s^{-2} *downward.*

2. (a) 1.5π rad s^{-1} **(b)** 0.18π m s^{-1}

3. (a) 5.0 rad s^{-1}; 25 N

(b) 30 N: 20 N

4. (i) 4/3 N   **(ii)** $\sqrt{3\pi/5}$ s

5. 1.01×10^3 N; 1.70×10^2 N; 1.85×10^3 N

6. (a) 44 rad s^{-1} **(b)** 11 m s^{-1} **(c)** 22 m s^{-1} **(d)** 87 J
7. 1.9×10^{-2} N m
8. 8 s
9. (b) 4.0×10^{-4} kg m^2
10. 9.83 N
11. $v_B/v_C = y/x$
12. 24 hours
13. 7.71×10^{22} kg; 1.15×10^3 m
14. 3.83×10^5 km
15. (a) (i) 0.12 N **(ii)** 0.10 N **(c)** 7.0×10^5 J

Chapter 8. Mechanical oscillations

1. (a) 75 cm s^{-1} **(b)** 1.4×10^4 cm s^{-2}
2. 14 cm s^{-1}
3. 1.04
4. (a) 31 cm s^{-1} **(b)** 50 cm s^{-2} **(c)** 0.33 s
   (d) 5.0×10^{-3} J at A; zero at limits of motion
   (e) 5.0×10^{-3} J
5. An s.h.m. of period $\pi/\sqrt{50}$ s : maximum velocity $= 0.71$ m s^{-1}; maximum acceleration $=$
   10 m s^{-2}
6. 24(.5) N kg^{-1} m^{-1}
7. 16 Hz
8. 1.3×10^2 cm s^{-1}; 1.6 cm

Chapter 9. Fluids at rest

1. 1.0×10^4 N; 5.0×10^2 N
2. (b) 9 N
3. Silver 4.90 g   Gold 30.3 g
4. 2.4 cm
5. 757.3 mm; 0.2736 mm; 760.3 mm
6. 118 (The general expression is $p_n = p[V/(V + v)]^n$ where p is the initial pressure of the air
   in the vessel, V is the volume of the vessel, v is the volume swept out by the piston of the
   pump per stroke and n is the number of strokes to achieve pressure p_n. The expression is
   obtained by repeated application of Boyle's law to the mass of air in the vessel at the start of
   each stroke.)
8. 7.1×10^{-2} N m^{-1}
9. 49 mm; the water does not overflow (and violate conservation of energy) but it remains at the
   top of the tube (i.e. $h = 30$ mm) with an angle of contact of 52.2°. The weight of the raised
   column of water is then supported by the vertical components of the surface tension forces.
10. 76.93 cm

Chapter 10. Fluids in motion

3. 8.7×10^{-4} m s^{-1}
4. 0.26 mm^2 (Air and petrol do not have the same velocities; Bernoulli's equation must be used)
5. Yes. (Hint: apply Bernoulli's equation using

$$\frac{\text{velocity of water in barrel}}{\text{velocity of water from jet}} = \frac{\text{area of jet}}{\text{area of barrel}}$$

to find velocity of water from jet. Find the time taken by a drop of water to travel 3.5 m (assuming its horizontal velocity is constant) and then calculate how far it falls in this time).

Objective-type questions

Multiple choice

1.1. B   **1.2.** A
2.1. C   **2.2.** C   **2.3.** B   **2.4.** A, E, D
3.1. A   **3.2.** D   **3.3.** C   **3.4.** E   **3.5.** D   **3.6.** B
3.7. C (This is a *balanced* Wheatstone bridge) **3.8.** C   **3.9.** D
4.1. B   **4.2.** C   **4.3.** E   **4.4.** B
5.1. B   **5.2.** A   **5.3.** D
6.1. B   **6.2.** C   **6.3.** E   **6.4.** D   **6.5.** B
7.1. A   **7.2.** D   **7.3.** D   **7.4.** C
8.1. E   **8.2.** B
9.1. C
10.1. E

Multiple selection

3.10. E   **5.4.** A   **6.6.** C   **8.3.** C

Assertion-reason

5.5. C   **7.5.** C   **8.4.** B   **9.2.** A

Index

413

INDEX

INDEX

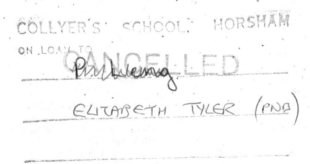